KU-279-895

# Natural inorganic hydrochemistry in relation to groundwater

# Natural inorganic hydrochemistry in relation to groundwater

## An introduction

For hydrogeologists, civil engineers, and chemists involved in the application of groundwater chemistry to water resources studies

J. W. LLOYD
*Reader in Hydrogeology*

J. A. HEATHCOTE
*Research Fellow in Hydrogeology*

*Department of Geological Sciences, University of Birmingham, UK*

CLARENDON PRESS · OXFORD
1985

*Oxford University Press, Walton Street, Oxford OX2 6DP*
*London New York Toronto*
*Delhi Bombay Calcutta Madras Karachi*
*Kuala Lumpur Singapore Hong Kong Tokyo*
*Nairobi Dar es Salaam Cape Town*
*Melbourne Auckland*
*and associated companies in*
*Beirut Berlin Ibadan Mexico City Nicosia*

*Oxford is a trade mark of Oxford University Press*



*Published in the United States*
*by Oxford University Press, New York*



*British Library Cataloguing in Publication Data*
*Lloyd, J. W. (John William)*
*Natural inorganic hydrochemistry in relation to groundwater.*
*1. Water, Underground—Composition*
*I. Title II. Heathcote, J. A.*
*ISBN 0-19-854422-7*

*Library of Congress Cataloging in Publication Data*
*Lloyd, J. W.*
*Natural inorganic hydrochemistry in relation to groundwater.*
*Bibliography: p.*
*Includes index.*
*1. Water chemistry. 2. Hydrogeology. I. Heathcote, J. A. II. Title.*
*GB855.L54 1984 551.49 84.16696*
*ISBN 0-19-854422-7*

*Typeset by Joshua Associates, Oxford*
*Printed in Great Britain by*
*St Edmundsbury Press,*
*Bury St Edmunds, Suffolk*

# Acknowledgement

The authors would like to thank Dr J. H. Tellam of the Department of Geological Sciences of the University of Birmingham, for his considerable help in reading the text before publication.

# Contents

# 1 Introduction

Groundwater chemistry was initially studied in relation to the suitability of groundwater for use purposes. As the overall knowledge of groundwater has increased the hydrochemical aspects have been interpreted with respect to chemical evolutionary processes and quality criteria. Man's impact upon groundwater systems has created many environmental problems so that hydrochemical studies related to pollution have become very important. In a relatively short period information about groundwater chemistry has multiplied by orders of magnitude and from professional necessity much of the work carried out has been and still is, quite rightly, the responsibility of chemists.

Traditionally groundwater has been the domain of the civil engineer and the hydrogeologist, who with the best will in the world frequently have difficulty in understanding each other; now with the advent of the chemist a further discipline has been added to make life even more difficult for the engineer and hydrogeologist. Although the chemists may be providing excellent data and reliable pollution control it is probable that much of the information made available is not fully appreciated by engineers and hydrogeologists; also it may well be that the chemists have problems in viewing their data in a hydrogeological context. Clearly an integrated understanding at a reasonable professional level is to the benefit of everybody.

In the field of groundwater studies some excellent textbooks are available on the hydraulic and engineering aspects of the subject (Walton 1970; Campbell and Lehr 1973). In water chemistry very comprehensive texts by, for example, Garrels and Christ (1965) and Stumm and Morgan (1981) provide essential reading for those well versed in chemistry. To some extent Freeze and Cherry (1979) and Fetter (1980) have tried to bridge the gap; however, it is felt that a more comprehensive but simple text explaining groundwater chemistry for the benefit of engineers and hydrogeologists is necessary, and this is the objective of this book.

The emphasis of the book is primarily towards the engineer and hydrogeologist involved in groundwater resources and other related work, and it is believed that more integration of hydrochemical data into overall aquifer assessment is essential and very worth while. It is hoped, however, that the book will provide information to the chemist in providing an explanation of the context in which hydrochemical data can be used. Only the inorganic aspects of unpolluted groundwaters have been considered as these provide the grounding to the understanding of overall hydrochemistry.

The book is written as an introduction to the subject in terms of the various chemical parameters that can be studied to advantage, the type of hydrochemical processes that occur in aquifers, and the methods of interpretation. Most of the basic chemistry required to understand the hydrochemistry is described; however, the reader is referred to the standard texts of Lewis and Waller (1982) and Atkins (1982). The book is not intended as a review of current inorganic groundwater chemistry research.

In addition to this introductory chapter, ten chapters are included. Chapters 2 and 3 describe the fundamental terminology used and the physical chemistry of aqueous solutions. In Chapters 4 and 5 the parameters that can be measured and calculated are discussed together with certain back-up features such as geophysical logging. Because so many parameters are available for interpretation in hydrochemistry today, their representation and the classification of hydrochemical types poses a problem; this is considered in Chapter 6. Most groundwater studies relate to good quality or low salinity waters and it is primarily these types of waters that are dealt with in Chapters 2 to 6; however, saline groundwaters can have a serious impact on groundwater resources so that these are discussed in Chapter 7 to indicate how they can be classified and related to flow conditions.

Isotopes are in common use in hydrochemical studies today; their value in resources problems is open to debate but they can be of use in integrated hydrochemical interpretation. The more commonly analysed isotopes are discussed in Chapter 8. Case studies are discussed in both Chapter 8 and Chapter 9 in which integrated hydrochemical studies from differing hydrogeological environments are considered. In conclusion Chapters 10 and 11 outline some of the criteria and interpretations adopted for the assessment of groundwater suitability for use. Throughout the book an attempt has been made to use hydrochemical examples from a wide variety of environments and a number of different countries.

## References

Atkins, P. W. (1982). *Physical chemistry*. Clarendon Press, Oxford.

Campbell, M. D. and Lehr, J. H. (1973). *Water well technology*. McGraw-Hill, New York.

Fetter, C. W. (1980). *Applied hydrogeology*. Merrill, Columbus, OH.

Freeze, R. A. and Cherry, J. A. (1979). *Groundwater*, Prentice-Hall, Englewood Cliffs, NJ.

Garrels, R. M. and Christ, C. L. (1965). *Solutions, minerals and equilibria*. Harper and Row, New York.

Lewis, M. and Waller, G. (1982). *Advancing chemistry*. Clarendon Press, Oxford.

Stumm, W. and Morgan, J. J. (1981). *Aquatic chemistry* (2nd edn.). John Wiley, New York.

Walton, W. C. (1970). *Groundwater resource evaluation*. McGraw-Hill, New York.

# 2 Chemistry of groundwater

## 2.1. Introduction

This chapter aims to discuss in qualitative terms the types of chemical reactions that occur in groundwater. Groundwaters, like other natural waters, are solutions of a variety of substances in the solvent water. The chemical properties of water as a solvent profoundly influence the chemical reactions that take place in aqueous solution; thus it is pertinent to discuss the chemistry of water itself before describing the chemical reactions that occur in groundwater.

## 2.2. Structure

The water molecule consists of two hydrogen atoms joined by single covalent bonds to an oxygen atom. The electronic arrangement of this molecule can be represented by Fig. 2.1(a), from which it can be seen that the outer shell of the oxygen atom contains four electron pairs. These are distributed approximately tetrahedrally as shown in Fig. 2.1(b). The electron distribution in the O–H bonds is uneven, being biased towards the oxygen atom, which leads to the water molecule having the distribution of partial charges shown in Fig. 2.1(b). As a consequence of this structure individual water molecules have a large electric dipole moment.

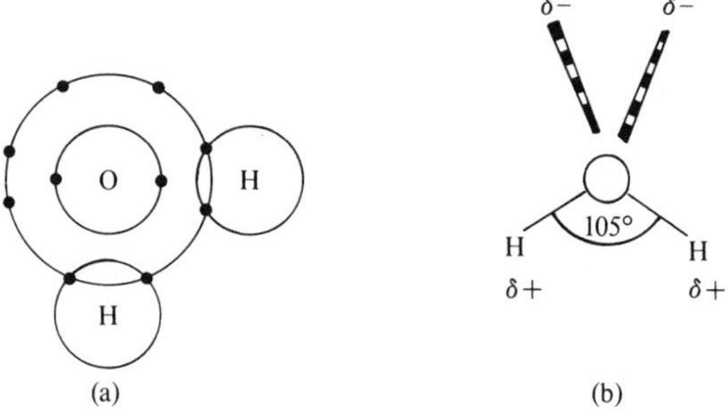


**Fig. 2.1.** Structure of the water molecule.

The unlike charges of water molecules attract each other, forming aggregates of molecules (Fig. 2.2). The bond shown as O----H is a hydrogen bond which is considerably weaker than the O–H bond (about 34 kJ $mol^{-1}$ compared with 458 kJ $mol^{-1}$). There appears to be some electron overlap between the two water molecules and this leads to transfer of hydrogen atoms.

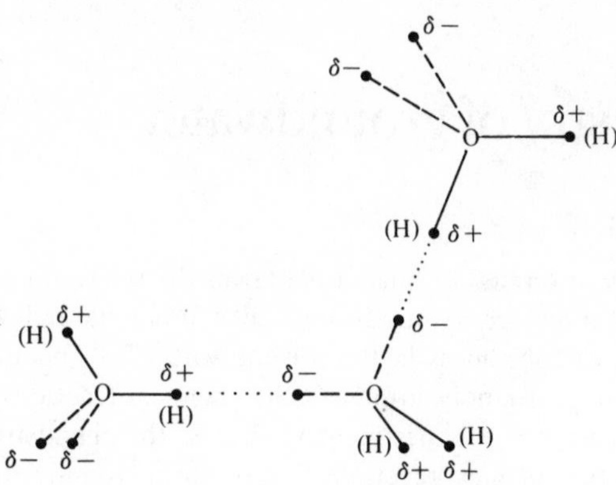


**Fig. 2.2.** Aggregate of water molecules.

On a larger scale, the spatial distribution of the partial charges on the water molecule imposes geometrical constraints on how the molecules may be linked and thus imposes a degree of order upon the structure of liquid water. Open-structured aggregates of up to 100 molecules are formed at normal temperature. The aggregates are very transitory, having a lifetime of the order of $10^{-11}$ s. With decreasing temperature the size of the clusters increases causing the structure of the liquid to become more open until the very open structure of ice is obtained (Fig. 2.3). This ordering is apparent from the expansion of water as it is cooled from 4°C to 0°C and its further expansion on freezing. The extensive association by hydrogen bonding of water molecules is also responsible for the relatively high melting and boiling points of water and for its high latent

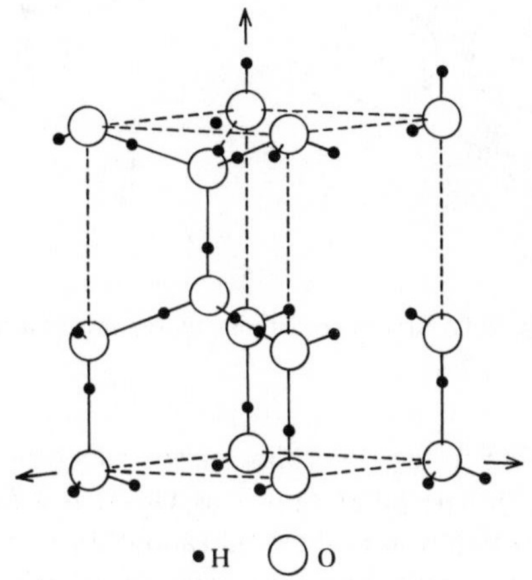


**Fig. 2.3.** Structure of ice. (After Evans 1964.)

heat of vaporization, since both melting and boiling require the breaking of hydrogen bonds.

The structure of water is affected by adding a solute. An ionic solute in solution dissociates into positively and negatively charged ions. Each ion then becomes surrounded by water molecules oriented oxygen inwards around cations and hydrogen inwards around anions, as shown in Fig. 2.4(a). This

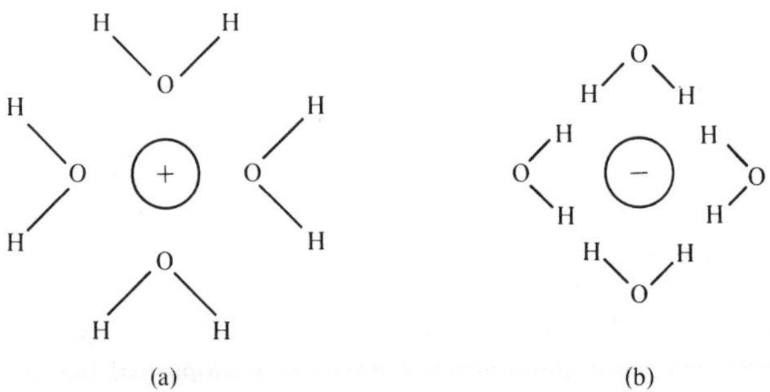


**Fig. 2.4.** Hydration of ions.

hydration of the ions stabilizes the solution by reducing the attractive forces between oppositely charged ions. The degree of hydration depends on the electrostatic field strength around the ion and thus hydration increases with increasing charge and decreasing ionic radius. For example, the $Li^+$ ion has a radius of 0.06 nm but the radius of the hydrated ion is 0.34 nm, whilst the chemically similar $Cs^+$ ion, with a radius of 0.169 nm, has a hydrated radius of 0.228 nm. As anions are almost invariably larger than cations, cations are usually more strongly hydrated than anions. Hydration of cations must be distinguished from complexation of cations by water, although there is inevitably overlap between the categories. Many transition metal cations chemically bind with water molecules forming complex ions of fixed composition and considerable stability; for example the exchange time for water molecules in $[Cr(H_2O)_6]^{3+}$ is measured in hours. In contrast, water molecules in hydration sheaths are only loosely associated, with exchange times of the order $10^{-11}$ s, and no fixed number of molecules is involved. The presence of a hydration sheath increases the size of an ion thus reducing its mobility and altering rates of chemical reaction. The hydration sheath is particularly important in crystallization, which is often hindered by the need to dehydrate cations.

There is an electrostatic attraction between cations and ions in solution because of their opposite charge, and therefore the anion concentration is locally increased around each cation and vice versa. Thermal agitation tends to prevent the arrangement becoming ordered. Because each ion is surrounded by an atmosphere of oppositely charged ions the effective concentration of each ionic species is reduced, an effect discussed quantitatively in Section 3.3. At high solute concentrations oppositely charged ions form transient ion pairs, thus reducing the effective concentration even further. Since there is an element of chemical interaction in ion pair formation, the formation of constants of ion pairs depend strongly on the participating ions.

Non-ionic solutes alter the structure of water in a variety of ways. Small non-polar molecules, such as methane and the noble gases, can fit into the gaps in the open water structure and thus stabilize it. At high pressures such molecules produce stable solids called *clathrates*. Methane clathrate has been reported from deep sea drilling. Polar non-ionic molecules are accepted into the hydrogen-bonded water structure, their effect depending on molecular size. For example, ammonia gas is highly soluble in water as a consequence of the similar size of its molecule and its ability to form hydrogen bonds. Large polar molecules, e.g. starch $(C_6H_{10}O_5)_n$, disperse in water to form stable colloids. Large non-polar molecules such as hydrocarbons display little solubility in water as they cannot enter into hydrogen bonding and therefore disrupt the structure of water. Similarly, hydrocarbons do not form stable colloids in water. Interesting properties arise when non-polar and polar groups form part of the same molecule, as in soaps and detergents. Thus molecules preferentially migrate to water surfaces, whose behaviour they greatly modify.

## 2.3. Chemical reactions in water—the equilibrium concept

Chemical reactions in groundwater can be summarized as follows: acid–base reactions, redox reactions and solid phase interactions (solution–precipitation and adsorption including ion exchange). Reactions of all these types are reversible and can be described in terms of equilibria:

$$A + B \rightleftharpoons C + D. \qquad (2.1)$$

A mixture of A + D will react to give some C + D, but conversely a mixture of C + D will react to give some A + B. The reaction reaches an equilibrium position where A, B, C, and D are all present regardless of from which end the reaction was started. The qualitative behaviour of a chemical system at equilibrium is described by *le Chatelier's principle*: a chemical system in equilibrium, when subjected to a stress that disturbs the equilibrium, will adjust its position of equilibrium in the direction that tends to relieve the stress. For example, carbon dioxide gas dissolves *exothermically* (evolving heat) in water to produce carbonic acid:

$$CO_2(g) + H_2O\,(\ell) \rightleftharpoons H_2CO_3 + \text{heat}. \qquad (2.2)$$

If the concentration of carbon dioxide is increased by raising its partial pressure, the reaction proceeds to the right to establish a new equilibrium, thus consuming carbon dioxide. If the temperature of the solution is raised, the equilibrium moves to the left, thus absorbing heat. Equilibrium conditions are susceptible to a rigorous theoretical treatment and this forms the basis of the quantitative hydrochemistry discussed in Chapter 3.

## 2.4. Acid-base reactions

Acid and bases are normally defined in terms of proton transfer (the Brønsted-Lowry concept). The electric field around bare protons is so strong that they cannot exist in a solvent, but react with solvent molecules to form complexes, e.g. in water

$$H^+ + H_2O \longrightarrow H_3O^+. \tag{2.3}$$

The resulting hydronium ion is itself hydrated as is usual for a small cation in aqueous solution. The formation of the hydronium ion is so universal that it is normally omitted in writing chemical equations; thus in aqueous solutions $H^+$ and $H_3O^+$ are synonymous.

According to the Brønsted-Lowry definition, acids are proton donors and bases are proton acceptors. Implicit in this definition is the idea of an acid-base pair. The behaviour of a solution of hydrogen chloride in water (hydrochloric acid) can be represented

$$\underset{\text{acid 1}}{HCl} + \underset{\text{base 2}}{H_2O} \rightleftharpoons \underset{\text{acid 2}}{H_3O^+} + \underset{\text{base 1}}{Cl^-}. \tag{2.4}$$

Hydrogen chloride behaves as an acid by donating a proton to water, which behaves as a base by accepting it. The reaction is reversible; thus the hydronium ion can behave as an acid by donating a proton to the chloride ion behaving as a base. Hydrogen chloride and chloride ion form a *conjugate acid-base pair*. In this particular case the equilibrium concentration of hydrogen chloride is negligible: hydrogen chloride is a strong acid and chloride ion is a weak base. The behaviour of a solution of acetate ions in water can be represented as

$$\underset{\text{base 1}}{Ac^-} + \underset{\text{acid 2}}{H_2O} \rightleftharpoons \underset{\text{acid 1}}{HAc} + \underset{\text{base 2}}{OH^-}. \tag{2.5}$$

In this case the concentration of undissociated acetic acid is appreciable: acetic acid is a weak acid and acetate ion is a strong base. A strong acid always has a weak conjugate base and vice versa.

In the examples above water has been seen behaving as a weak acid and a weak base. These reactions can be summarized by

$$H_2O + H_2O \rightleftharpoons H_3O^+ + OH^-. \tag{2.6}$$

This equation shows that pure water must contain equal numbers (about $10^{-7}$ mol $l^{-1}$) of hydronium and hydroxyl ions. Pure water contains neither acid nor alkali and is therefore neutral. A solution is described as *neutral* if it contains equal numbers of hydronium and hydroxyl ions. If hydronium ions are in excess the solution is *acid*, and if hydroxyl ions are in excess the solution is *alkaline*.

Acid–base pairs commonly present in groundwater are those associated with carbonic acid and water itself. Boric, orthophosphoric, and humic acids are minor constituents of groundwaters and are seldom important in controlling the acid–base chemistry. Large amounts of carbonic acid are produced in groundwater by the solution of carbon dioxide in the atmosphere and soil air and therefore the equilibria associated with this acid, which is considerably stronger than water, dominate the acid–base chemistry of most groundwaters. Many aquifers of sedimentary origin contain significant amounts of solid carbonate such as calcite ($CaCO_3$) or dolomite ($CaMg(CO_3)_2$) which participate in the carbonic acid equilibria via the reactions

$$CO_2 + H_2O \rightleftharpoons H_2CO_3 \quad (2.7)$$

$$H_2CO_3 \rightleftharpoons H^+ + HCO_3^- \quad (2.8)$$

$$HCO_3^- \rightleftharpoons H^+ + CO_3^{2-} \quad (2.9)$$

$$CaCO_3 \rightleftharpoons Ca^{2+} + CO_3^{2-}. \quad (2.10)$$

The carbonate chemistry of many aquifers is almost completely described by these equations, which will be analysed further in Chapter 3.

Acid–base chemistry affects many types of reactions in addition to those limited to acids and bases. In the example above, the dissolution of calcite contributes carbonate ion, a fairly strong base, to the solution and therefore reduces the hydrogen ion concentration, thus rendering the solution more alkaline. Further examples of this kind of behaviour will be seen later. All acid–base reactions encountered in natural hydrochemistry are very fast, and therefore acid–base systems are always in equilibrium in the solution although the solution may react only slowly with solid and gaseous phases.

## 2.5. Redox reactions

Many reactions in groundwater involve the transfer of electrons between dissolved, gaseous and solid constituents. Electron loss results in oxidation and electron gain results in reduction, but since free electrons do not exist in solution, oxidation and reduction occur simultaneously and the overall reduction is called a redox reaction. Oxidation increases oxidation number and reduction decreases oxidation number. For example, in the reaction of iron with water

$$Fe + 2H^+ \longrightarrow Fe^{2+} + H_2 \quad (2.11)$$

iron is oxidized to $Fe^{2+}$ and protons are simultaneously reduced to hydrogen. It is convenient to write redox reactions in terms of two half reactions, one oxidation and one reduction, which can then be added so that the electrons cancel:

$$Fe \longrightarrow Fe^{2+} + 2e^- \qquad \text{oxidation} \quad (2.12)$$

$$2(H^+ + e^- \longrightarrow \tfrac{1}{2}H_2) \qquad \text{reduction} \quad (2.13)$$

$$2H^+ + Fe \longrightarrow Fe^{2+} + H_2 \qquad \text{complete reaction} \quad (2.14)$$

**Table 2.1** *Hydrochemically important redox reactions*

| | | | |
|---|---|---|---|
| (1) | $O_2 + 4H^+ + 4e^-$ | $\rightarrow$ | $2H_2O$ |
| (2) | $2H^+ + 2e^-$ | $\rightarrow$ | $H_2$ |
| (3) | $2NO_3^{2-} + 12H^+ + 10e^-$ | $\rightarrow$ | $N_2 + 6H_2O$ |
| (4) | $CH_2O + 4H^+ + 4e^-$ | $\rightarrow$ | $CH_4 + H_2O$ |
| (5) | $CO_2 + 4H^+ + 4e^-$ | $\rightarrow$ | $CH_2O + H_2O$ |
| (6) | $CO_2 + 8H^+ + 8e^-$ | $\rightarrow$ | $CH_4 + 2H_2O$ |
| (7) | $SO_4^{2-} + 8H^+ + 6e^-$ | $\rightarrow$ | $S + 4H_2O$ |
| (8) | $SO_4^{2-} + 10H^+ + 8e^-$ | $\rightarrow$ | $H_2S + 4H_2O$ |
| (9) | $S + 2e^-$ | $\rightarrow$ | $S^{2-}$ |
| (10) | $Fe^{3+} + e^-$ | $\rightarrow$ | $Fe^{2+}$ |
| (11) | $Fe(OH)_3 + 3H^+ + e^-$ | $\rightarrow$ | $Fe^{2+} + 3H_2O$ |
| (12) | $Fe(OH)_3 + H^+ + e^-$ | $\rightarrow$ | $Fe(OH)_2 + H_2O$ |

A list of hydrochemically important half-reactions is presented in Table 2.1. All the reactions in this table are written as reductions, but these can easily be converted to oxidations by reversing the arrow.

Criteria for assessing whether or not any given redox reaction will occur are discussed in Section 3.5. It can be seen from Table 2.1 that water may participate in redox reactions in four ways. It may be an inert solvent, in which case it does not appear in the reaction. It may fulfil an acid-base role, in which case water appears on one side of the equation and $H^+$ or $OH^-$ on the other side, or it may be oxidized (reaction (1)) or reduced (reaction (2)). In the reaction discussed above, water is reduced since the hydrogen ions come from the reaction

$$H_2O \rightarrow H^+ + OH^- \qquad (2.15)$$

This is an example of an acid-base reaction occurring simultaneously with another reaction. As the reaction proceeds, the solution becomes progressively more alkaline because of the loss of protons as hydrogen.

The most important oxidizing agents in groundwater are dissolved oxygen, oxo-anions such as nitrate and sulphate, and water itself. Reducing agents in groundwater include a wide variety of organic compounds such as carbohydrates, humic substances, and hydrocarbons, inorganic sulphides such as pyrite, and iron(II) silicates. Most aquifers contain at least one of these groups of reducing agents, while most water entering aquifers contains oxidizing agents such as dissolved oxygen and nitrate. As groundwater passes through an aquifer, oxidizing agents in the water are consumed progressively by reaction with reducing agents in the aquifer, the most powerful oxidizing agents reacting first. Dissolved oxygen is the most powerful oxidizing agent encountered naturally and this is normally consumed within a short distance from the recharge area. Nitrate reduction follows the consumption of oxygen and this may in turn be followed by sulphate reduction. As a corollary, the presence of oxidizing species may indicate recent groundwater recharge.

Many of the important redox reactions that occur in groundwater are catalysed by bacteria and other microorganisms. Although the reactions must be thermodynamically possible, without the intervention of bacteria the rate at which the reaction proceeds is negligible. Most bacteria derive the energy required for their metabolism from the oxidation of organic matter, and this oxidation must be accompanied by a simultaneous reduction. If oxygen is available, i.e. the water is *aerobic*, the prefered reduction is that of oxygen to water, this being the metabolic pathway used by all higher organisms. In the absence of oxygen, i.e. in *anaerobic* conditions, certain specialized bacteria can utilize other reduction reactions, for example the reduction of nitrate to nitrogen (reaction (3)) by *Pseudomonas* and the reduction of sulphate to sulphide (reaction (8)) by *Desulphovibrio* and *Desulphotomaculum*. Bacteria can thrive in a wide variety of conditions: pressures of several hundred bars, pH between 1 and 10, temperatures up to 75°C, and a very high salinities. Given favourable physical conditions, the other requirement for significant bacterial activity is an adequate supply of nutrients which include nitrogen, phosphorus, and sulphur compounds, trace elements, organic carbon, and a suitable oxidizing agent. The persistence of oxyanions such as sulphate in reducing waters probably indicates a lack of adequate nutrient.

## 2.6. Solution-precipitation

Because of the powerful solvent properties of water, solution-precipitation reactions with the aquifer matrix are frequently important in controlling groundwater chemistry. The dissolution of a mineral in water can be described by an equilibrium relationship: when water is brought in contact with an excess of mineral, the concentration of the solution increases to a maximum for given physical conditions whereupon the solution is said to be *saturated*. The concentration of the saturated solution is the *solubility* of the mineral and in general it depends on temperature and pressure and sometimes upon external chemical factors. Table 2.2 gives the solubilities of several commonly encountered sedimentary minerals under surface conditions. It can be seen that solubilities vary considerably.

The solution process of calcite, an extremely common mineral in aquifers of sedimentary origin, merits further study. It is described by the reaction

$$CaCO_3(s) \rightleftharpoons Ca^{2+} + CO_3^{2-}. \qquad (2.16)$$

According to le Chatelier's principle, the solubility can be controlled by altering the concentration of carbonate ion. The carbonate ion concentration is controlled by the complex acid-base equilibrium of carbonic acid, and therefore the solubility of calcite depends on the hydrogen ion concentration of the solution and on the partial pressure of carbon dioxide in contact with the solution. The solubility of calcite at 25 °C and $10^{-3}$ bar partial pressure of carbon dioxide is 90 mg $l^{-1}$ and rises to 480 mg $l^{-1}$ at $10^{-1}$ bar partial pressure of carbon

**Table 2.2** *Solubilities of minerals that dissolve congruently in water at 25 °C and 1 bar total pressure*

| Mineral | Formula | Solubility at pH 7 (mg $\ell^{-1}$) |
|---|---|---|
| Gibbsite | $Al_2O_3 \cdot 2H_2O$ | 0.001 |
| Quartz | $SiO_2$ | 12 |
| Hydroxyapatite | $Ca_5(OH)(PO_4)_3$ | 30 |
| Amorphous silica | $SiO_2$ | 120 |
| Fluorite | $CaF_2$ | 160 |
| Dolomite | $CaMg(CO_3)_2$ | 90[a] |
| Calcite | $CaCO_3$ | 100[a] |
| Gypsum | $CaSO_4 \cdot 2H_2O$ | 2100 |
| Sylvite | KCl | 264 000 |
| Epsomite | $MgSO_4 \cdot 7H_2O$ | 267 000 |
| Mirabillite | $NaSO_4 \cdot 10H_2O$ | 280 000 |
| Halite | NaCl | 360 000 |

[a] $CO_2$ of partial pressure $10^{-3}$ bar.
Data from Seidell (1958).

dioxide. This is an example of *conditional solubility*. A full discussion of the carbonate system is contained in Chapter 3.

All the minerals listed in Table 2.2 dissolve *congruently* in pure water: no solid phases are produced as a result of the dissolution reaction. In contrast, most rock-forming silicates, with the exception of quartz, dissolve *incongruently*, producing a solid phase during the dissolution reaction. For example, orthoclase, a common mineral in granites and in sandstones derived from granites, may dissolve in water according to the reaction

$$11H_2O + \underset{\text{orthoclase}}{2KAlSi_3O_8(s)} \rightleftharpoons 2K^+ + 2OH^- + 4Si(OH)_4 + \underset{\text{kaolinite}}{Al_2Si_2O_5(OH)_3(s)}. \tag{2.17}$$

Kaolinite is left as a solid phase, although it too may dissolve via its own dissolution reaction which is also incongruent. Since hydroxyl ions are produced in the reaction above, it can be deduced that the solubility is conditional on pH. Incongruent dissolution processes may also arise when two or more minerals, each of which dissolves congruently in pure water, dissolve successively in the same water, a common example being the dissolution of calcite and gypsum. Consider the dissolution of gypsum in a solution which is already saturated with calcite. The dissolution of gypsum contributes calcium ions according to

$$CaSO_4 \cdot 2H_2O \rightleftharpoons Ca^{2+} + SO_4^{2-} + 2H_2O. \tag{2.18}$$

If the calcium ion concentration is increased, reference to eqn (2.18) and le Chatelier's principle shows that the solubility of calcite must decrease. Thus in these circumstances gypsum dissolution is incongruent since it is accompanied by calcite precipitation.

When the products of a dissolution reaction are present in a solution at a higher concentration than would occur in equilibrium with undissolved solid the solution is said to be *supersaturated.* Under these conditions precipitation may occur. However, mineral precipitation, unlike mineral dissolution, is not well described by equilibrium chemistry. Certain minerals, particularly calcite, dolomite, and most silicates, are reluctant to precipitate from solution despite considerable supersaturation. Concentrations may eventually fall to equilibrium values but the time taken may be significant even on the geological time scale. For example, the oceans are supersaturated with respect to both calcite and dolomite but show no tendency to precipitate either mineral. The precipitation of calcite is prevented by the presence of magnesium ions in solution, probably because the magnesium ions, which would otherwise be incorporated in the calcite lattice, are too highly hydrated. Seawater can precipitate aragonite, a more soluble calcium carbonate polymorph, when calcium and carbonate concentrations become sufficiently high because magnesium ions will not fit in the aragonite lattice and therefore do not inhibit its precipitation. However, because seawater is supersaturated with respect to calcite it cannot dissolve additional calcite. Ordered dolomite has never been synthesized at surface temperatures and hence it is not suprising that it does not crystallize from seawater. It appears that the necessary order of calcium and magnesium ions in the dolomite cannot be achieved. Similar ordering considerations prevent the crystallization of most silicates. Thus caution must be exercised in using solution precipitation equilibria in groundwater chemistry—equilibrium theory is useful in predicting when a mineral may dissolve but it is not necessarily useful in predicting when it will precipitate.

## 2.7. Adsorption

Some dissolved constituents in water are concentrated at the interface between solid and solution; they are said to be *adsorbed* on the solid (the term *adsorption* describes a purely surface phenomenon, whereas *absorption* implies permeation of the bulk material). Such surface phenomena are of particular relevance in groundwater chemistry because several minerals present in aquifers, of which clay minerals, and iron and manganese oxides are the most common, are very fine grained and thus have a large specific area. Specific areas of clay minerals are of the order of $10^3$ $m^2$ $g^{-1}$ which contrasts with a clean sand which may have a specific area as low as $10^{-3}$ $m^2$ $g^{-1}$. Particles with equivalent spherical diameters less than 1 or 2 $\mu$m, which are so small that surface effects dominate, are termed *colloidal* particles.

At the boundary of a crystal there must be atoms with unsatisfied valencies as shown in Fig. 2.5. Consequently the crystal boundary is characterized by

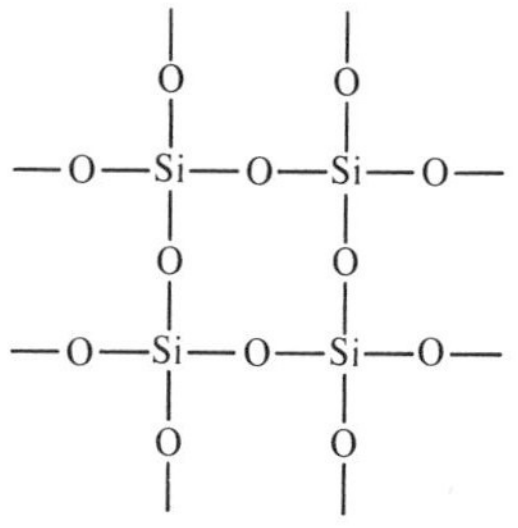

```
        |       |
        O       O
        |       |
—O—Si—O—Si—O—
        |       |
        O       O
        |       |
—O—Si—O—Si—O—
        |       |
        O       O
        |       |
```

**Fig. 2.5.** Unsatisfied valencies at a silicate surface.

unsatisfied charge. In oxides and silicates, which together constitute most hydrogeologically important colloids, these boundary atoms are oxide ions, which may react with water. The predominant species is determined by the hydrogen ion concentration of the solution via the equilibria

$$\equiv Si{-}O^- + H^+ \rightleftharpoons \equiv Si{-}OH + H^+ \rightleftharpoons \equiv Si{-}O^+H_2 \qquad (2.19)$$

Thus the surface charge is pH dependent, being positive in acid solution and negative in alkaline solutions. At some intermediate pH, known as the zero point of charge, the surface is uncharged. This pH varies between 9 for basic oxides such as $Al_2O_3$ to around 2 for clay minerals and feldspars which therefore behave as weak acids. From these data it can be deduced that the charge on clay minerals in most natural waters is negative. Iron and manganese oxides, which usually have a point of zero charge near pH 7, can be either positively or negatively charged in natural waters.

For clay mineral grains there is an additional charge which arises from atomic substitution within the crystal lattice. In all clay minerals there is limited substitution of quadrivalent silicon by trivalent aluminium and, except in kandites such as kaolinite and serpentine, there is also extensive substitution of divalent magnesium or iron for trivalent aluminium. The result of these substitutions is to leave the crystal lattice with a permanent negative charge.

In contact with ionic solutions, charged colloid particles are surrounded by oppositely charged *counterions* in a diffuse adsorbed layer, the number of ions in this layer being just sufficient to balance the charge on the colloid particle. In many cases these adsorbed ions are relatively mobile because of the long-range nature of the electrostatic forces involved and therefore the counterions are readily exchangeable, subject to the constraint of charge balance being maintained. Since colloidal clays are negatively charged at normal pH values the exchangeable ions are normally cations and therefore ion exchange phenomena are dominated by cation exchange. The cation exchange capacity of a clay is limited by the density of negative charge, which is much higher for smectites and vermiculites than for other types of clay mineral. Typical cation exchange capacities (CEC) are given in Table 2.3.

**Table 2.3** *Cation exchange capacities of clay minerals*

| Clay mineral | CEC (meq (100 g)$^{-1}$) |
|---|---|
| Kaolinite | 3–15 |
| Smectites | 60–120 |
| Illite | 10–40 |
| Vermiculite | 100–160 |
| Chlorite | 10–40 |

Cation exchange is ion selective: because of their higher charge, divalent ions such as $Ca^{2+}$ are absorbed more strongly than monovalent ions such as $Na^+$; for ions with the same charge the ion with the smallest hydrated radius is most strongly adsorbed and therefore $Ca^{2+}$ is favoured over $Mg^{2+}$.

Under suitable circumstances ion exchange phenomena may have a pronounced effect on groundwater chemistry. In sedimentary aquifers of marine origin the clays are sodium saturated because of the high sodium-to-calcium ratio of seawater, but the cation chemistry of recharge waters is usually dominated by calcium and magnesium. Recharge waters entering such a sodium-saturated aquifer have their calcium and magnesium almost completely replaced by sodium until the exchange capacity of the aquifer is exhausted. The removal of calcium by this process disturbs the calcite solubility equilibrium and thus may change the chemistry of the water profoundly. Low calcium water is preferred for many domestic uses, but high sodium water can be disastrous for irrigation and therefore the occurrence of ion exchange has important consequences for water resources. Ion exchange in the reverse direction, i.e. the replacement of sodium in water by calcium, may take place when saline water percolates through terrigenous clays deposited in river estuaries.

Some cations are absorbed irreversibly on clay minerals and are therefore non-exchangeable, or *fixed*; fixation occurs with large unhydrated ions such as $K^+$ and $NH_4^+$ and with small highly hydrated ions such as $Li^+$ and to some extent $Mg^{2+}$ which become part of the clay mineral lattice. Irreversible adsorption tends to occur when there is strong chemical interaction between the adsorbent and the adsorbate. Phosphate ions are adsorbed by positively charged iron(III) oxide colloids and the adsorption becomes irreversible through the formation of insoluble iron(III) phosphates. Irreversible adsorption also occurs in *coprecipitation* in which ions adsorbed on actively growing colloidal particles become buried in the structure and are therefore no longer in contact with the bulk solution. Iron(III) hydroxides coprecipitate many other transition metal ions by this process.

## 2.8. Chemistry of groundwater

In any aquifer situation, one or more of the above processes may be occurring, modifying the chemistry of the water. Which process dominates at any time

**Table 2.4** *Processes controlling the occurrences of some constituents of groundwater*

| | Physical | | Geochemical | | | | | | | | Biochemical |
|---|---|---|---|---|---|---|---|---|---|---|---|
| | Dispersion | Filtration | Geochemical abundance | Complex formation | Ionic strength | Acid-base | Oxidation-reduction | Solution-precipitation | Adsorption-desorption | Membrane filtration | |
| $Ca^{2+}$ | x | | | (x) | x | | | (x) | x | (x) | |
| $Mg^{2+}$ | x | | | (x) | x | | | (x) | x | (x) | |
| $Na^{+}$ | x | | | | | | | | x | (x) | |
| $K^{+}$ | x | | | | | | | | x | x | |
| $Sr^{2+}$ | x | | (x) | | x | | | (x) | x | x | |
| $Ba^{2+}$ | x | | (x) | | x | | | (x) | x | x | |
| $Li^{+}$ | x | | x | | | | | | (x) | (x) | |
| $Rb^{+}$ | x | | x | | | | | | x | x | |
| $Cs^{+}$ | x | | x | | | | | | x | x | |
| $NH_4^{+}$ | x | | | (x) | | x | x | | x | x | x |
| As | x | | x | x | x | x | x | x | x | | |
| Fe | x | x | | x | x | x | x | x | x | | |
| Mn | x | x | (x) | x | (x) | x | x | x | x | | |
| Cu | x | (x) | x | x | (x) | x | x | x | x | | |
| Zn | x | (x) | x | x | (x) | x | x | x | x | | |
| Pb | x | (x) | x | x | (x) | x | x | x | x | | |
| Ni | x | (x) | x | x | (x) | x | x | x | x | | |
| Co | x | (x) | x | x | (x) | x | x | x | x | | |
| Cd | x | (x) | x | x | (x) | x | x | x | x | | |
| $HCO_3^{-}$ | x | | | (x) | | x | | x | | (x) | (x) |
| $SO_4^{2-}$ | x | | | (x) | (x) | (x) | x | (x) | (x) | x | (x) |
| $Cl^{-}$ | x | | | (x) | | | | | | x | |
| $NO_3^{-}$ | x | | | | | (x) | x | | | | x |
| $HPO_4^{-}$ | x | | (x) | x | x | x | | x | x | | x |
| $F^{-}$ | x | | (x) | x | | | | x | x | | |
| $H_3BO_3$ | x | | (x) | (x) | (x) | | | | | x | |
| $Br^{-}$ | x | | (x) | | | | | | | x | |
| $I^{-}$ | x | | x | | | | | | | x | |
| $SiO_2$ | x | (x) | | x | | x | | x | | | |

The parentheses indicate minor controls.
After Edmunds 1977.

depends on the mineralogy of the aquifer, the hydrogeological environment, and the history of groundwater movement. In this dependence lies the usefulness of hydrochemistry: if the inverse problem can be solved, information about aquifer environment and history can be deduced.

These changes will be illustrated by describing a typical evolutionary sequence as an example. Rainwater arriving at the soil surface is a very dilute solution, predominantly containing sodium chloride of marine origin and dissolved carbon dioxide. In its passage through the soil zone it gains more carbon dioxide from the soil atmosphere and may also leach nitrate, phosphate, and chloride together with sodium and potassium originally applied to the soil as fertilizers. In the aquifer this $CO_2$ charged water dissolves calcium and magnesium carbonates, almost ubiquitous in aquifer lithologies, thus becoming a calcium-magnesium-bicarbonate solution via a sequence of acid-base reactions. In the absence of carbonates, silica may be contributed by silicate minerals and sulphate by the oxidation of sulphides such as pyrite. Bacterial reduction of dissolved organic carbon in the water consumes dissolved oxygen, followed by oxyanions such as nitrate and sulphate. As the water becomes more reducing it may leach iron oxides, or conversely may precipitate transition metal sulphides, of which iron as $FeS_2$ is usually most important. Many aquifers are of marine sedimentary origin, or have otherwise once contained dominantly sodium-chloride water. As the groundwater encounters the sodium clays produced by this saline water, ion exchange occurs, changing the calcium bicarbonate water to a sodium bicarbonate water. At the end of the flow path some saline water may remain, producing a sodium chloride water by mixing. This classic sequence, observed in many aquifers, has been described by Chebotarev (1955). Nearly static water at the end of the flow path actively participates in the ongoing diagenesis of the aquifer. Elements incorporated in minerals by organic intervention are liberated to the porewater, which therefore acquires high concentrations of certain trace elements.

The common elements involved in hydrogeological processes, and their controls, are summarized in Table 2.4. The full implications of these controls are discussed in later chapters.

## References

Chebotarev, I. I. (1955). Metamorphism of natural waters in the crust of weathering. *Geochim. cosmochim. Acta* 8, 22–48, 137–70, 198–212.

Edmunds, W. M. (1977). Groundwater geochemistry—controls and processes. *Proc. Conf. on Groundwater Quality—Measurement, Prediction and Protection*. pp. 115–47. *Water Research Centre* Medmenham, England.

Evans, R. C. (1964). *An introduction to crystal chemistry*. Cambridge University Press, Cambridge.

Seidell, A. (1958). *Solubilities, I* (4th edn.). American Chemical Society–Van Nostrand, Princeton, NJ.

# 3 Physical chemistry of aqueous solutions

## 3.1. Introduction

Physical chemistry deals with many of the quantitative aspects of chemical interaction and is therefore important in hydrochemistry, where quantitative aspects often dominate. Much physical chemistry concerns equilibrium conditions between 'ideal' substances, for reasons of mathematical tractability. Fortunately, many hydrochemical systems approach chemical equilibrium, or can usefully be discussed in terms of equilibrium. Various approaches are available to cope with the fact that the real world is not 'ideal'.

The way in which the theory of chemical equilibrium can be applied in hydrochemistry and the ways of coping with non-ideality are discussed in this chapter. The material is basically a quantitative treatment of that discussed qualitatively in Chapter 2.

## 3.2. Chemical equilibrium

A discussion of chemical equilibrium can be approached from either a kinetic or an energetic viewpoint; both produce the same results but the kinetic approach may be easier to understand while the energetic approach produces more extensive conclusions. The kinetic approach will be discussed first.

The rate at which substances react together is described by the *law of mass action*. This states that the rate of a reaction is proportional to the effective concentrations of the reacting substances. For example, the reaction

$$A + 2B \rightarrow C \qquad (3.1)$$

proceeds at a rate given by

$$\text{rate} \propto [A]\,[B]^2$$

since eqn. (3.1) can be written as

$$A + B + B \rightarrow C \qquad (3.2)$$

for which

$$\text{rate} \propto [A]\,[B]\,[B].$$

Alternatively

$$\text{rate} = k[A]\,[B]^2.$$

The value of the proportionality constant clearly depends on the concentration units used (usually molality for solutions)†. Pure solids have effectively constant concentration. It should be noted that the law of mass action applies only to the reacting species, which may differ from those in the equation for the reaction because of the formation of reaction intermediates. For example, suppose the reaction

$$\mathrm{A + B + C \rightarrow D} \tag{3.3}$$

proceeds as

$$\begin{aligned} &\mathrm{A + B \rightarrow E} \\ &\mathrm{C + E \rightarrow D.} \end{aligned} \tag{3.4}$$

If the second reaction in (3.4) is the slower of the two, the rate will be proportional to [C] alone. By measurement of the effect of concentration on reaction rate, many reaction pathways have been elucidated.

Rate equations for a reversible reaction

$$a\mathrm{A} + b\mathrm{B} \leftrightharpoons c\mathrm{C} + d\mathrm{D} \tag{3.5}$$

can be written for both directions:

$$\begin{aligned} \text{forward rate} &= k_1\ [\mathrm{A}]^a\ [\mathrm{B}]^b \\ \text{backward rate} &= k_2\ [\mathrm{C}]^c\ [\mathrm{D}]^d. \end{aligned} \tag{3.6}$$

At equilibrium the concentrations of all the species remain constant, implying that the forward and backward rates are equal. The equilibrium is dynamic and therefore neither rate is zero. From (3.6) we have

$$K = \frac{k_1}{k_2} = \frac{[\mathrm{C}]^c\ [\mathrm{D}]^d}{[\mathrm{A}]^a\ [\mathrm{B}]^b}. \tag{3.7}$$

Thus the concentrations of the reactants at equilibrium can be expressed in terms of the equilibrium constant $K$. The numerical value of $K$ depends on the units used, usually molalities for solutions, and on the temperature. Conventionally the value of $K$ also incorporates constant concentrations such as pure solids and solvent water.

The ability to calculate reactant concentrations from equilibrium constants is often required. It should be noted that eqn (3.7) applies only to equilibrium situations, but this is not a severe restriction as many reactions of hydrochemical interest are sufficiently rapid to reach equilibrium. The principles are best illustrated by example.

*Example*. The solubility product of silver chloride is given by $\log K = -9.8$. What is the concentration of silver ions in a saturated solution of silver chloride in distilled water and in seawater?
(The equilibrium constant of a dissolution reaction is known as the solubility

† Molality (m) and molarity (M) are defined in Section 5.1.

product. Because of the wide range of values for equilibrium constants and the mathematics involved in their calculation, they are normally reported as logarithms.)

The dissolution reaction is

$$AgCl \rightleftharpoons Ag^+ + Cl^- \quad (3.8)$$

i.e.

$$K = [Ag^+]\,[Cl^-].$$

(By convention the concentration of solid AgCl is contained in the equilibrium constant.)

Dissolution of AgCl in pure water produces equal quantities of $Ag^+$ and $Cl^-$, i.e.

$$[Ag^+] = [Cl^-].$$

Therefore at equilibrium

$$\begin{aligned} K &= [Ag^+]^2 \\ [Ag^+] &= \sqrt{K} \\ &= \sqrt{10^{-9.8}} \\ &= 1.26 \times 10^{-5} \text{ mol } Ag^+ \text{ kg}^{-1} \\ &= 1.26 \times 10^{-5} \times 108 \times 10^3 \text{ mg } Ag^+ \text{ kg}^{-1} \\ &= 1.36 \text{mg } Ag^+ \text{kg}^{-1} \text{ in distilled water.} \end{aligned}$$

Seawater already contains 0.535 mol $Cl^-$ $kg^{-1}$, which is so large that the additional contribution from silver chloride dissolution can be neglected. From (3.8)

$$[Ag^+] = \frac{K}{[Cl^-]}.$$

Substituting the $[Cl^-]$ value of seawater gives

$$\begin{aligned} [Ag^+] &= \frac{10^{-9.8}}{0.535} \\ &= 2.9 \times 10^{-10} \text{ mol } Ag^+ \text{ kg}^{-1} \\ &= 3.2 \times 10^{-5} \text{ mg } Ag^+ \text{ kg}^{-1} \text{ in seawater.} \end{aligned}$$

This considerable reduction in solubility illustrates the common ion effect: if two salts share a common ion, they will mutually reduce each other's solubility (the effect on sodium chloride in seawater is negligible in this case).

Returning to the theoretical discussion of chemical equilibrium, chemical equilibrium can be defined in terms of energy. By analogy with mechanical systems, the most stable composition of a mixture of reactants is the composition having the lowest energy. It is therefore necessary to define the energy of chemical systems. Many chemical reactions produce obvious energy changes—for example the combustion of fuel in an engine produces mechanical work and some heat.

A rigorous thermodynamic treatment of energy involves three parameters:

*enthalpy*, defined as the heat content at constant pressure, *entropy*, a measure of the disorder of a system, and *Gibbs' free energy*, defined by

$$G = H - TS \tag{3.9}$$

where $G$ is the free energy (kJ mol$^{-1}$), $H$ is the enthalpy (kJ mol$^{-1}$), $T$ is the absolute temperature, and $S$ is the entropy (kJ K$^{-1}$ mol$^{-1}$). A derivation of this equation is beyond the scope of this book.

It is not actually possible to measure enthalpy and free energy values, but changes during a reaction can be measured. Therefore, by convention, elements in their *standard states* (their most stable form at 25 °C and 1 atm pressure) are assigned enthalpy and free energy values of zero. By using this approach values can be measured for most substances; those of hydrochemical interest are presented in Appendix I.

Free energies, enthalpies, and entropies can be added and subtracted to derive the changes for any reaction.

*Example.* Evaluate the enthalpy change in the reaction

$$\underset{\text{gas}}{CO_2} + \underset{\text{calcite}}{CaCO_3} + H_2O \rightarrow \underset{\text{aq}}{Ca^{2+}} + \underset{\text{aq}}{2HCO_3^-}$$

(Note that the states of the reactants are specified.) The enthalpies of each side of the equation are totalled, taking into account stoichiometric coefficients:

$$\underset{(-393)}{CO_2} + \underset{(-1205)}{CaCO_3} + \underset{(-286)}{H_2O} \rightarrow \underset{(-543)}{Ca^{2+}} + \underset{2(-690)}{2HCO_3^-}$$

$$-1884 \qquad\qquad -1923$$

$$\begin{aligned}\Delta H &= H\text{ products} - H\text{ reactants}\\ &= -1923 + 1884\\ &= -39\text{ kJ.}\end{aligned}$$

At constant temperature

$$\Delta G = \Delta H - T\Delta S. \tag{3.10}$$

At equilibrium a chemical system has minimum free energy, which leads to the relation

$$\Delta G = -RT\ln K \tag{3.11}$$

where $R$ is the gas constant (8.3143 J K$^{-1}$ mol$^{-1}$) and $K$ is the equilibrium constant defined by eqn (3.7). This is a useful result since standard thermodynamic data are more readily available than equilibrium constants.

The temperature dependence of an equilibrium constant is related to the enthalpy change by the van 't Hoff equation

$$\ln\left(\frac{K_{T_2}}{K_{T_1}}\right) = \frac{-\Delta H^\circ}{R}\left(\frac{1}{T_2} - \frac{1}{T_1}\right) \tag{3.12}$$

assuming $\Delta H$ is not a function of temperature.† More generally

$$\ln K_T = \frac{\Delta H^\circ}{RT} + A\ln T + BT + CT^2 \ldots + I. \tag{3.13}$$

where $A, B, C \ldots I$ are constants.

The constants in equations of this type are obtained by fitting experimental data. This has been done for several of the more important equilibrium constants in hydrochemistry. If such an equation is available, it can be expected to be more accurate than calculations based on basic thermodynamic data.

## 3.3. Activity in solution

Reference has already been made to the fact that the effective concentrations of ions in solution differ from their actual concentration (Section 2.2). The effective concentration of a dissolved substance is referred to as its *activity*. Activity $a$ and molality $m$ are related by an activity coefficient‡

$$a = \gamma\, m. \tag{3.14}$$

The activity of a substance is the correct quantity to use in a thermodynamic relationship, but it will be seen that the actual concentration is often needed in addition. Throughout this book activity and concentration will be distinguished by the use of round and square brackets respectively: eqn (3.14) can be written

$$(\mathrm{A}) = \gamma_\mathrm{A}\,[\mathrm{A}].$$

*Example.* Calculate the equilibrium constant for the conversion of calcite to dolomite by magnesium ions. At what magnesium-to-calcium activity ratio should the transition take place at 25 °C and at 200 °C. Relevant data are as follows.

| | $\Delta H_f^\circ$ (kJ mol$^{-1}$) | $\Delta S^\circ$ (J K$^{-1}$ mol$^{-1}$) |
|---|---|---|
| $Ca^{2+}$ | −542.83 | −53 |
| $CaCO_3$ | −1207.4 | 91.7 |
| $Mg^{2+}$ | −466.8 | −138 |
| $CaMg(CO_3)_2$ | −2324.5 | 155.2 |

$\Delta H_f^\circ$ is standard enthalpy of formation.

† $\Delta H^\circ$ is $\Delta H$ at standard state (25 °C, one atmosphere and one mole of compound).

‡ Thermodynamic equations are strictly valid only for molalities, but under the physical conditions involved in natural hydrochemistry the differences between molality $m$ and molarity M (related by the density of water) are usually negligible. It is experimentally more convenient to use molarities, which will be used in this text.

$\Delta G$ is calculated from eqn (3.10). We need $\Delta H$ and $\Delta S$, which are calculated from the equation for the reaction

$$2CaCO_3 + Mg^{2+} \rightleftharpoons CaMg(CO_3)_2 + Ca^{2+}.$$

$H$ 2 (−1207.4) (−446.8) (−2324.5) (−542.83) $\Delta H = 14.27$ kJ

$S$ 2 ( 91.7) (−138) ( 155.2) ( −53) $\Delta S = 56.8$ J K$^{-1}$

$$\begin{aligned}\Delta G &= \Delta H - T\Delta S \\ &= 14.27 \times 10^3 - 298 \times 56.8 \\ &= -2.66 \text{ kJ}.\end{aligned}$$

Using eqn (3.11)

$$\begin{aligned}\ln K &= \frac{-\Delta G}{RT} \\ &= \frac{2.66 \times 10^3}{8.314 \times 298} \\ &= 1.074\end{aligned}$$

$$K = 2.96 \text{ and } \log K = 0.466 \text{ at } 25\ ^\circ\text{C}.$$

We also need $K$ at 200 °C. This can be obtained either from eqn (3.10) or from the van't Hoff equation (eqn (3.12)).

Using eqn (3.10)

$$\begin{aligned}\text{at } 473\text{ K} \quad \Delta G &= 14.27 \times 10^3 - 473 \times 56.8 \\ &= -12.596 \text{ kJ} \\ \ln K &= \frac{-\Delta G}{RT} \\ &= \frac{12.596 \times 10^3}{8.314 \times 473} \\ &= 3.203.\end{aligned}$$

$$K = 24.607 \text{ and } \log K = 1.391 \text{ at } 473 \text{ K}$$

Using eqn (3.12)

$$\begin{aligned}\ln K_{473} &= \frac{-14.27 \times 10^3}{8.314}\left(\frac{1}{473} - \frac{1}{298}\right) + 1.074 \\ &= 3.204\end{aligned}$$

$$K = 24.631 \text{ and } \log \text{K} = 1.391 \text{ at } 473 \text{ K}$$

From the reaction equation

$$K = \frac{(Ca^{2+})}{(Mg^{2+})}.$$

Thus the $Ca^{2+}$-to-$Mg^{2+}$ activity ratio at the transition is 2.9 at 25 °C and 24.6 at

200 °C. The transition of calcite to dolomite is therefore favoured by higher temperature. It is interesting to note that the molar Ca-to-Mg ratio of seawater is 0.185 and the activity ratio is 0.165. Dolomitization of calcite by seawater is therefore thermodynamically favoured even at 25 °C.

From the discussion in Section 2.2, it is expected that the activity coefficient of an ion depends on the other ions in solution. The appropriate measure of the concentration of ions is the *ionic strength*, defined by

$$I = \tfrac{1}{2} \sum_i m_i z_i^2 . \tag{3.15}$$

The activity of an ion approaches its concentration, i.e. $\gamma$ tends to unity, as the solution becomes more dilute. The activity equals the concentration at infinite dilution. At normal concentrations, activity coefficients of ions are less than unity. The activity coefficient of an ion is related to the ionic strength by the extended Debye-Hückel equation

$$\log \gamma_i = -Az_i^2 \left( \frac{\sqrt{I}}{1 + Ba_i\sqrt{I}} \right) \quad I < 0.1. \tag{3.16}$$

$A$ and $B$ are temperature-dependent parameters involving universal constants and the physical properties of water. Values are given in Table 3.1. The parameter

**Table 3.1** *Constants for the extended Debye–Hückel equation*

Variation of A and B with temperature

| $t$ (°C) | $A$ | $B$ | $t$ (°C) | $A$ | $B$ |
|---|---|---|---|---|---|
| 0 | 0.4917 | 0.3248 | 22 | 0.5083 | 0.3282 |
| 2 | 0.4931 | 0.3251 | 24 | 0.5100 | 0.3285 |
| 4 | 0.4945 | 0.3254 | 26 | 0.5117 | 0.3288 |
| 6 | 0.4959 | 0.3257 | 28 | 0.5134 | 0.3291 |
| 8 | 0.4973 | 0.3260 | 30 | 0.5152 | 0.3294 |
| 10 | 0.4988 | 0.3263 | 32 | 0.5170 | 0.3297 |
| 12 | 0.5003 | 0.3266 | 34 | 0.5188 | 0.3301 |
| 14 | 0.5019 | 0.3269 | 36 | 0.5207 | 0.3304 |
| 16 | 0.5034 | 0.3272 | 38 | 0.5226 | 0.3307 |
| 18 | 0.5050 | 0.3276 | 40 | 0.5245 | 0.3310 |
| 20 | 0.5067 | 0.3279 | | | |

*Ionic size parameter $a_i$*

| $a_i$ | Ion | $a_i$ | Ion |
|---|---|---|---|
| 9 | $H^+$ | 5 | $Ba^{2+}$, $Sr^{2+}$, $Pb^{2+}$, $CO_3^{2-}$ |
| | $Al^{3+}$, $Fe^{3+}$, $La^{3+}$, $Ce^{3+}$ | 4 | $Na^+$, $HCO_3^-$, $SO_4^{2-}$ |
| 8 | $Mg^{2+}$, $Be^{2+}$ | 3 | $K^+$, $Ag^+$, $NH_4^+$, $OH^-$, $Cl^-$, $I^-$ |
| 6 | $Ca^{2+}$, $Zn^{2+}$, $Cu^{2+}$, $Sn^{2+}$, $Mn^{2+}$, $Sn^{2+}$ | | |

After Kielland 1937.

$a_i$ is the Debye-Hückel radius of the ion, which differs from its actual radius, and $z_i$ is its charge. These parameters are tabulated in Table 3.1. The empirical Davies equation (eqn (3.17)) gives a better result at higher ionic strengths:

$$\log \gamma_i = -Az_i^2 \left( \frac{\sqrt{I}}{1 + Ba_i\sqrt{I}} \right) + b_i I \quad I<0.5. \tag{3.17}$$

$A$ and $B$ are the same as in eqn (3.16) and the parameters $a_i$ and $b_i$ are given in Table 3.2. Example curves produced by the Davies equation are shown in

**Table 3.2** *Constants for the Davies equation*

| Ion | $a_i$ | $b_i$ |
|---|---|---|
| $Ca^{2+}$ | 5.0 | 0.165 |
| $Mg^{2+}$ | 5.5 | 0.2 |
| $Na^+$ | 4.0 | 0.075 |
| $K^+$ | 3.5 | 0.015 |
| $Cl^-$ | 3.5 | 0.015 |
| $SO_4^{2-}$ | 5.0 | −0.04 |
| $HCO_3^-$ | 5.4 | 0.0 |
| $CO_3^{2-}$ | 5.4 | 0.0 |

Fig. 3.1. The Davies equation is the better of the two although data for more ions are available for the Debye-Hückel equation. For most everyday work a value taken from the curves of Fig. 3.1 is sufficiently accurate.

The activities of uncharged species are also affected by dissolved ions because polar water molecules are involved in hydration sheaths around ions. This water, although not chemically bound, is less available for chemical reaction.

The activity of water is approximated by

$$(H_2O) = 1 - 0.017 \sum_i m_i. \tag{3.18}$$

Conversely, the activity of other uncharged species is increased by ionic strength:

$$\log \gamma_i = 0.1\, I. \tag{3.19}$$

*Example*. Evaluate the activities of calcium and water in this analysis (concentrations in mg l$^{-1}$): $Ca^{2+}$, 167; $Mg^{2+}$, 5.6; $Na^+$, 20; $K^+$, 2.5; $HCO_3^-$, 365; $SO_4^{2-}$, 53; $Cl^-$, 33; temperature, 10 °C; conductivity, 640 $\mu$S cm$^{-1}$.

It is first necessary to convert the concentrations to molalities. Since the density of the water is not given and the water is quite dilute, it can be assumed that molality equals molarity.

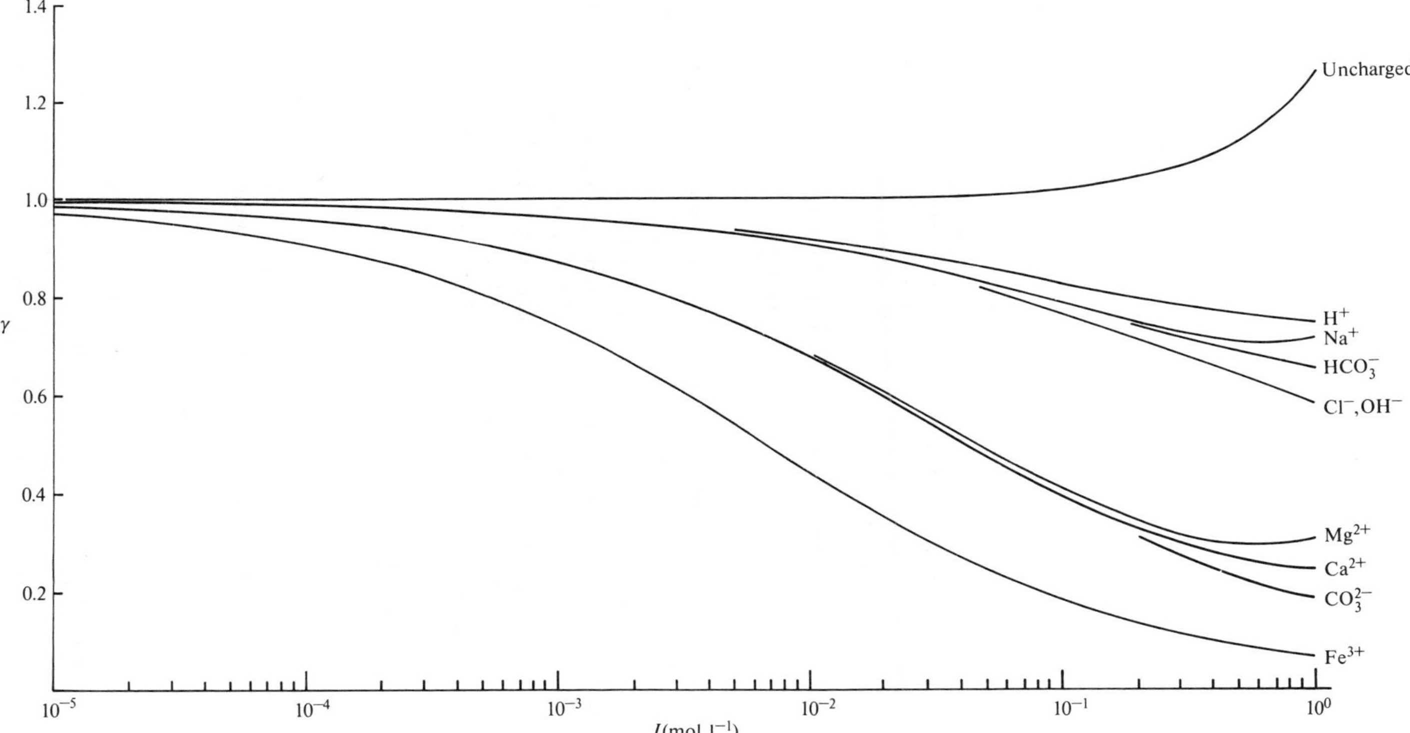


**Fig. 3.1.** Activity coefficients as a function of ionic strength.

The calculation is easily carried out in tabular form.

| Ion | mg $l^{-1}$ | Molecular weight | mol $kg^{-1}$ | $z$ | $mz^2$ |
|---|---|---|---|---|---|
| Ca | 167 | 40 | $4.175 \times 10^{-1}$ | +2 | $1.670 \times 10^{-2}$ |
| Mg | 5.6 | 24 | $2.333 \times 10^{-4}$ | +2 | $9.333 \times 10^{-4}$ |
| Na | 20 | 23 | $8.696 \times 10^{-4}$ | +1 | $8.696 \times 10^{-4}$ |
| K | 2.5 | 39 | $6.410 \times 10^{-5}$ | +1 | $6.410 \times 10^{-5}$ |
| $HCO_3^-$ | 365 | 61 | $5.984 \times 10^{-3}$ | −1 | $5.984 \times 10^{-3}$ |
| $SO_4^{2-}$ | 53 | 96 | $5.521 \times 10^{-4}$ | −2 | $2.208 \times 10^{-3}$ |
| $Cl^-$ | 33 | 35.5 | $9.26 \times 10^{-4}$ | −1 | $9.296 \times 10^{-4}$ |
| | | | $\Sigma m_i = 1.281 \times 10^{-2}$ | | $\frac{1}{2}\Sigma mz^2 = 1.384 \times 10^{-2}$ |

Since $I < 0.1$, eqn (3.16) is used. Substituting values for $Ca^{2+}$ from Table 3.1:

$$\log \gamma_{Ca} = -0.4988 \times (2)^2 \times \frac{\sqrt{(1.384 \times 10^{-2})}}{1 + 0.3263 \times 6 \times \sqrt{(1.384 \times 10^{-2})}}$$

$$\gamma = 0.64.$$

The activity of water is given by eqn (3.17). On substituting we obtain;

$$(H_2O) = 1 - 0.017 \times 1.281 \times 10^{-2}$$
$$= 0.9978.$$

N.B. This small difference from unity confirms the assumption that $(H_2O) = 1$ for most natural fresh waters.

The ionic strength of a water can be related empirically to its electrical conductivity. The relationship depends on the type of chemistry the water exhibits, and therefore it is best to deduce the relationship for a particular water type from several complete analyses before extrapolating to partial analyses. The technique can be useful if analytical time is at a premium or for historic analyses. For the example above, $I = 2.16 \times 10^{-3} \times$ conductivity ($\mu$S $cm^{-1}$).

## 3.4. Acid–base equilibria

The general concepts of acid–base equilibria were covered in Section 2.4 where the fundamental role of the hydrogen (hydronium) ion was explained. Because hydrogen ion concentrations vary over many orders of magnitude the p notation (eqn (3.20)) is universally used and the hydrogen ion concentration becomes pH:

$$pX = -\log X. \tag{3.20}$$

In the original definition of pH, the quantity was p[H], i.e. the actual concentration of hydrogen ions was used. However, with the advent of electrodes for the measurement of pH, which actually measure p(H), i.e. the hydrogen ion activity, the convention that pH represents p(H) has become widespread in hydrochemistry. Other conventions in use are described by Stumm and Morgan (1981) (Section 3.4).

The ionization of an acid in water is represented by the equation

$$HB + H_2O \rightleftharpoons H_3O^+ + B^- \tag{3.21}$$

for which an *acidity constant* $K_A$ can be defined:

$$K_A = \frac{(H^+)(B^-)}{(HB)}. \tag{3.22}$$

By convention, $(H^+) = (H_3O^+)$, which is equivalent to stating that $\Delta G$ for the reaction

$$H^+ + H_2O \rightleftharpoons H_3O^+$$

is zero. In dilute aqueous solution the concentration of water (55.5 mol kg$^{-1}$) is effectively constant and is conventionally absorbed in the equilibrium constant. The acidity constant $K_A$ is often expressed as $pK_A$:

$$pK_A = -\log K_A = pH + \log \frac{(HB)}{(B^-)}. \tag{3.23}$$

From eqn (3.23) it can be seen that when the solution pH equals $pK_A$, then (HB) and $(B^-)$ are equal. Acids with $pK_A < 0$ are known as strong acids; they are almost completely dissociated in aqueous solution. The conjugate bases of acids with $pK_A > 14$ are known as strong bases; their aqueous solutions contain almost no hydrogen ions.

The self-ionization of water is represented by

$$H_2O \rightleftharpoons H^+ + OH^- : K_w = (H^+)(OH^-). \tag{3.24}$$

The variation of $K_w$ with temperature is tabulated in Table 3.3. In pure water the ionic strength is very low and hence $(H^+) = [H^+]$ and $(OH^-) = [OH^-]$. Since $(H^+) = (OH^-)$ at neutrality, it is easy to show that at the neutral point $pH = pK_w/2$.

**Table 3.3** *Variation of* $pK_w$ *with temperature*

| $T$(°C) | $pK_w$ | pH of neutral solution |
|---|---|---|
| 0 | 14.93 | 7.465 |
| 5 | 14.73 | 7.365 |
| 10 | 14.53 | 7.265 |
| 15 | 14.35 | 7.175 |
| 20 | 14.17 | 7.085 |
| 25 | 14.00 | 7.000 |
| 30 | 13.83 | 6.915 |
| 50 | 13.26 | 6.630 |

After Harned and Owen 1958.

To gain a better understanding of the behaviour of acids and bases, it will be useful to explore the equilibrium of a hypothetical weak acid HB, for which $pK_A = 6$. Consider a $10^{-3}$ M solution of this acid in water. The following equations describe the composition of the system:

ionization of acid $\quad HB \rightleftharpoons H^+ + B^- : K_A = \dfrac{(H^+)(B^-)}{(HB)}$ (3.25)

ionization of water $\quad H_2O \rightleftharpoons H^+ + OH^- : K_w = (H^+)(OH^-)$ (3.24)

mass balance of acid $\quad C_B = [HB] + [B^-] = 10^{-3}$ M (3.26)

proton condition $\quad [H^+] = [B^-] + [OH^-]$. (3.27)

These four equations allow the four unknowns $(H^+)$, $(OH^-)$, (HB), and $(B^-)$ to be evaluated, if the simplifying assumption that all activity coefficients are unity is made.

Eqns. (3.24) and (3.27) give

$$[B^-] = [H^+] - \frac{K_w}{[H^+]}. \tag{3.28}$$

Eqns. (3.25) and (3.26) give

$$[B^-] = \frac{K_A(C_B - [B^-])}{[H^+]}. \tag{3.29}$$

Substituting (3.28) in (3.29) gives

$$[H^+] - \frac{K_w}{[H^+]} = \frac{K_A}{[H^+]}\left(C_B + \frac{K_w}{[H^+]} - [H^+]\right) \tag{3.30}$$

or

$$[H^+]^3 + K_A[H^+]^2 - [H^+](K_w + K_A C_B) - K_A K_w = 0.$$

Substituting numerical values gives

$$[H^+]^3 + 10^{-6}[H^+]^2 - 10^{-9}[H^+] - 10^{-20} = 0.$$

This cubic equation is most readily solved either by plotting $f([H^+])$ against $[H^+]$ to locate the value of $[H^+]$ for which $f([H^+])$ is zero or by using more sophisticated computer techniques. These methods give the following results:

$[H^+] = 3.1 \times 10^{-5}$ M, pH = 4.51 $\quad [OH^-] = 3.2 \times 10^{-10}$ M

$[HB] = 9.6 \times 10^{-4}$ M $\quad [B^-] = 3.1 \times 10^{-5}$ M.

Inspection of these values shows that an approximate solution could have been obtained more readily, since $[OH^-] \ll [B^-]$. Using this approximation, Eqn (3.27) becomes

$$[H^+] = [B^-]$$

Since [HB] $\gg$ [B$^-$], eqn (3.26) becomes

$$C_B = [HB]$$

Substituting in (3.25) gives $[H^+]^2 = K_A C_B$, whence $[H^+] = 3.16 \times 10^{-5}$M, pH = 4.5. The error of 2 per cent is within the experimental error of normal hydrochemical measurements.

A particularly interesting situation arises when the solution contains significant quantities of both HB and B$^-$, for example when NaB, which dissociates completely, is added to a solution of HB. From eqn (3.23)

$$\mathrm{pH} = \mathrm{p}K_A - \log\left(\frac{[\mathrm{HB}]}{[\mathrm{B}^-]}\right). \tag{3.31}$$

The pH of the solution is defined by the ratio of the species. This is clearly invariant as the concentration of the solution is changed as long as [B$^-$] remains large compared with [OH$^-$] produced by dissociation of water. The pH of the solution is also relatively unchanged by the addition of strong acid or base, provided that the quantity added is small compared with $C_B$. Suppose a solution contains $10^{-3}$ M HB and $10^{-3}$ M NaB. The concentration of B$^-$ produced by dissociation of HB can be neglected by comparison with B$^-$ produced by complete dissociation of NaB. Hence the pH of the solution from eqn (3.31) is

$$\mathrm{pH} = 6 - \log\left(\frac{[10^{-3}]}{[10^{-3}]}\right)$$
$$= 6.$$

If $10^{-4}$ mol l$^{-1}$ of a strong acid such as HCl is added, the effective reaction is

$$H^+ + B^- \rightarrow HB.$$

Thus (HB) becomes $1.1 \times 10^{-3}$ M and (B$^-$) becomes $0.9 \times 10^{-3}$ M. The new pH is 5.91.

This is a pH change of less than 0.1 units. If the same quantity of HCl had been added to pure water the pH would have changed from 7 to 4, a change of 3 units. The presence of the weak acid and its conjugate base *buffers* the solution against changes in pH. Solutions are also buffered if they contain a weak base and its conjugate acid and these solutions are alkaline. A *buffered solution* is a solution for which the pH is resistant to change. A little thought shows that the maximum buffering behaviour of a solution occurs when the acid and base are present in equal concentrations. More precisely, it can be shown that the buffer intensity $\beta$ of a solution is given by

$$\beta = \frac{\mathrm{d}C_A}{\mathrm{dpH}} = 2.303\left([\mathrm{H}^+] + [\mathrm{OH}^-] + \frac{[\mathrm{HB}]\,[\mathrm{B}^-]}{[\mathrm{HB}] + [\mathrm{B}^-]}\right) \tag{3.32}$$

where dpH is the pH change produced by adding $\mathrm{d}C_A$ mol l$^{-1}$ of strong acid.

This equation shows that the buffer capacity of a solution also becomes large at extreme pH values.

The most important acid-base system in the hydrochemistry of most natural waters is the carbonate system and it will therefore be described in some detail as a practical example of the theory discussed above. Carbonic acid is produced by the reaction of water with carbon dioxide, which comprises 0.03 volume per cent of normal air rising to several volume per cent in soil air. Carbonic acid is a dibasic acid, i.e. it is capable of donating two protons, and most groundwater systems contain solid carbonate, usually calcium carbonate as calcite. Thus the carbonate system is quite complex.

Carbon dioxide dissolves in water largely without reaction, carbonic acid being formed in only small quantities:

$$CO_2(g) \rightleftharpoons CO_2(aq) : K_{CO_2} = \frac{(CO_2(aq))}{(CO_2(g))} \tag{3.33}$$

$$CO_2(aq) + H_2O \rightleftharpoons H_2CO_3 : K = \frac{(H_2CO_3)}{(CO_2(aq))}. \tag{3.34}$$

The concentration of $H_2CO_3$ is difficult to measure analytically and therefore a composite parameter, total dissolved $CO_2$ or 'free $CO_2$' defined by eqn (3.35) is used:

$$[H_2CO_3^*] = [CO_2(aq)] + [H_2CO_3]. \tag{3.35}$$

At normal temperatures and pressures $[H_2CO_3^*] \approx [CO_2(aq)]$. Instead of using the carbon dioxide concentration as in eqn (3.33), it is more convenient to use the partial pressure of the gas. A new equilibrium constant, Henry's Law constant $K_H$ can be defined using the measured parameters:

$$K_H = \frac{(H_2CO_3^*)}{P_{CO_2}} \tag{3.36}$$

The ionization of dissolved $CO_2$ is described by eqns (3.37) and (3.38)

$$H_2CO_3^* \rightleftharpoons H^+ + HCO_3^- \quad K_1 = \frac{(H^+)(HCO_3^-)}{(H_2CO_3^*)} \tag{3.37}$$

$$HCO_3^- \rightleftharpoons H^+ + CO_3^{2-} \quad K_2 = \frac{(H^+)(CO_3^{2-})}{(HCO_3^-)}. \tag{3.38}$$

Values for these constants are given in Table 3.4. The relative distribution of the three carbonate species, $H_2CO_3^*$, $HCO_3^-$, and $CO_3^{2-}$ are shown in Fig. 3.2. A logarithmic composition diagram† for a total inorganic carbon content of $\Sigma$ of $10^{-3}$ M is given in Fig. 3.3. In this figure it is assumed that the system has no carbon dioxide gas phase, i.e. that the system is *closed*. Alternatively a system can be envisaged where the partial pressure of carbon dioxide is kept constant;

† See Stumm and Morgan (1981) for a description of logarithmic composition diagrams.

**Table 3.4** *Equilibrium constants of the carbonate system*

| $t$ (°C) | $-\log K_H$ | $-\log K_1$ | $-\log K_2$ |
|---|---|---|---|
| 0 | 1.112 | 6.579 | 10.628 |
| 5 | 1.193 | 6.518 | 10.556 |
| 10 | 1.269 | 6.466 | 10.490 |
| 15 | 1.339 | 6.421 | 10.431 |
| 20 | 1.404 | 6.383 | 10.378 |
| 25 | 1.464 | 6.352 | 10.331 |
| 30 | 1.520 | 6.327 | 10.289 |
| 40 | 1.619 | 6.296 | 10.221 |
| 50 | 1.702 | 6.287 | 10.172 |

$-\log K_H = 14.0184 - 0.015264\,T - 2385.73/T$
$-\log K_1 = 14.8435 + 0.032786T + 3404.71/T$
$-\log K_2 = 6.498 + 0.02379T + 2902.39/T$
$T\,K = 273.16 + t\,°C.$

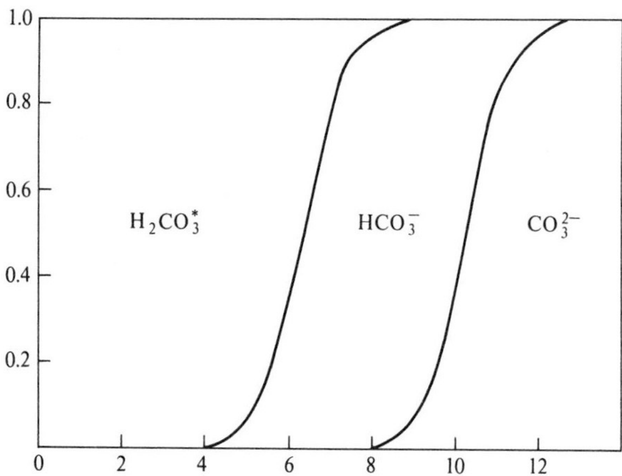


**Fig. 3.2.** Distribution of major inorganic carbon species.

the logarithmic composition diagram for this *open* system is given in Fig. 3.4. As the pH is increased (by the addition of strong alkali) $[H_2CO_3^*]$ remains constant, since this is determined by equilibrium with the gas phase, but $\Sigma$ increases because of the contribution from $[HCO_3^-]$ and $[CO_3^{2-}]$.

The analytical methods used to determine the carbonate system illustrate some of the uses of the composition diagram. The carbonate system contains at least five species: $H_2CO_3^*$, $HCO_3^-$, $CO_3^{2-}$, $OH^-$, $H^+$, and perhaps a metal cation. Because these species are not all independent the system can be specified by any two of them. The hydrogen ion can be measured by electrochemical methods,

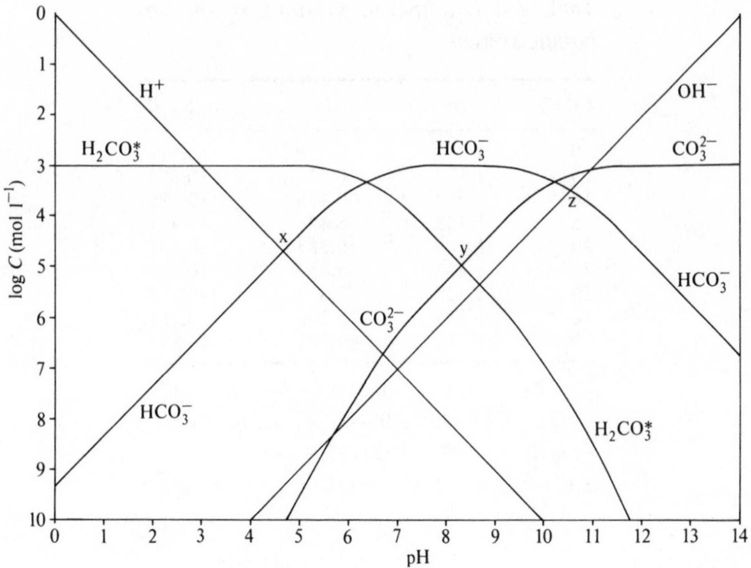


Fig. 3.3. Log composition diagram for $H_2CO_3^*$ (closed system; $\Sigma = 10^{-3}$ M).

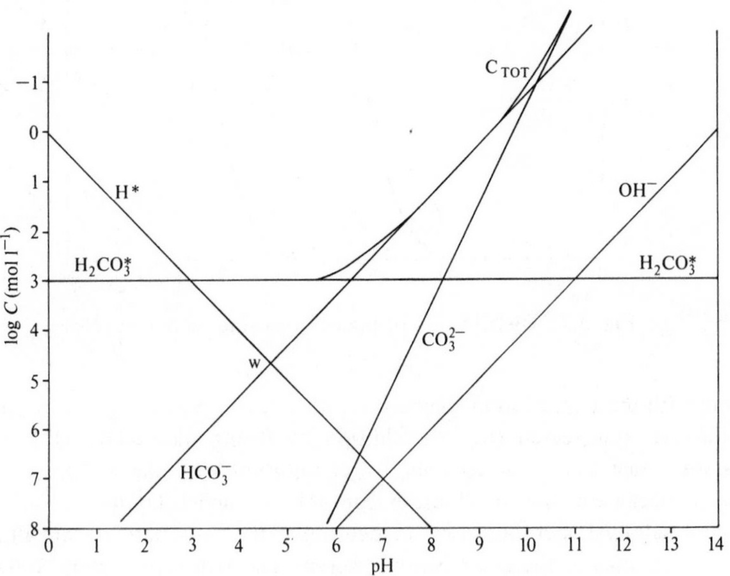


Fig. 3.4. Log composition diagram for $H_2CO_3^*$ in contact with soil air ($P_{CO_2} = 10^{-1.5}$).

but none of the other species can be determined directly. The usual technique is to titrate the solution with either strong acid or strong alkali and to deduce the species from the titration results. Reaction of atmospheric carbon dioxide with the solution is quite slow and therefore the reaction approximates to closed-system conditions (Fig. 3.3) provided that the solution is not agitated violently.

Consider first the titration of an alkaline ($pH>11$) solution with strong acid: as hydrogen ions are added $OH^-$ ions initially react to produce water until the pH determined by point z in Fig. 3.3 is reached. This corresponds to the proton condition for a pure carbonate solution such as $Na_2CO_3$:

$$2[H_2CO_3^*] + [HCO_3^-] + [H^+] = [OH^-]. \qquad (3.39)$$

The quantity of acid required to achieve this condition is known as the caustic alkalinity of the solution. Unfortunately, at this pH ($\approx$10.3) the buffer capacity of most natural waters is too high for the end point to be detected by indicators or potentiometrically.

Addition of further acid results mainly in the reaction of $CO_3^{2-}$ ions to produce $HCO_3^-$. At the pH determined by point y in Fig. 3.3 the proton condition corresponding to a pure bicarbonate solution is satisfied:

$$[H_2CO_3^*] + [H^+] = [OH^-] + [CO_3^{2-}]. \qquad (3.40)$$

The rapid change of pH with quantity of acid at this point is commonly determined using the indicator phenolphthalein, and hence the total quantity of acid required is known as phenolphthalein alkalinity (p-alkalinity).

Further addition of acid results in the reaction of $HCO_3^-$ to produce $H_2CO_3^*$. At the pH determined by point x the proton condition corresponding to a pure $CO_2$ solution is satisfied:

$$[H^+] = [HCO_3^-] + 2[CO_3^{2-}] + [OH^-]. \qquad (3.41)$$

This end point is commonly determined with the indicator methyl orange. The total quantity of acid required to reach this point is known as the (total) alkalinity of the solution. At this point the total quantity of acid added equals the quantity of strong alkali that would have to be added to a $CO_2$-$H_2O$ solution to produce the original solution, hence the term alkalinity. The proton level specified by $CO_2$-$H_2O$ is taken as a reference level. Addition of further strong acid beyond this point produces mineral acidity.

Titration with strong alkali reverses all the above reactions. The point x corresponds to the reaction of protons produced by stronger acids than the reference $CO_2$-$H_2O$; this is mineral acidity. The point y is reached when $H_2CO_3^*$ has reacted to produce $HCO_3^-$ and the additional alkali required is therefore known as the $CO_3^{2-}$ acidity. The point z corresponds to the reaction of $HCO_3^-$ to produce $CO_3^{2-}$ and the additional alkali required is therefore the bicarbonate acidity, although this point is not experimentally determinable as mentioned above. Further addition of alkali produces caustic alkalinity. It is clear that a solution possessing caustic alkalinity has no acidity, and that a solution possessing mineral acidity has no alkalinity.

The discussion above shows that the only parameters of the carbonate system unequivocably determinable by titration are $CO_2$-acidity, and the difference between the p-alkalinity and total alkalinity (p-alkalinity may include caustic alkalinity). If the pH of the solution is known, the concentrations of the individual species can be calculated from either of the two measurable parameters with a knowledge of the equilibrium constants. Examination of Fig. 3.3 shows that the total alkalinity is almost identical with $[HCO_3^-]$ for pH $<9$ and similarly that $CO_2$ acidity is almost identical with $[H_2CO_3^*]$ for $5.7 < \text{pH} < 7.6$ and therefore more elaborate calculations are unnecessary for many natural waters. If caustic alkalinity is absent, i.e. pH $<10$, p-alkalinity is almost identical with $[CO_3^{2-}]$. It should be noted that for solutions of original pH in the range 7.6 to 9.0, the determinations of $CO_2$ acidity and p-alkalinity are experimentally imprecise.

Whether alkalinity or $CO_2$ acidity is measured is determined by practical considerations. Because $CO_2$-$H_2O$ is the reference point for the alkalinity titration, alkalinity is unaffected by gains or losses of carbon dioxide. Because most groundwaters are in equilibrium with pressures of carbon dioxide greater than atmospheric, loss of carbon dioxide between sampling and analysis is a real possibility. The use of the alkalinity titration circumvents this, provided that the original pH is known and that no precipitation of solid carbonate occurs. In the latter case, slow titration in the original bottle will allow the carbonate to redissolve. For low pH waters (pH $<5.7$), the $CO_2$-acidity titration is experimentally more precise than the alkalinity titration and is therefore preferred. However, such waters are particularly likely to lose $CO_2$ on storage and are therefore best analysed in the field.

Groundwaters containing dissolved inorganic carbon are usually produced by the reaction of water and carbon dioxide with solid carbonates. This is obviously important in limestone and dolomite aquifers, and many sandstones contain either carbonate cement or detrital carbonate—even basalts may have calcite-filled vesicles. The reactions with calcite (rhombohedral $CaCO_3$), the most common carbonate, can be considered typical.

The addition of calcite to the $CO_2$-$H_2O$ system adds two equations, the dissociation equation (eqn 3.42) and the charge balance equation (eqn 3.43):

$$CaCO_3 \rightleftharpoons Ca^{2+} + CO_3^{2-} : K_C = (Ca^{2+})(CO_3^{2-}) \tag{3.42}$$

$$2[Ca^{2+}] + [H^+] = [OH^-] + [HCO_3^-] + 2[CO_3^{2-}]. \tag{3.43}$$

Two reaction schemes are of particular interest to the groundwater hydrochemist: firstly, the reaction of water with soil zone carbon dioxide in the absence of calcite followed by the solution achieving equilibrium with calcite out of contact with the carbon dioxide reservoir, and secondly the three-phase equilibrium between solution, soil zone carbon dioxide, and soil zone calcite. Most natural systems fall somewhere between these two schemes.

The first stage of scheme 1 is illustrated by the open-system composition diagram given in Fig. 3.4. The dissolution of carbon dioxide results in a pure $CO_2$ solution, for which the proton condition is given by eqn (3.41). This is satisfied by point w in Fig. 3.4. This diagram has been constructed using a typical carbon dioxide partial pressure for soil air in temperate climates, $P_{CO_2} = 10^{-1.5}$ or, using p notation, $pP_{CO_2} = 1.5$. The solution produced by this process is quite acid (pH $\approx$ 4.7). Subsequent dissolution of calcite out of contact with carbon dioxide gas can be represented by

$$H_2CO_3^* + CaCO_3 \rightarrow Ca^{2+} + 2HCO_3^-. \tag{3.44}$$

Since there is now no $CO_2$ reservoir the system is closed and therefore resembles the system shown in Fig. 3.3. Reaction (3.44) corresponds to region x-y of Fig. 3.3, where it can be seen that the reaction causes an increase in pH. Simultaneously, the diagram shows that the equilibrium partial pressure of carbon dioxide falls and that $[CO_3^{2-}]$ rises. Because of the latter the system reaches a point, which cannot be deduced from the diagram, where the saturation condition for calcite (eqn (3.42)) is satisfied and the reaction ceases. The behaviour of the system is more clearly illustrated by the family of curves for various initial conditions (Fig. 3.5). Because the actual dissolution of calcite takes place under closed-system conditions, this reaction scheme is usually described as *closed-system dissolution*, although the complete reaction is a two-stage process involving both open and closed steps.

Scheme 2 is entirely represented by the open-system composition diagram shown in Fig. 3.4. The reaction under these conditions is represented by

$$H_2O + CO_2 + CaCO_3 \rightarrow Ca^{2+} + 2HCO_3^-. \tag{3.45}$$

This represents a steady addition of bicarbonate ion and therefore the pH increased steadily. By definition $pP_{CO_2}$ remains constant and therefore $\Sigma$ increases. As $[HCO_3^-]$ and pH increase, so does $[CO_3^{2-}]$ and therefore the reaction stops when the saturation condition for calcite (eqn (3.42)) is satisfied. The behaviour of this reaction scheme, usually described as *open-system dissolution*, can be compared with closed-system dissolution in Fig. 3.5. It is readily apparent that for any fixed $pP_{CO_2}$ the open system reaches saturation at a lower pH and a higher $[HCO_3^-]$ than the corresponding closed system does. It should also be noted that the equilibrium $pP_{CO_2}$ of closed-system dissolution is always less than the starting value and therefore it is not easy to deduce the soil-zone $P_{CO_2}$ from a study of groundwater chemistry in this case.

An alternative approach to the open-system case is a direct numerical solution of the defining eqns. (3.24), (3.36), (3.37), (3.38), (3.42) and (3.43). If the simplifying assumption that all activity coefficients are unity is made, simultaneous solution of these equations yields.

$$[H^+]^4 \frac{K_C}{P_{CO_2} K_H K_1 K_2} + [H^+]^3 - [H^+] \{(P_{CO_2} K_H K_1) + K_w\} - 2P_{CO_2} K_H K_1 K_2 = 0. \quad (3.46)$$

It is clear that in general the solution of carbonate equilibria by this approach is inconvenient.

The discussion above has centred on pure calcium carbonate solutions and in general activity coefficients have been neglected. In real groundwaters other ions are present, particularly $Na^+$, $K^+$, $Mg^{2+}$, $Cl^-$, and $SO_4^{2-}$, and activity coefficients are not negligible. In addition ion pairing reduces the concentration of free ions (Section 2.2). The result is that in the system $Ca^{2+}$-$Mg^{2+}$-$Na^+$-$K^+$-$Cl^-$-$SO_4^{2-}$-$CO_2$-$H_2O$, the bare minimum that can represent most groundwaters, 37 ionic species may be necessary to describe the system. The solution of this system therefore requires the solution of 37 simultaneous equations which is impracticable and usually analytically impossible. For this reason iterative solution techniques using computers have been developed to study the chemistry of natural water systems. A good introduction to the mathematics of these models is given by Wigley (1977). One of the most useful programs for investigating chemical processes in natural waters is MIX2 (Plummer, Parkhurst, and Kosiur 1975), which allows chemical reactions and saturation equilibria in the above system to be explored.

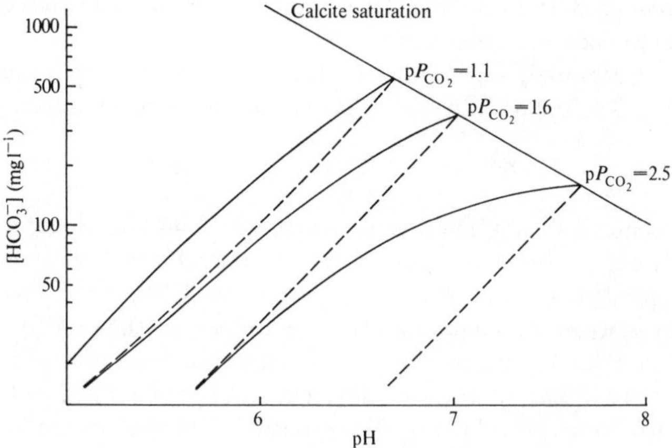


**Fig. 3.5.** Composition of a solution during calcite solution by closed-system (——) and open-system (– – –) reaction schemes.

## 3.5. Redox processes

The general nature of redox processes has been described in Section 2.5 where it was shown that redox reactions can be treated as electron transfer reactions in much the same way as acid-base reactions are proton transfer reactions. The main difference is that electrons, either free or solvated, do not exist in aqueous solution. However, by using the concept of hypothetical electron activity, redox reactions can be described by equilibrium theory. Because electron transfer is involved in redox processes, these processes can manifest themselves by the production of electric potential differences in electrochemical cells. The measurement of cell potentials is a valuable method of studying redox processes and is the basis of several analytical techniques, including the pH electrode.

The equilibrium theory relevant to redox processes will be developed by reference to an example, the oxidation of iron(II) solutions by atmospheric oxygen. From this example the general case should be clear without the necessity for a rigorous proof. The oxidation of iron(II) solutions by oxygen is described by the overall equation

$$4Fe^{2+} + 4H^+ + O_2 \rightarrow 4Fe^{3+} + 2H_2O \tag{3.47}$$

This can be represented as the sum of two half-reactions (conventionally written as reductions)

$$Fe^{3+} + e^- \rightarrow Fe^{2+} \tag{3.48}$$

$$O_2 + 4H^+ + 4e^- \rightarrow 2H_2O. \tag{3.49}$$

For each half-reaction, a formal equilibrium constant incorporating the hypothetical electron activity can be defined. For example, for eqn (3.49)

$$K = \frac{(H_2O)^2}{(O_2)(H^+)^4(e^-)^4}. \tag{3.50}$$

The electron activity is usually expressed as p*e*, where p*e* = —log(*e*). The p*e* value can be defined in terms of the equilibrium constant by rearranging eqn (3.50);

$$pe = \tfrac{1}{4}\log K + \tfrac{1}{4}\log\left(\frac{(O_2)(H^+)^4}{(H_2O)^2}\right). \tag{3.51}$$

It is useful to define a standard p*e* value, p$e^\circ$, as the (relative) electron activity when all other reactants are at unit activity. Equation (3.51) then becomes

$$pe = pe^\circ + \tfrac{1}{4}\log\left(\frac{(O_2)(H^+)^4}{(H_2O)^2}\right). \tag{3.52}$$

Because the electron activities are hypothetical no absolute values are measurable. An arbitrary standard has been agreed by setting p$e^\circ$ = 0 for the reduction of aqueous hydrogen ions to gaseous hydrogen:

$$H^+(aq) + e^- = \tfrac{1}{2}H_2(g). \tag{3.53}$$

The general expression for eqn (3.52) is given by

$$pe = pe^\circ + \frac{1}{n}\log\left(\frac{\Pi(\text{oxidized})}{\Pi(\text{reduced})}\right) \tag{3.54}$$

where $n$ is the number of electrons transferred.

The transfer of electrons in redox reactions can be realized as an external electric current in an electrochemical cell. Consider a rod of zinc metal immersed in a solution containing $Zn^{2+}$ ions at unit activity (Fig. 3.6). There is a tendency for the zinc rod to dissolve, expressed by the equilibrium

$$Zn \rightleftharpoons Zn^{2+} + 2e^-. \tag{3.55}$$

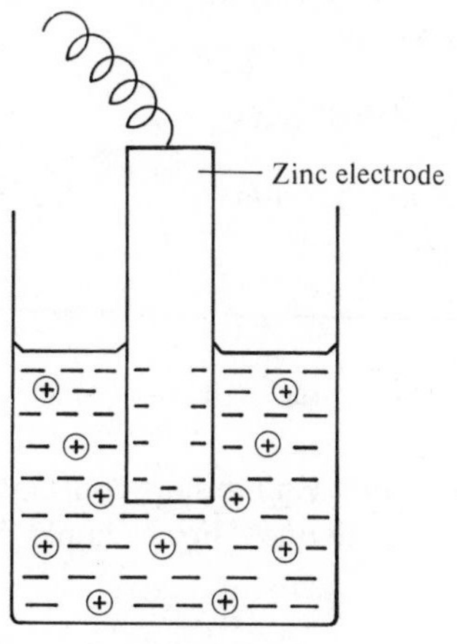


**Fig. 3.6.** Zinc electrode.

This process contributes positive $Zn^{2+}$ ions to the solution and leaves negative electrons in the zinc rod, as shown in Fig. 3.6, and thus produces a difference in electrical potential between the rod and the solution. Any electrode immersed in the solution will do this to a greater or lesser degree and therefore it is impossible to measure the potential difference absolutely. This problem has been resolved by defining the potential of the standard hydrogen electrode (Fig. 3.7) as zero. The reaction taking place at this electrode is

$$H^+ + e^- \rightleftharpoons \tfrac{1}{2}H_2. \tag{3.56}$$

This electrode is connected to the system of interest by a salt bridge, i.e. a tube filled with saturated potassium chloride solution, whose purpose is to form an electrical connection between the two solutions without introducing further potential differences. Under these circumstances, the observed potential difference between the zinc and hydrogen electrodes is 0.763 V with the zinc electrode negative, denoted by $E_H = -0.763$ V ($E_H$ signifies a measurement relative to the standard hydrogen electrode).

The hydrogen and zinc electrodes together form an electrochemical cell. Electrons produced at the zinc electrode travel through the external circuit, in which they do electrical work, to the hydrogen electrode, where they combine with hydrogen ions to produce hydrogen, the overall reaction being

$$Zn + 2H^+ \longrightarrow Zn^{2+} + H_2. \tag{3.57}$$

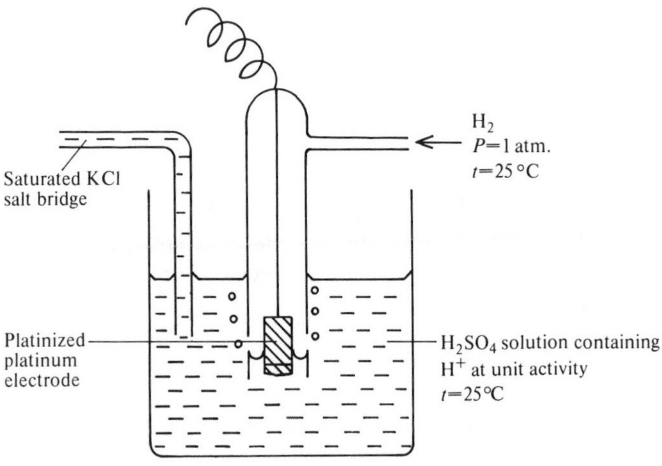


**Fig. 3.7.** Standard hydrogen electrode.

A standard electrode potential can be defined for any redox half-reaction—it is the $E_H$ of an electrode when all reactants are present in their standard states at 25 °C and 1 atm pressure (for dissolved species, the standard state is unit activity). By the International Union of Pure and Applied Chemistry (IUPAC) 1953 convention, standard electrode potentials are quoted for reactions written as reductions and the sign of the potential is the sign of the electrode when connected to a standard hydrogen electrode. (Certain lists of standard electrode potentials use another convention, in which the signs are reversed.) Examples of standard electrode potentials are

$$2H^+ + 2e^- \rightarrow H_2(g) \qquad E_H^\circ = 0.000 \text{ V (defined)}$$

$$Zn^{2+} + 2e^- \rightarrow Zn(s) \qquad E_H^\circ = -0.763 \text{ V}$$

$$Cu^{2+} + 2e^- \rightarrow Cu(s) \qquad E_H^\circ = 0.337 \text{ V}$$

$$Fe^{3+}(aq) + e^- \rightarrow Fe^{2+}(aq) \quad E_H^\circ = 0.771 \text{ V}$$

An extensive list of electrode potentials is given in Appendix II.

Standard electrode potentials enable the assessment of relative reducing and oxidizing power. Good reducing agents are poor oxidizing agents and vice versa. Increasingly negative values of standard electrode potential indicate the increasing reducing power of the species on the right-hand side of the equation. Increasingly positive potentials indicate the increasing oxidizing power of the speices on the left-hand side of the equation. Reducing agents can only reduce species with more positive potentials and oxidizing agents can only oxidize species with more negative potentials. Thus the data above indicate that zinc may reduce hydrogen ions to hydrogen, but that copper cannot. Note that electrode potentials indicate only the feasibility of a reaction, not whether it will occur. The electromotive force (e.m.f.) of a cell is readily calculated as the difference between the two electrode potentials:

$$E_{\text{cell}} = E_{\text{cathode}} - E_{\text{anode}}. \tag{3.58}$$

The cathode is the electrode at which reduction takes place (positive electrode in the case of a cell supplying current).

A consideration of the work done by the electron in the external circuit supplied by a cell leads to a relationship between the cell e.m.f. and the free energy change of the cell reaction:

$$\Delta G = -nFE \tag{3.59}$$

where $n$ is the number of electrons transferred and $F$ is Faraday's constant ($9.64867 \times 10^4$ C mol$^{-1}$). Electrode potentials and free energies can be used interchangeably in evaluating the possibility of chemical reactions. It is sometimes more convenient to use electrode potentials if these are directly measurable. Eqn (3.59) also applies to the half-reactions of electrodes and by substituting in eqn (3.11) gives a relationship between $E$ and the equilibrium constant:

$$E = \frac{RT}{nF}\ln K. \tag{3.60}$$

The standard potential $E^\circ$ is defined by all reactants (other than electrons) having unit activity. Substituting this condition in eqn (3.60) gives, by analogy with the discussion of p$e$,

$$E = E^\circ + \frac{RT}{nF}\ln\left\{\frac{\Pi\,(\text{oxidized})}{\Pi\,(\text{reduced})}\right\} \tag{3.61}$$

or

$$E = E^\circ + 2.303\frac{RT}{nF}\log\left\{\frac{\Pi\,(\text{oxidized})}{\Pi\,(\text{reduced})}\right\}$$

(note: $2.303RT/F = 0.059$ V at 25 °C). This is the Nernst equation, which allows electrode potentials to be calculated for non-standard conditions.

Comparison with eqn (3.54) shows that

$$pe = \frac{F}{2.303RT} E_H. \tag{3.62}$$

Thus the $pe$ and $E_H$ scales are effectively equivalent, differing only by a numerical factor.

*Example.* The standard free energy of formation of water is −237.18 kJ mol$^{-1}$. Calculate $E^\circ$ for the $O_2|H_2O$ couple. Is atmospheric oxygen capable of oxidizing aqueous $Fe^{2+}$ at pH 2 and 25 °C?

The equation for the reduction of oxygen to water is given by eqn (3.49):

$$O_2 + 4H^+ + 4e^- \rightarrow 2H_2O.$$

First it is necessary to evaluate $\Delta G$. The reactants on the left-hand side of the equation all have $G_f^\circ = 0$ by definition; thus $\Delta G = -474.36$ kJ.

Substituting in eqn (3.59) gives

$$E = \frac{-\Delta G}{nF}$$

$$E^\circ = 1.230 \text{ V}$$

and from (3.62)

$$pe = 20.85.$$

To answer the second part of the question, the potential of the couple is determined under the stated conditions from the Nernst equation ($P_{O_2} = 0.2$ atm for the atmosphere). The appropriate equation is

$$E = E^\circ + \frac{RT}{4F} \ln \left( \frac{(O_2)\,(H^+)^4}{(H_2O)^2} \right).$$

Substituting gives

$$E = 1.23 + \frac{0.059}{4} \log \left( \frac{(0.2)\,(10^{-2})^4}{1} \right)$$

$$= 1.102\text{V}.$$

This is compared with $E^\circ$ of 0.771 V for the $Fe^{3+}|Fe^{2+}$ reaction given earlier. The value for the oxygen couple is greater and therefore $Fe^{2+}$ will reduce oxygen to water, which is equivalent to saying that atmospheric oxygen can oxidize $Fe^{2+}$ to $Fe^{3+}$.

Problems of stability, like the example above, are common in hydrochemistry. Of the various factors determining the species in which an element will exist, $E_H$ and pH are usually the most important. Stability problems are often best tackled graphically using a stability field diagram, where stability fields are plotted in terms of $E_H$ and pH. Other diagrams may be useful in certain situations, and in complex systems governed by more than two variables a whole series of diagrams may be necessary. The techniques for constructing an $E_H$-pH diagram are outlined below for the iron-water-air system. Other diagrams can be constructed in a similar way.

Before considering the distribution of the iron species, it is useful to plot the stability field of water itself. The stability of water is restricted by reduction to hydrogen or oxidation to oxygen. The former is described by

$$2H^+ + 2e^- \rightleftharpoons H_2 \quad E^\circ = 0.00 \text{ V}. \tag{3.63}$$

Substituting in the Nernst equation gives

$$E_H = E^\circ + \frac{0.059}{2}\log\frac{(H^+)^2}{(H_2)} \tag{3.64}$$

Since pH = $-\log(H^+)$,

$$E_H = 0.00 - 0.059\text{pH} - 0.029\log(H_2). \tag{3.65}$$

It is necessary to make an assumption concerning the partial pressure of hydrogen gas in contact with the solution. In surface environments, $P_{H_2}$ is unlikely to exceed 1 atm and therefore eqn (3.65) becomes

$$E_H = -0.059\text{pH}. \tag{3.66}$$

This line defines the lower limit of water stability. In more reducing environments water can be reduced to gaseous hydrogen but it is possible for such environments to exist metastably.

The oxidation of water is described by

$$O_2 + 4H^+ + 4e^- \rightarrow 2H_2O. \tag{3.67}$$

The Nernst equation becomes

$$E_H = E^\circ + \frac{0.059}{4}\log\left(\frac{(O_2)(H^+)^4}{(H_2O)^2}\right) \tag{3.68}$$

In most groundwaters $(H_2O)$ is near to unity, and it can be assumed that $P_{O_2}$ = 0.2 atm in contact with the atmosphere. On substituting these values, eqn (3.68) becomes

$$E_H = 1.22 - 0.059\text{pH}. \tag{3.69}$$

Plotting eqns (3.66) and (3.69) on the $E_H$–pH diagram (Fig. 3.8) defines the stability field of water for the conditions assumed. No substance defined by $E_H$-pH conditions outside this field is stable in the presence of water and atmospheric oxygen.

Returning to the iron-water-air system, it is necessary to specify the compounds that exist in the system. It can be shown that in the $E_H$-pH diagram, single substances are represented by areas, reactions between two substances are represented by lines, and reactions between three substances are represented by a point. Four substances cannot coexist. Thus the diagram can be constructed by deriving the lines for coexisting pairs, and can be checked by affirming that equations represented by the intersection points are correct.

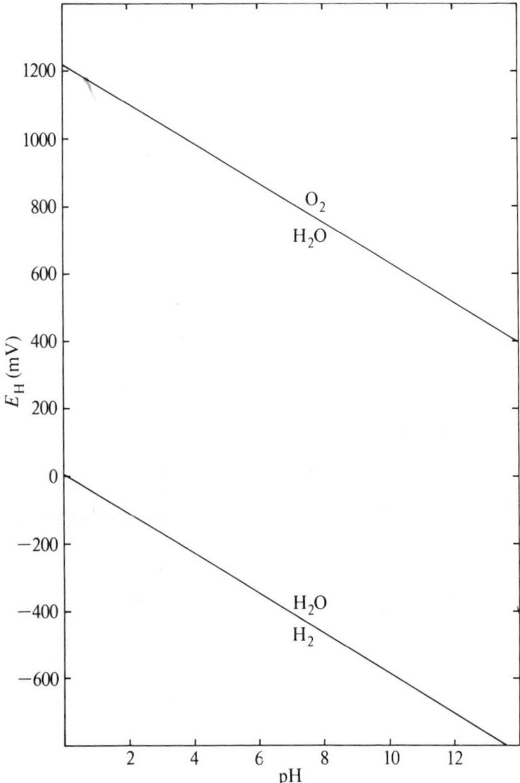


**Fig. 3.8.** $E_H$–pH diagram showing the stability field of water.

The following species can form in the iron–water–air system (ignoring ion pairs):

| Fe | $Fe^{2+}$ | $Fe^{3+}$ |
|---|---|---|
| | $Fe(OH)_2$ | $Fe(OH)_3$ |

Which pairs of species can coexist may not be fully known until the diagram is drawn; thus it is necessary to proceed stepwise.

Consider first the formation of the simple ions:

$$Fe^{2+} + 2e^- \longrightarrow Fe \qquad E^\circ = -0.44 \text{ V} \tag{3.70}$$

$$Fe^{3+} + e^- \longrightarrow Fe^{2+} \qquad E^\circ = 0.77 \text{ V}. \tag{3.71}$$

These potentials indicate that Fe is capable of reducing $Fe^{3+}$ to $Fe^{2+}$, and hence $Fe^{3+}$ and Fe do not coexist at equilibrium. Since the reactions are independent of pH, they are easily plotted on the $E_H$-pH diagram as horizontal lines. For the reaction of eqn (3.70), the Nernst equation is

$$E_H = E^\circ + \frac{0.059}{2} \log (Fe^{2+}). \tag{3.72}$$

It is therefore necessary to specify ($Fe^{2+}$). This can be done by specifying a total dissolved iron activity; a value of $10^{-5}$ M will be used in this example. Equation (3.72) then becomes

$$E_H = -0.58 \text{ V}. \tag{3.73}$$

This is plotted in line a in Fig. 3.9.

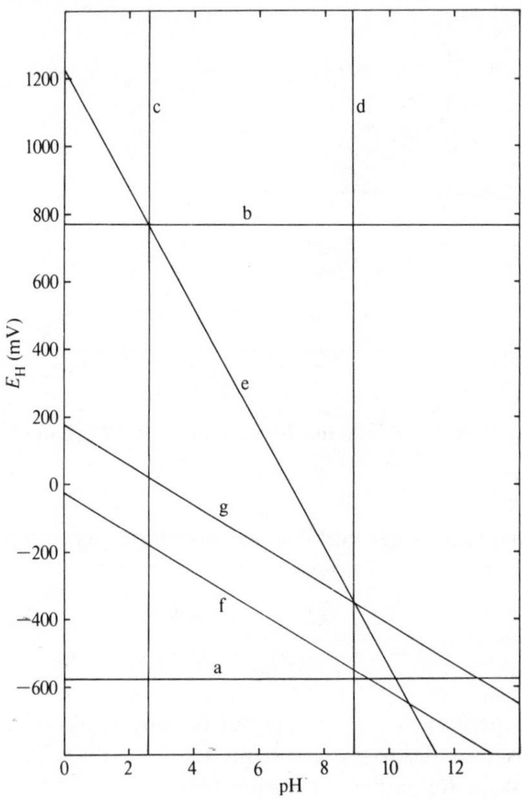


**Fig. 3.9.** Construction of an $E_H$–pH diagram for an Fe–$H_2O$–$O_2$ system.

For the reaction of eqn (3.71), the Nernst equation is

$$E_H = E^\circ + 0.059 \log \left( \frac{(Fe^{3+})}{(Fe^{2+})} \right).$$

By convention, this line is drawn where the two activities are equal, and therefore

$$E_H = 0.77\ \text{V}. \tag{3.74}$$

This is shown by line b, Fig. 3.9. Note that for this reaction, the $E_H$ is determined by the ratio of the two ions and not by their concentration. This is analogous to the situation in a buffered solution (Section 3.3). A solution such as this whose $E_H$ is relatively invariant to change is said to be *poised*.

The formations of the hydroxides are described by

$$Fe(OH)_2 \rightleftharpoons Fe^{2+} + 2OH^- \quad \log K = -15.1 \tag{3.75}$$

$$Fe(OH)_3 \rightleftharpoons Fe^{3+} + 3OH^- \quad \log K = -39.1. \tag{3.76}$$

Equation (3.75) gives

$$\log K = \log (Fe^{2+}) + 2 \log (OH^-). \tag{3.77}$$

Since the diagram uses pH as a variable, $(OH^-)$ must be expressed in terms of pH. From eqn (3.24)

$$\log (OH^-) = \text{pH} + \log K_W. \tag{3.78}$$

Substituting and rearranging gives

$$\text{pH} = \frac{\log K - \log (Fe^{2+})}{2} - \log K_W. \tag{3.79}$$

Substituting the total dissolved iron activity produces line c:

$$\text{pH} = 8.95. \tag{3.80}$$

Proceeding similarly for eqn (3.76) gives line d:

$$\text{pH} = 2.63. \tag{3.81}$$

From Fig. 3.9 it can now be seen that the reaction of eqn (3.76) cannot take place at pH values greater than 2.63. Above this pH, the reaction is

$$Fe(OH)_3 + 3H^+ + e^- \rightarrow Fe^{2+} + 3H_2O \quad E^\circ = 0.941\ \text{V}. \tag{3.82}$$

The Nernst equation is

$$E_H = E^\circ + 0.059 \log \left( \frac{(H^+)^3}{(Fe^{2+})} \right) \tag{3.83}$$

from which

$$E_H = 1.24 - 0.177\text{pH}. \tag{3.84}$$

This reaction is represented by line e.

Above pH 8.95, reactions involving $Fe^{2+}$ are replaced by reactions involving $Fe(OH)_2$, namely

$$Fe(OH)_2 + 2H^+ + 2e^- \rightarrow Fe + 2H_2O \qquad E^\circ = -0.03\ \text{V} \tag{3.85}$$

$$Fe(OH)_3 + H^+ + e^- \rightarrow Fe(OH)_2 + H_2O \quad E^\circ = 0.17\ \text{V}. \tag{3.86}$$

Equation (3.85) is represented by line f and

$$E_H = -0.03 - 0.059\text{pH} \tag{3.87}$$

Equation (3.86) is represented by line g and

$$E_H = 0.17 - 0.059\text{pH}. \tag{3.88}$$

The lines representing all these reactions are shown by Fig.3.9. It is then necessary to decide which portions of the intersecting lines shown can exist. For example, line d shows that $Fe^{3+}$ does not exist at pH > 2.63; hence the continuation of the $Fe^{3+} \mid Fe^{2+}$ line, b, beyond this is meaningless. Application of similar criteria until an internally consistent result is achieved results in Fig. 3.10. Certain features may be noticed. The lines describing $Fe^{3+} \mid Fe(OH)_3$, $Fe^{3+} \mid Fe^{2+}$, and $Fe(OH)_3 \mid Fe^{2+}$ meet at a point as they must do for consistency. Two other points are similarly defined and therefore the diagram can in fact be constructred with less calculation than above by appreciating that certain reactions are independent of $E_H$ or pH. However, this presumes that the data from which the diagram is constructed are themselves self-consistent

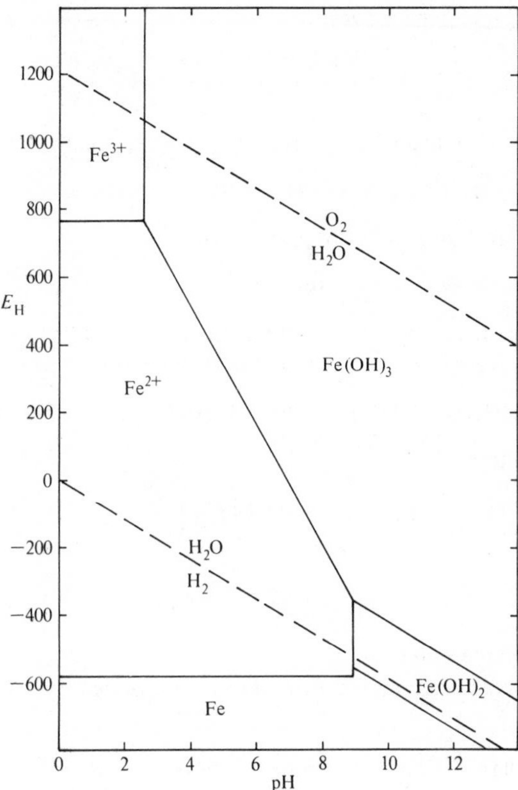


**Fig. 3.10.** $E_H$–pH diagram for an Fe–$H_2O$–$O_2$ system ($Fe_{tot} = 10^{-5}$ M).

and this is often not the case. Many solid precipitates in particular are rather poorly defined and this makes exact measurement of thermodynamic variables impossible. In this example it should be noted that the $Fe|Fe^{2+}|Fe(OH)_2$ point shows a triangle of error.

Various conclusions about the iron-water-air system can be drawn from Fig. 3.10. Firstly, metallic iron is never stable in contact with water. In contact with the atmosphere the usual product is $Fe(OH)_3$, represented by rust. In aqueous environments buffered near pH 7 by the carbonate system, the stable iron species is soluble $Fe^{2+}$ or insoluble $Fe(OH)_3$, depending on $E_H$, and therefore $E_H$ has a profound effect upon iron mobility. The high $E_H$ required for $Fe^{3+}$ stability limits it to well-aerated aerobic environments; the low pH required restricts its occurrence in geological environments since carbonates and some clay minerals are unstable at this pH. Further information about geological systems can be obtained by combining this diagram with those of other elements, e.g. sulphur.

## 3.6. Transport

### *3.6.1. Diffusion*

From eqn (3.11) it can be deduced that the free energy of a compound in solution depends on its activity in the solution. Thus a free-energy difference exists between two solutions of the same solute in contact, which provides a driving force for the spontaneous mixing of the solutions. The minimum total free energy occurs when the solution has a uniform concentration. Equilibrium thermodynamics can say nothing about the rate of the spontaneous mixing, only that it will occur. The process by which solutes migrate along concentration gradients is known as *diffusion*.

The rate of diffusion is described empirically by Fick's first law

$$f = -D\frac{\mathrm{d}C}{\mathrm{d}x} \tag{3.89}$$

where $f$ is the flux per unit area, $\mathrm{d}C/\mathrm{d}x$ is the concentration gradient, and $D$ is the diffusion coefficient. The minus sign indicates that the flux is in the opposite direction to the concentration gradient. If conservation of the solute is assumed, it is possible to derive Fick's second law

$$D\left(\frac{\partial^2 C}{\partial x^2}\right)_t = \left(\frac{\partial C}{\partial t}\right)_x. \tag{3.90}$$

This is known as the diffusion equation. Similar equations describe the flow of heat and the flow of groundwater; hence a large number of solutions for various boundary conditions are available (see Carslaw and Jaeger (1959) and Crank (1956)).

In hydrochemistry it is usually diffusion of ionic solutes that is of interest. Diffusion of ions clearly implies a movement of electric charge and therefore diffusion coefficients can be related to ionic conductances, which are readily determined from conductivity measurements:

$$D^\circ = \frac{RT}{F^2\,|z|}\,\lambda^\circ \tag{3.91}$$

where $D^\circ$ is the diffusion coefficient at infinite dilution and $\lambda^\circ$ is the equivalent conductance of the ion at infinite dilution. A list of individual ion diffusion coefficients is given in Table 3.5. It is noticeable that ions of the same charge have similar diffusion coefficients, with the exception of $H^+$ and $OH^-$. The diffusion of a salt involves the simultaneous migration of two oppositely charged ions in order to preserve electoneutrality. The combined diffusion coefficient $D^\circ_{12}$ for ions of charge $z_1$ and $z_2$ is

$$D^\circ_{12} = \left(\frac{1}{|z_1|} + \frac{1}{|z_2|}\right)\left(\frac{D^\circ_1\,D^\circ_2}{D^\circ_1 + D^\circ_2}\right). \tag{3.92}$$

**Table 3.5** *Tracer diffusion coefficients of ions at infinite dilution in water*

| Cation | $D^\circ$ ($\times 10^{-10}$ m$^2$ s$^{-1}$) | | | Anion | $D^\circ$ ($\times 10^{-10}$ m$^2$ s$^{-1}$) | | |
|---|---|---|---|---|---|---|---|
| | 0 °C | 18 °C | 25 °C | | 0 °C | 18 °C | 25 °C |
| $H^+$ | 56.1 | 81.7 | 93.1 | $OH^-$ | 25.6 | 44.9 | 52.7 |
| $Na^+$ | 6.27 | 11.3 | 13.3 | $F^-$ | — | 12.1 | 14.6 |
| $K^+$ | 9.86 | 16.7 | 19.6 | $Cl^-$ | 10.1 | 17.1 | 20.3 |
| $NH_4^+$ | 9.80 | 16.8 | 19.8 | $Br^-$ | 10.5 | 17.6 | 20.1 |
| $Mg^{2+}$ | 3.56 | 5.94 | 7.05 | $SO_4^{2-}$ | 5.00 | 8.90 | 10.7 |
| $Ca^{2+}$ | 3.73 | 6.73 | 7.93 | $NO_3^-$ | 7.78 | 16.1 | 19.0 |
| $Sr^{2+}$ | 3.72 | 6.70 | 7.94 | $HCO_3^-$ | — | — | 11.8 |
| $Fe^{2+}$ | 3.41 | 5.82 | 7.19 | $CO_3^{2-}$ | 4.39 | 7.80 | 9.55 |
| $Pb^{2+}$ | 4.56 | 7.95 | 9.45 | | | | |
| $Fe^{3+}$ | — | 5.28 | 6.07 | | | | |
| $Al^{3+}$ | 2.36 | 3.46 | 5.59 | | | | |

Data from Li and Gregory (1974)

Because of the complex effects described in Section 2.2, diffusion coefficients vary with ionic strength but the effect, which is illustrated in Table 3.6, is small enough to be neglected in most hydrogeological work. The variation of diffusion coefficients with temperature is approximately described by

$$D^\circ_{T_1} = D^\circ_0\,(1 + \alpha\,T_1) \tag{3.93}$$

where $D^\circ_0$ is the value of 0 °C, $T_1$ is in degrees Celsius, and $\alpha$ is 0.048 for small ions and 0.04 for large ions (most anions and also $K^+$).

**Table 3.6** *Variation of diffusion coefficient for NaCl with concentration*

| Concentration (mol $l^{-1}$) | 0 | 0.05 | 0.1 | 0.2 | 0.5 | 1.0 | 2.0 | 3.0 | 4.0 | 5.0 |
|---|---|---|---|---|---|---|---|---|---|---|
| $D$ at 25 °C ($10^{-9}$ $m^2$ $s^{-1}$) | 1.610 | 1.507 | 1.483 | 1.475 | 1.474 | 1.484 | 1.516 | 1.565 | 1.594 | 1.690 |

Data from Lerman and Weiler (1970)

Diffusion through the pore space of sediments is of considerable interest to the groundwater chemist. The factors affecting the rate of diffusion are the porosity $\phi$ of the sediment and the tortuosity $\theta$ of the pore space. On theoretical grounds, the diffusion coefficient in a porous medium has been shown by Lerman (1979) to be given by

$$D' = D\frac{\phi}{\theta}. \tag{3.94}$$

Empirical measurements have given the result

$$D' = D\phi^{n} \tag{3.95}$$

where $n = 1.2$-$5.4$. For general calculations, an assumption of $n = 2$ is reasonable. Since electrical conduction in saturated porous media proceeds primarily by ion transport within the pore space, the resistivity of the porous formation and diffusion coefficient are phenomenologically related. Thus the effective value of the diffusion coefficient can be calculated from the resistivity formation factor $F'$, where $F'$ is the ratio of the electrical resistance of a brine-filled porous medium to the electrical resistance of the brine occupying the same volume as the porous medium:

$$D' = \frac{D}{F'}. \tag{3.96}$$

The formation factor is easily determined by direct measurement. For very small pore sizes (less than 10 nm) these relationships break down because of increased interference between the hydration sheaths of the ions and the matrix.

To gain an impression of the significance of diffusion processes, a numerical idea of the rate of the process is necessary. Solutions of the diffusion equation (eqn 3.90) are characterized by a relaxation time

$$\tau = \frac{l^2}{GD} \tag{3.97}$$

where $l$ is the dimension of the system and $G$ is a geometrical factor ($G$ is 4 for a point source and $\pi^2$ for plane parallel sources). The amount of change that takes place in time $\tau$ depends on the geometrical configuration, but is typically 50 per cent of the total change (see Crank 1956). Substitution in eqn (3.97) shows that for $l = 0.1$ m (typical fissure spacing), $\tau = 37$ days, and for $l = 100$ m (typical aquifer thickness), $\tau = 100\,000$ years. Thus diffusion is significant only in very small systems or when large amounts of time are available.

### *3.6.2. Advection and dispersion*

Because diffusion, i.e. transport induced by chemical gradients, is so slow, transport in most groundwater systems is dominated by *advection*, i.e. bulk movement of water induced by hydraulic gradients. The solute flux $f$ is related to the seepage velocity $v$ by

$$f = Cv \tag{3.98}$$

where $C$ is the concentration of the solute.

The seepage velocity is related by the porosity $\phi$ to the Darcy velocity $Q$:

$$v = \frac{Q}{\phi} \tag{3.99}$$

These equations are strictly true only if every water molecule moves with the same velocity, a situation that does not obtain in a porous medium because of the velocity gradients in each microscopic flow channel. This produces a spreading of any moving solute front, known as *dispersion*. The effect of dispersion is the same as that of diffusion although on a different scale. The process is described by the coefficient of hydrodynamic dispersion $D_h$ which in general depends on the grain-size distribution of the sediment and also on the seepage velocity; at zero $v$, $D_h$ reduces to $D'$. The combined equation for one-dimensional advection and dispersion is

$$\frac{\partial C}{\partial t} = D_h \frac{\partial^2 C}{\partial x^2} - v \frac{\partial C}{\partial x}. \tag{3.100}$$

Note that this reduces to the diffusion equation (eqn 3.90) at zero $v$ when $D_h = D'$ for a porous medium.

The general form of processes described by eqn (3.100) can be seen by solving it for a step input at the beginning of a long column. Appropriate boundary conditions are:

$$\begin{array}{llll} t<0, -\infty<x<0 & C=C_0 & t>0, x=\pm\infty & \partial C/\partial x = 0 \\ \quad 0<x<+\infty & C=0 & \quad x=-\infty & C=C_0 \\ & & \quad x=+\infty & C=0. \end{array}$$

The solution is (Bear 1972)

$$C = \frac{C_0}{2} \operatorname{erfc}\left\{\frac{x - vt}{2(D_h t)^{\frac{1}{2}}}\right\} \tag{3.101}$$

where

$$\operatorname{erfc}(x) = \frac{2}{\sqrt{\pi}} \int_x^{\infty} \exp^{(-\alpha^2)}\, d\alpha.$$

From this it can be shown that the point of mean concentration $C_0/2$ travels with the mean velocity $v$, and that the variance of the dispersion front is $2D_h t$, i.e. it is proportional to the time taken. Note that this implies that in oscillatory motion, where the net displacement of the mean point is zero, the dispersion increases steadily. The shape of the dispersion profile produced by eqn (3.101) is shown in Fig. 3.11.

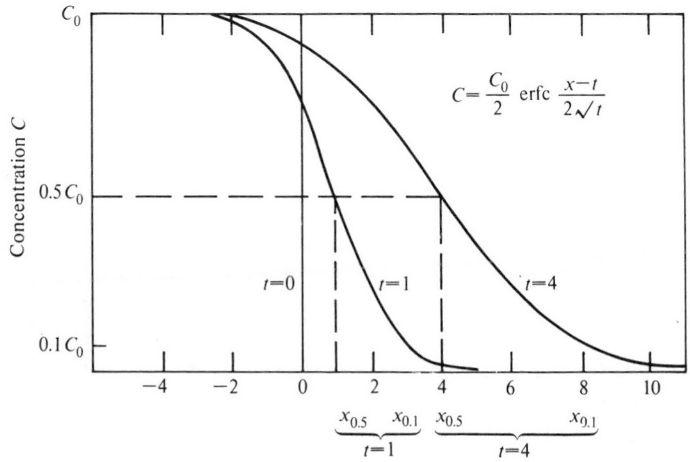


**Fig. 3.11.** Concentration of a dissolved substance in the presence of diffusion and flow in a one-dimensional system. Profiles are shown for three consecutive values of time $t$. The dispersal of concentration is shown on the horizontal axis for $t = 1$ and $t = 4$.

Solutions to the dispersion equation are limited to a few simple geometrical configurations (Al-Niami and Rushton 1978) and therefore numerical models have been developed to study solute transport in complex situations (Konikow and Bredehoeft 1978).

Once transport is considered in the chemical processes of an aquifer, the time dimension enters consideration of the chemical processes. Some reactions are so slow that equilibrium will not be achieved in the time available before conditions change. Slow reactions likely to be affected by such considerations are solution and precipitation reactions and redox reactions. As an example of the first, many waters do not saturate with calcite in the soil zone because of inadequate reaction opportunity, and consequently dissolved calcium only reaches the relatively low concentrations characteristic of closed-system dissolution. The persistence of undersaturation far into the aquifer is important in the development of secondary permeability. The mathematics of a solution process limited by reaction rate, chemical equilibrium, and solute transport have been discussed by Palciauskas and Domenico (1976).

## 3.7. Adsorption and ion exchange

### *3.7.1. Adsorption*

The interface between solution and solid, or solution and gas, causes a discontinuity in the structure of the solvent and therefore the energy levels of surface solvent molecules differ from those of the bulk solution. The effect is particularly marked in the case of water because of the strong hydrogen bonding. A consequence of the difference in energy between the bulk solution and its surface is that solute molecules may be concentrated or depleted at the surface, depending on the way in which they interact with the solvent. For a general discussion of the physical chemistry of adsorption, see Stumm and Morgan (1981). In natural groundwater chemistry, the most important aspect of adsorption is the adsorption of ions on mineral surfaces, which gives rise to the phenomenon of ion exchange.

### *3.7.2. Ion exchange*

The phenomenon of ion exchange arises from the adsorption behaviour of charged surfaces. The two modes of origin of charged surfaces relevant to geological materials have already been mentioned in Section 2.7. When a charged surface is present in a solution containing ions, the result is the formation of an electrical double layer; one layer is the fixed charge attributed to the surface, and the other is a diffuse layer within the solution. If it is assumed that the fixed layer is negatively charged, the diffuse layer will be characterized by an increased concentration of cations, which are attracted, and a decreased concentration of anions, which are repelled. This is the Gouy-Chapman model, illustrated in Fig. 3.12. The excess cations, which have total charge $\sigma_-$, comprise the exchangeable cations adsorbed on the surface. The negatively adsorbed anions have a total charge $\sigma_-$ and the sum $\sigma = \sigma_+ + \sigma_-$ equals the charge on the

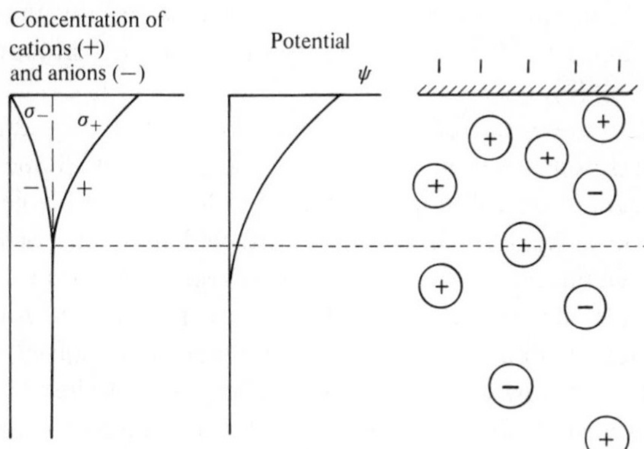


**Fig. 3.12.** Gouy–Chapman model of ions adsorbed on a negatively charged clay surface. (After Stumm and Morgan 1981.)

surface. Thus the cation exchange capacity is less than the surface charge. The unequal concentration of cations in the diffuse layer implies that there is an electric potential gradient through the diffuse layer and therefore that there is a potential difference $\psi$ between the surface and the bulk solution. The charge and the potential difference are related by the capacitance of the double layer:

$$C = \sigma / \psi. \tag{3.102}$$

The capacitance can be calculated from the mean layer thickness $1/\kappa$ by

$$C = \epsilon\kappa \tag{3.103}$$

where $\epsilon$ is the permittivity of the solution. The layer thickness can be related to the ionic strength $I$ (van Olphen 1977) by

$$\kappa = \left(\frac{e^2 I}{\epsilon RT}\right)^{\frac{1}{2}}. \tag{3.104}$$

When numerical values are substituted, the thickness of the diffuse layer varies from 5 to 20 nm for fresh water to 0.4 nm in seawater. This latter value is similar to the radii of hydrated ions and therefore the layer is no longer diffuse. A modification to the Gouy-Chapman model by Stern, to include the effects of ionic size, provides a better description at these ionic strengths.

Charged surfaces fall into two categories: surfaces of constant potential and surfaces of constant charge. Metal oxide particles, whose charge is determined by a pH-dependent equilibrium, approximate to surfaces of constant potential at constant pH. Clay particles, whose charge is determined mainly by atomic substitution within the lattice, approximate to surfaces of constant charge except at low pH values. As ionic strength is increased, which decreases the layer thickness and increases its capacitance (see eqns (3.103) and (3.104)), the two types of surface behave differently. For a constant potential surface, $\psi$ is constant and therefore $\sigma$ increases. It can be shown (van Olphen 1977) that the ratio $\sigma_+ / \sigma$ remains constant and thus the ion exchange capacity increases. For a constant charge surface, $\sigma$ remains constant so $\psi$ must fall. Under these circumstances $\sigma_+ / \sigma$ decreases and thus the ion exchange capacity is reduced.

The Gouy-Chapman theory predicts that ions of the same charge should be adsorbed equally, i.e. the ratio of the adsorbed ions should be the same as the activity ratio of the ions in solution. In practice this is found not to be so, ions being adsorbed in the order

$$Cs^+ > K^+ > Na^+ > Li^+$$

$$Ba^{2+} > Sr^{2+} > Ca^{2+} > Mg^{2+}.$$

These series are in order of increasing hydrated ionic radius, suggesting that large hydrated ions are less readily adsorbed than small ones of the same charge. Dehydration effects may also be important. The theory correctly predicts that divalent ions are much more strongly adsorbed than monovalent ions.

The exchange of different cations can be described by equilibrium equations, for example

$$2NaR + Ca^{2+} \rightleftharpoons CaR_2 + 2Na^+ \qquad (3.105)$$

$$Q_{(Na \rightarrow Ca)} = \frac{X_{Ca}}{X^2_{Na}} \frac{(Na^+)^2}{(Ca^{2+})}$$

where $X$ represents the equivalent fraction of the ion on the exchanger R (e.g. $X_{Ca} = 2\ [CaR_2]/\{2[CaR_2] + [NaR]\}$). The selectivity coefficient $Q$ describes the reaction in a semiquantitative way only since it varies with the ionic strength of the solution. Selectivity coefficients decrease markedly at high ionic strengths as can be deduced from Table 3.7.

**Table 3.7** *Ion exchange of clays with solutions containing $Ca^{2+}$ and $K^+$ at equal equivalent concentrations*

| Clay | Exchange capacity (meq $g^{-1}$) | $Ca^{2+}/K^+$ ratios on clay | | | |
|---|---|---|---|---|---|
| | | Concentration of solution (meq $l^{-1}$) | | | |
| | | 100 | 10 | 1 | 0.1 |
| Kaolinite | 0.023 | — | 1.8 | 5.0 | 11.1 |
| Illite | 0.162 | 1.1 | 3.4 | 8.1 | 12.3 |
| Montmorillonite | 0.810 | 1.5 | — | 22.1 | 38.8 |

This is important in considering the reaction of clays equilibrated with sea-water ($Na^+ \approx 500$ meq $l^{-1}$; $Ca^{2+} \approx 20$ meq $l^{-1}$) and groundwater.

## References


Al Niami, A. N. S. and Rushton, K. R. (1978). Analysis of flow against dispersion in porous media. *J. Hydrol.* **33**, 87–98.

Bear (1972).

Carslaw, H. S. and Jaeger, J. C. (1959). *Conduction of heat in solids*. Oxford University Press, London.

Crank, J. (1956). *Mathematics of diffusion*. Clarendon Press, Oxford.

International Union of Pure and Applied Chemistry (1969) *Division of Physical Chemistry Commission on Symbols, Terminology and Units. Manual of symbols and terminology for physiochemical quantities and units*. Butterworths, London.

Kielland, J. (1937). Individual activity coefficients of ions in aqueous solutions. *J. chem. Soc.* **59**, 1675–1678.

Konikow, L. F. and Bredehoeft, J. D. (1978). Computer model of two dimensional solute transport and dispersion in ground water. In *Techniques of water-resources investigations of the United States Geological Survey (Book 7): Automated data processing and computations*. U.S. Government Printing Office, Washington, DC.

Lerman, A. and Weiler, R. R. (1970). Diffusion and accumulation of chloride and sodium in Lake Ontario sediment. *Earth planet Sci. Lett.* **10**, 150–156.
— (1979). *Geochemical processes*. Wiley, New York.
Li, Y.-H. and Gregory, S. (1974). Diffusion of ions in sea water and in deep sea-sediments. *Geochim. cosmochim. Acta* **38**, 703–714.
Palciauskas, V. V. and Domenico, P. A. (1976). Solution chemistry, mass transfer and the approach to chemical equilibrium in porous carbonate rocks and sediments. *Geol. Soc. Am. Bull.* **87**, 207–214.
Plummer, L. N., Parkhurst, D., and Kosiur, D. R. (1975). MIX2, a computer program for modelling chemical reactions in natural waters. *U.S. Geol. Surv. Rep.* 61–75. (Water Resources Internal).
Stumm, W. and Morgan, J. J. (1981). *Aquatic chemistry. An introduction emphasizing chemical equilibria in natural waters*. Wiley-Interscience, New York.
Van Olphen, A. (1977). *An introduction to clay colloid chemistry*. Wiley, New York.
Wigley, T. M. L. (1977). Watspec: A computer program for determining the equilibrium speciation of aqueous solutions. *Brit. geomorph. res. tech. Bull.* **20**, 16.

# 4 Hydrochemical parameter measurement and sample collection

## 4.1. Introduction

The hydrochemical parameters requiring determination in a study will be dictated by the study objectives. From the discussion in Chapter 2 the potential number of parameters can be vast; nevertheless a reasonable consensus can be made of those parameters that normally are of value in interpretation and those that under the right hydrogeological circumstances can give information.

It is not possible to give rules about which parameters should be measured in a particular groundwater study. It is important that well-head chemistry as described in Section 4.2 be carried out, and clearly the major ion parameters should provide the opportunity for initial interpretation. Where no distinguishing major ions are present, as for example in some brackish and saline groundwaters, minor ions can be sampled for analysis and interpretation. Isotopes can be of considerable value but are expensive to analyse. Their real use lies in relative age assessments and palaeoclimate studies. Their use as trace parameters is limited in that major and minor ions frequently provide adequate data.

A hydrochemical data sheet is given on Fig. 4.1. The parameter list is comprehensive and covers most of the determinands used in groundwater studies. The parameters, measurements, and sampling procedures detailed below will be discussed with reference to Fig. 4.1.

## 4.2. Chemical parameters measured at the well head

The measurement of certain chemical parameters at the well head is carried out for the purposes of convenient rapid assessment and to provide control for laboratory measurements. Of the two objectives, control for laboratory measurements is the most important in that the physical conditions of a sample may change between the time of sampling and the laboratory measurements. The changes that occur most frequently effect the carbonate chemistry leading to carbonate mineral precipitation which can also induce coprecipitation of metals and other changes. Clearly also redox conditions can change considerably, resulting in metal oxide precipitation etc.

The field parameters normally measured are electrical conductivity, temperature, pH, $E_H$, and dissolved oxygen.

### *4.2.1. Electrical conductivity*

As ions principally occur in a dissociated form in waters the charged ions are able to move under the influence of an electrical potential. Therefore by imparting

LOCALITY .................... Grid ref: ....................

Sampled by .................... Date: ....................

SAMPLE INFORMATION

Sample No Filtered Acidified

Sampling details .................... Sample depth ....................

....................

WELL HEAD CHEMISTRY

Electrical conductivity ($\mu$S cm$^{-1}$) .......... Temperature (°C) .......... pH ..........

Redox potential (mV) .......... Dissolved oxygen (mg l$^{-1}$) ..........

GENERAL PARAMETERS

Laboratory pH .......... Laboratory E.C. ($\mu$S cm$^{-1}$) .......... Total hardness (mg l$^{-1}$ $CaCO_3$) ..........

Total dissolved solids (mg l$^{-1}$) .......... Alkalinity (mg l$^{-1}$ $CaCO_3$) ..........

| MAJOR IONS | mg l$^{-1}$ | meq l$^{-1}$ | | mg l$^{-1}$ | meq l$^{-1}$ |
|---|---|---|---|---|---|
| Sodium ($Na^+$) | | | Sulphate ($SO_4^{2-}$) | | |
| Potassium ($K^+$) | | | Chloride ($Cl^-$) | | |
| Calcium ($Ca^{2+}$) | | | Nitrate ($NO_3^-$) | | |
| Magnesium ($Mg^{2+}$) | | | Nitrite ($NO_3^-$) | | |
| Bicarbonate ($HCO_3^-$) | | | Ammonia ($NH_3$) | | |
| Carbonate ($CO_3^{2-}$) | | | Phosphate ($PO_4^{2-}$) | | |

| MINOR IONS | $\mu$g l$^{-1}$ | $\mu$eq l$^{-1}$ | | $\mu$g l$^{-1}$ | $\mu$eq l$^{-1}$ |
|---|---|---|---|---|---|
| Boron (B) | | | Fluoride ($F^-$) | | |
| Silica ($SiO_2$) | | | Bromide ($Br^-$) | | |
| Arsenic (As) | | | Iodide ($I^-$) | | |

| TRACE METALS | $\mu$g l$^{-1}$ | $\mu$eq l$^{-1}$ | | $\mu$g l$^{-1}$ | $\mu$eq l$^{-1}$ |
|---|---|---|---|---|---|
| Lithium ($Li^+$) | | | Manganese (Mn) | | |
| Rubidium ($Rb^+$) | | | Iron (Fe) | | |
| Strontium ($Sr^{2+}$) | | | Nickel (Ni) | | |
| Barium ($Ba^{2+}$) | | | Copper (Cu) | | |
| Aluminium ($Al^{3+}$) | | | Zinc (Zn) | | |
| Vanadium (V) | | | Cadmium (Cd) | | |
| Chromium (Cr) | | | Mercury (Hg) | | |
| Molybdenum (Mo) | | | Lead (Pb) | | |

ISOTOPES

Tritium (Tritium Units) .......... Carbon 13 ($\delta$‰) ..........

Deuterium ($\delta$‰) .......... Carbon 14 (Per cent modern carbon) ..........

Oxygen 18 ($\delta$‰) ..........

OTHER CONSTITUENTS

**Fig. 4.1.** Hydrochemical data sheet.

an electrical current to a solution the conductance of the solution can be determined. The ability of the solution to conduct the current is a function of the concentration and charge of the ions, and the rate at which the ions can move under the influence of the potential.

Electrical conductivity (EC) has the units of reciprocal ohms per metre or, in SI units, seimens per metre (S $m^{-1}$). Conductivities in groundwaters, however, are such that values are usually reported as micromhos per centimetre or microsiemens per centimetre ($\mu$S $cm^{-1}$). As ionic activity is affected by temperature, electrical conductivity increases with temperature and any measurement should be reported at a specific temperature, normally 25 °C. The electrical conductivity change with temperature for a 0.00702 N potassium chloride solution is shown in Fig. 4.2.

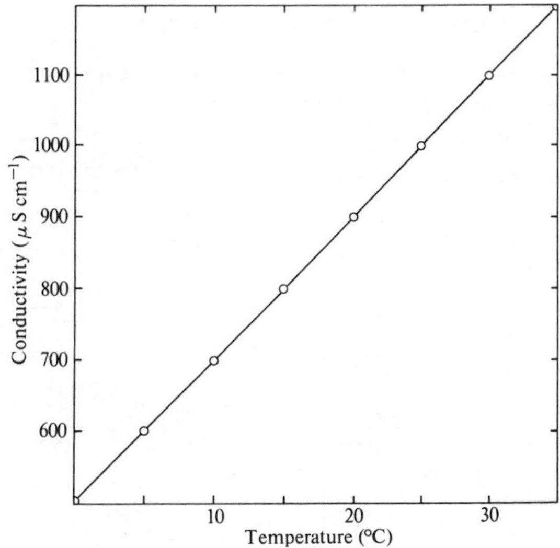


**Fig. 4.2.** Conductivity of a 0.00702 N potassium chloride solution at various temperatures. (After Wood 1976.)

As the ion concentration increases per unit volume of solution the rate of ionic activity increase of individual ions decreases because of interionic attraction; in consequence the relationship between ionic concentration and electrical conductivity is only linear for dilute solutions (below approximately 1000 $\mu$S $cm^{-1}$). Since natural waters contain a variety of both ionic and uncharged species in various amounts and proportions, conductivity determinations cannot be used to obtain accurate estimates of ion concentrations or total dissolved solids. An indication of the variation of conductivity for specific concentrations is shown in Fig. 4.3 for Lower Indus groundwaters (MacDonald and Partners 1965). The poor correlation between electrical conductivity and sulphate and bicarbonate, and the good correlation with chloride demonstrated by Fig. 4.3

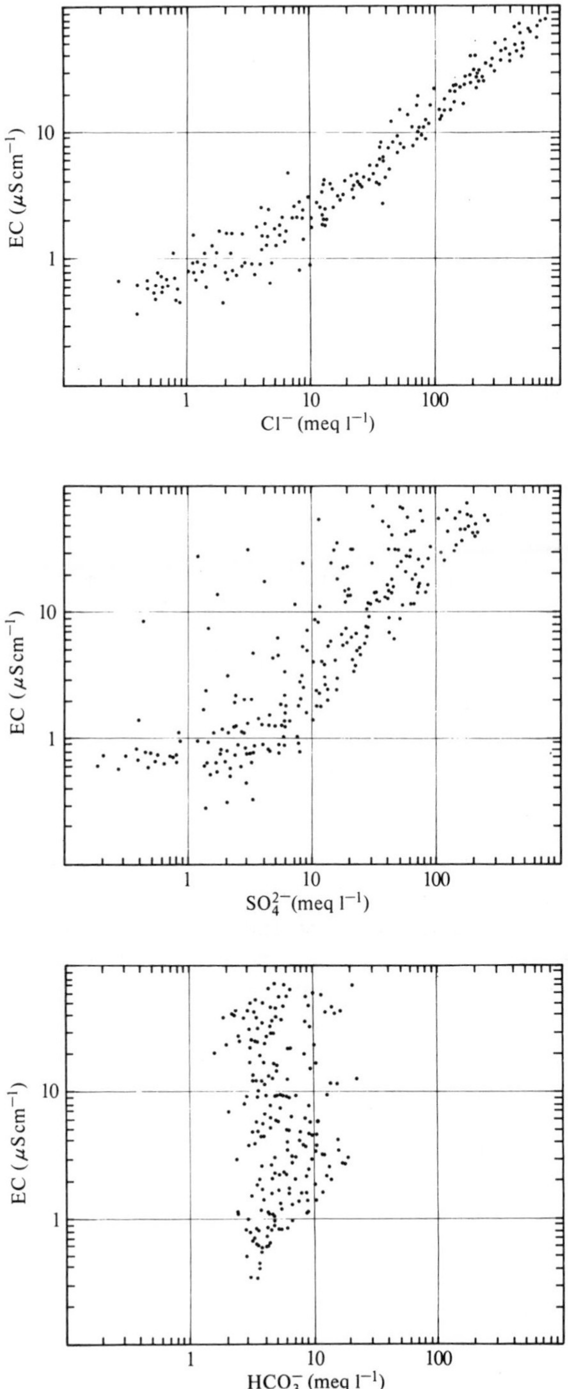


**Fig. 4.3.** Variation of EC for specific anions. (After McDonald and Partners 1965.)

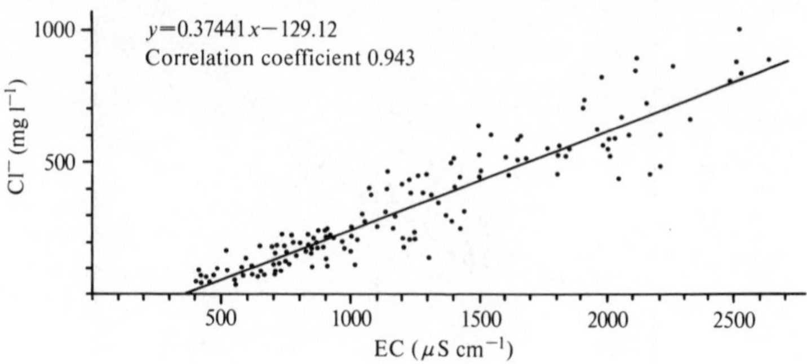


**Fig. 4.4.** Chloride and electrical conductivity correlation of groundwaters from Grand Cayman.

is a common feature of groundwaters. The latter correlation is frequently reliable as indicated by Fig.4.4 for groundwaters from the Cayman Islands (Bugg and Lloyd 1976).

Normally an approximate correlation between conductance and total dissolved solids (TDS) can be used:

$$\mathrm{TDS} = k_e \mathrm{EC} \tag{4.1}$$

where TDS is expressed in milligrams per litre and EC is in microseimens per centimetre at 25 °C. The correlation factor $k_e$ in eqn (4.1) varies between about 0.55 and 0.80 for groundwaters and needs to be determined specifically in each study.

Reliable equipment for either field or laboratory measurements of electrical conductivity is readily available. Samples are normally measured in an epoxy resin and carbon electrode assembly with measurement ranges from 0 to $10^5$ $\mu$S cm$^{-1}$. Units are battery operated and are usually automatically temperature corrected to 25 °C on all ranges at 2 per cent per degree centigrade.

The ease and rapidity of electrical conductivity measurements in the field allows it to be used as an excellent monitoring parameter both on an areal and a time basis. Conductivity measurements can aid in determining laboratory sample frequency and can be used to advantage to support hydraulic observations at the groundwater–surface water interface (i.e. effluent river conditions) or in pumping tests as shown in Fig. 4.5.

### *4.2.2. Temperature*

Temperature measurements (usually in degrees Celsius) at the well head are primarily of importance for thermodynamic calculations related to water chemistry, although as shown in Fig. 4.5 they can also support hydraulic and other chemical observations. Measurements are made using a standard mercury thermometer and should be to an accuracy of ±0.1 °C. When measurements are taken the thermometer reading should be given time to equilibrate. They can conveniently be taken in a cell as shown on Fig. 4.6.

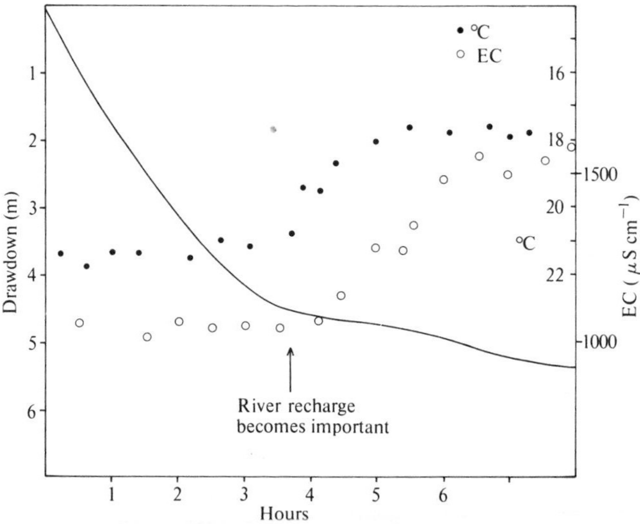


**Fig. 4.5.** Temperature (●) and electrical conductivity (EC) (○) indicating the effects of river recharge during a pumping test.

### *4.2.3. pH*

The pH of a solution is the negative logarithm of the hydrogen ion activity in moles per litre (see Section 3.4). When groundwater is removed from an aquifer the physical controls governing the hydrogen ion activity are changed and thus the pH changes. As water moves in a well pH changes will occur so that exact measurements are unlikely. However carefully water is sampled, once it is allowed to stand the pH will probably change so that laboratory measurements have little relevance and it is vital that it be measured at the well head.

pH is determined using a glass electrode compared with a reference electrode of known potential by means of a pH meter or other potential-measuring device with a very high impedance (Wood 1976). Details of pH electrodes and measurements are given by Barnes (1964) and Langmuir (1971).

Good quality field pH meters that are battery operated and will read from 0 to 14 pH units are available with either dial or digital displays. Measurements are made with respect to standard calibrating buffers which for precision should be within ±1 °C of the sample solution as pH is temperature sensitive. pH measurements in the field can realistically be made to ±0.01 of a unit; however, difficulties of equilibrium do occur particularly in hot climates and some 10–20 min may be required to obtain a dependable reading (see Fig. 4.7).

### *4.2.4. Redox potential*

As discussed in Section 3.5 the standard oxidation potential of a half-reaction is $E^{\circ}$. When the activities of species differ from unity the observed potential is

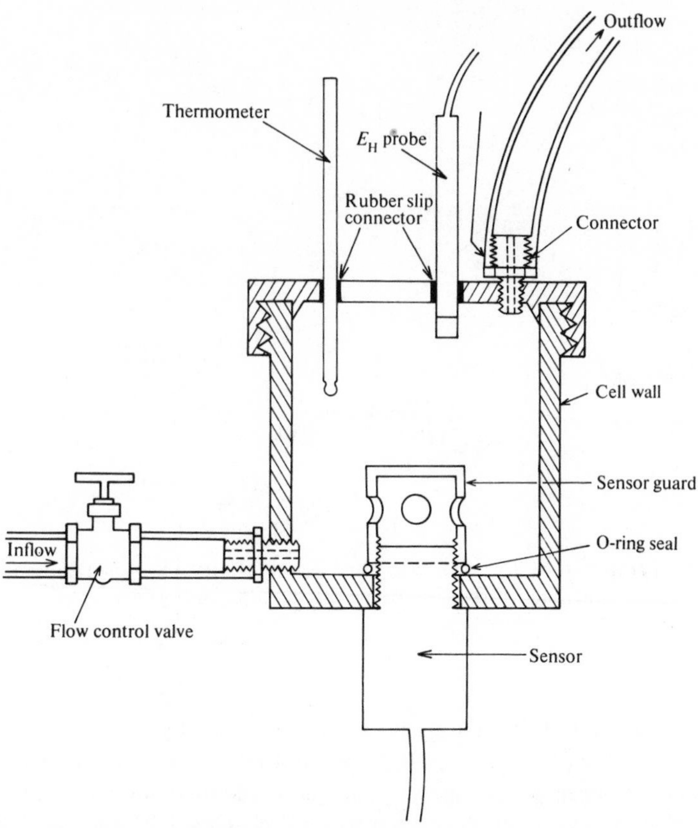


**Fig. 4.6.** Isolation cell for the measurement of $E_H$, temperature and DO. (Modified from Wood 1976.)

termed the redox potential $E_H$(oxidation-reduction potential). $E_H$ is a qualitative measurement but can be of value in understanding the nature of metal species in solutions and also the possible corrosive effects of groundwater upon well materials (see Section 11.2.5).

$E_H$ is measured using dual or combination probes consisting of a platinum redox and reference electrode in one unit. The sensing element is platinum foil or wire which in either case must be kept polished if meaningful and stable $E_H$ readings are to be obtained. In addition, in the presence of high concentrations of sulphide the platinum surface can become temporarily 'poisoned', leading to erroneous results. The reference electrode is usually of the Ag–AgCl type, and its filling solution is KCl saturated with AgCl. The potential generated by the reference measured relative to a standard hydrogen electrode (SHE) will vary according to the concentration of the KCl electrolyte as given by

$$E_H = E_0 + C_{SHE} \tag{4.2}$$

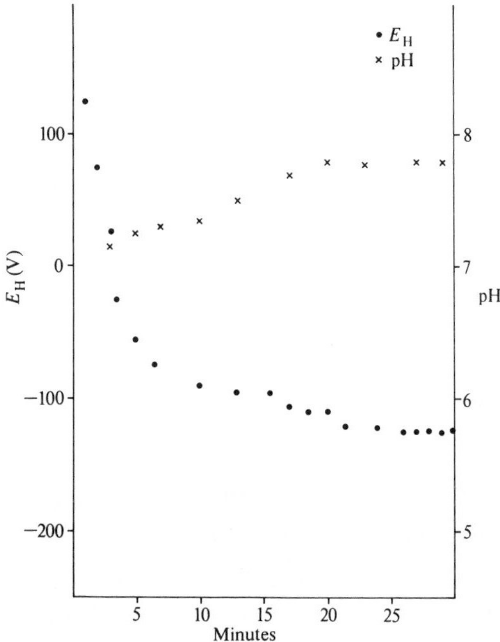


**Fig. 4.7.** Time stabilization of pH and $E_H$ measurements.

where $E_H$ is the redox potential of the sample relative to the normal hydrogen electrode, $E_0$ is the potential developed by the platinum electrode in the sample, and $C_{SHE}$ is the potential developed by the reference electrode relative to the normal hydrogen electrode.

The electrolyte used should have a known concentration (usually 4 M) which ensures that the potential ($C_{SHE}$) generated by the reference can be obtained from one of the various tables given in the literature (Carver 1971). This potential is also dictated by temperature which should be measured when $E_H$ readings are taken.

As the oxidation-reduction state of a solution will change on exposure to the atmosphere $E_H$ measurements have to be carried out in a sealed cell. For convenience at the well head, it is preferable to carry out $E_H$, and temperature measurements in the same cell as shown diagrammatically in Fig. 4.6. The cell is attached to a 'bleed line' on the discharge pipe of the well and water is allowed to flow through the cell until stable $E_H$ conditions are obtained. The time required for stability may be up to 30 min as shown in Fig. 4.7.

### *4.2.5. Dissolved oxygen*

The presence of dissolved oxygen (DO) in groundwaters normally indicates previous and fairly recent exposure of the water to atmospheric influences such as recharge. Once a sample is taken, reoxygenation can occur quite rapidly so that, as with $E_H$, dependable measurements can only be obtained from a sample which is sealed from the atmosphere in the type of cell shown in Fig. 4.6.

The electrode used for the measurements of dissolved oxygen is usually of the MacKereth type. This consists of a central lead anode covered with polythene and surrounded by a perforated coaxial silver cathode which is itself coated with an oxygen-permeable polythene membrane. The annular space between the two electrodes is filled with an alkaline electrolyte. The electrode relies upon the diffusion of oxygen through the membrane and the subsequent reduction of the oxygen at the cathode to give a current proportional to the oxygen partial pressure. Water in the DO cell flows past the electrode and dissolved oxygen diffuses through the membrane into the cell where it is consumed. Minimum flow conditions are critical since if the flow is too low the sample water around the electrode may be depleted of oxygen. A flow rate of about 10 cm $s^{-1}$ is usually recommended.

## 4.3. Sampling and the influence of well conditions

Because wells disturb the natural ground conditions the chemistry of well waters is unlikely to represent true aquifer hydrochemistry; in particular they often provide a mixture of waters from several horizons. In sampling, therefore, account must be taken of the source situation to ensure that minimal changes in the aquifer hydrochemistry have occurred. Changes at a source are dependent upon well construction, construction materials, aquifer penetration, and piezometric level. Spring sources can also be considered in this discussion.

### *4.3.1. Non-flowing wells and springs*

Pumping wells and springs normally provide the easiest sources for groundwater samples. As with any sampling situation clean easily sealed bottles should be used for sampling. Bottles should be totally filled with minimal exposure of the sample to the atmosphere. A litre sample should be adequate for analyses of the major and minor ions listed in Fig. 4.1. Half of the sample should preferably be acidified for cation analysis. 10 ml of concentrated hydrochloric acid of analytical grade quality should suffice.

Hydrochemical stratification is a common feature of aquifers so that samples taken from pumping wells will inevitably have mixed chemistry. Nevertheless, in practical terms they may be the only means of obtaining information. Well-head measurements can easily be performed on pumping wells provided that care is taken to restrict contact between the discharge and the atmosphere. A 'bleed line' from the pump discharge is usually a convenient method of sampling, with measurements carried out in a cell (Fig. 4.6). $E_H$ and DO measurements can be suspect in pump discharges where old rising mains draw air into the discharge through poor joints.

Notwithstanding mixing, samples from pumping wells are most reliable chemically if pumps are in regular operation. Samples from wells that are rarely pumped or are not pumped at all can be very suspect owing to the chemical changes that take place during water stagnation in a well. This is a particular problem within cased sections in a well in which reduction can radically alter groundwater chemistry. Some of the types of changes that can occur are discussed in Section 4.3.2 with respect to flowing wells where they can be more easily recognized. It is emphasized, however, that the same type of changes are particularly significant in unpumped non-flowing wells.

Springs flowing from unconfined aquifers theoretically provide excellent sampling points; unfortunately they are frequently polluted at the source. Where flow is from confined aquifer conditions, aeration at the discharge point completely changes the hydrochemistry so that it becomes totally unrepresentative. To overcome the problems at spring sources it is advisable to drive a sampling tube as deeply into the bedrock as possible. Perforated well points form convenient tubes.

### *4.3.2. Flowing wells*

Irrespective of groundwaters mixing as they flow up a well, it is probable that flowing-well hydrochemical data are more dependable than pumping-well data. Problems can occur, however, and it is particularly important, as explained below, that a well be allowed to flow for some time before well-head measurements and samples are taken.

Marsh and Lloyd (1980), in an investigation of flowing-well chemistry in limestones, examined hydrochemical changes in water stored in casing in wells with a natural flow potential by allowing a small constant flow to occur after a well had been standing for some time until the cased solution was completely flushed. The distance travelled by a particle of water up the well is defined as

$$d_t = \frac{Qt}{\pi r^2} \tag{4.3}$$

where $d_t$ is the distance travelled by a particle after time $t$, $Q$ is the discharge rate, $t$ is the time, and $r$ is the diameter of the internal casing.

The type of changes encountered in well-head chemistry are illustrated in Fig. 4.8 for wells with mild steel casing in their upper sections. The chemical parameters are seen to change rapidly, with stability not being established until some time after a volume of groundwater in excess of the complete well storage has been removed. The modifications of the groundwaters are the result of reactions with the casing by acid attack due to carbonic acid or hydrogen sulphide:

$$Fe + 2H^+ + CO_3^{2-} \rightleftharpoons FeCO_3 + H_2 \tag{4.4}$$

$$Fe + 2H^+ + 2HS^- \rightleftharpoons FeS_2 + 2H_2\,. \tag{4.5}$$

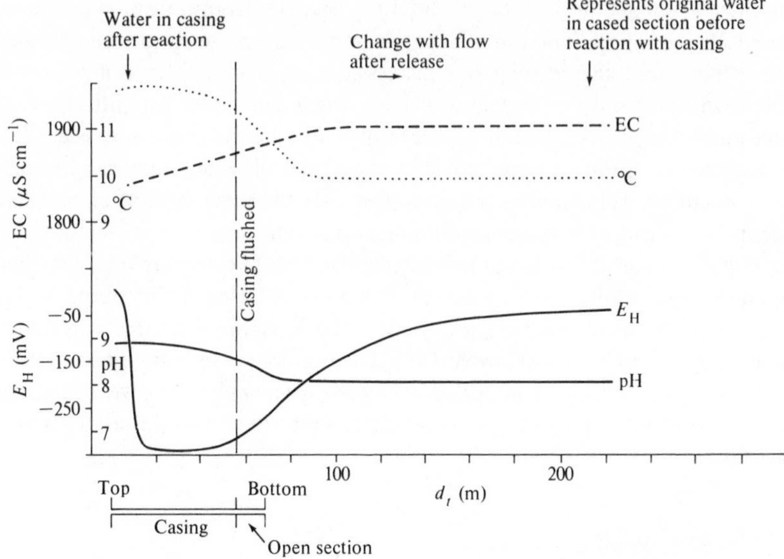


**Fig. 4.8.** Changes in flowing-well chemistry after the release of the well following a period of stagnation. The well is partially cased with mild steel.

Both of these reactions result in the removal of hydrogen ions from solution with a consequent increase in pH. The progress of these reactions will be governed by the rate of removal of the molecular hydrogen. In the presence of suitable bacteria this might occur very quickly. The continued supply of molecular hydrogen for oxidation and the precipitation of sulphides causes the reduction of $E_H$ shown in Fig. 4.8. It is probable that these reactions are aided by bacterial action. The lower $E_H$ and the high pH within the casing is therefore attributed to corrosion and precipitation reactions. The low electrical conductivity in the casing shown in Fig. 4.8 is due to the high pH which causes carbonate mineral precipitation and thus a depletion in dissolved solids in the stored water.

In contrast with the hydrochemical effects outlined above, the same study showed that in wells lined with non-metal casings such effects were very slight.

The inferences in the changes discussed are probably more important for non-flowing unpumped wells in that samples from within the steel-cased sections are unlikely to have any validity in terms of aquifer hydrochemistry.

### *4.3.3. Hydrochemical sampling and monitoring programmes*

Sampling and monitoring will be unique to each hydrogeological environment. While statistical methods have been applied to determine representative well patterns under average conditions for sampling or monitoring purposes, such techniques have limited values and programmes must be established to suit local requirements. Programmes should be flexible and take account of time changes in groundwater chemistry. Unfortunately most programmes are constrained by finance so that the advisable practice is to select sampling points for full sampling and analyses (i.e. major and minor ions) with other support sampling points at which only indicator parameters are measured such as electrical conductivity. In monitoring programmes sequential sampling is often only performed twice annually for full analyses where groundwater conditions undergo significant seasonal effects. Depending upon local conditions, indicator parameter surveys may be more frequent. Hydrochemical monitoring on a monthly basis or less is usually only necessary where pollution or saline groundwater intrusion is a problem.

Sampling and monitoring programmes are usually carried out from surface sources although depth sampling and logging (Section 4.4) are frequently included. Sequential sampling from layered systems can be carried out by depth sampling, but in order to avoid mixing of groundwaters *in situ* samplers can be installed as shown in Fig. 4.9. In the type of sampler shown the formational water that enters through the intake chamber is periodically removed by displacement with nitrogen.

## 4.4. Hydrochemical logging and depth sampling

### *4.4.1. Temperature and electrical conductivity logging of well fluids*

Profile logging using combination temperature and electrical conductivity probes is of particular value in hydrochemical studies. Measurements can be made in absolute terms or in differential terms using a double-probe system or surface module differentiator. This type of logging provides much hydrogeological data and control of the depth-sampling location. Logging can be carried out in most types of well and is often used below the pump base in pumping wells. The best results are normally obtained from observation wells adjacent to pumping wells that are discharging at a low rate. Temperature and conductivity profiles of wells in which there is no groundwater movement can prove misleading.

Temperature is measured by a thermistor or thermistor pair. For reliable measurements the thermistor head should be well exposed to water in the well, although protected. Absolute measurements can be made with laboratory-calibrated thermistors to a practical precision of $\pm 0.1$ °C. Such temperature measurements can then be used in thermodynamic equations for hydrochemical interpretation as they will be more reliable than measurements made at the surface from depth samples. Differential temperature measurements can be made to a precision of $\pm 0.01$ °C and are very useful in understanding well hydraulics. They are, however, of little direct hydrochemical use.

Electrical conductivity in logging probes is measured in the same manner as in surface sampling (Section 4.2). Absolute values are obtained that can be corrected to a standard temperature by correlation with the temperature log.

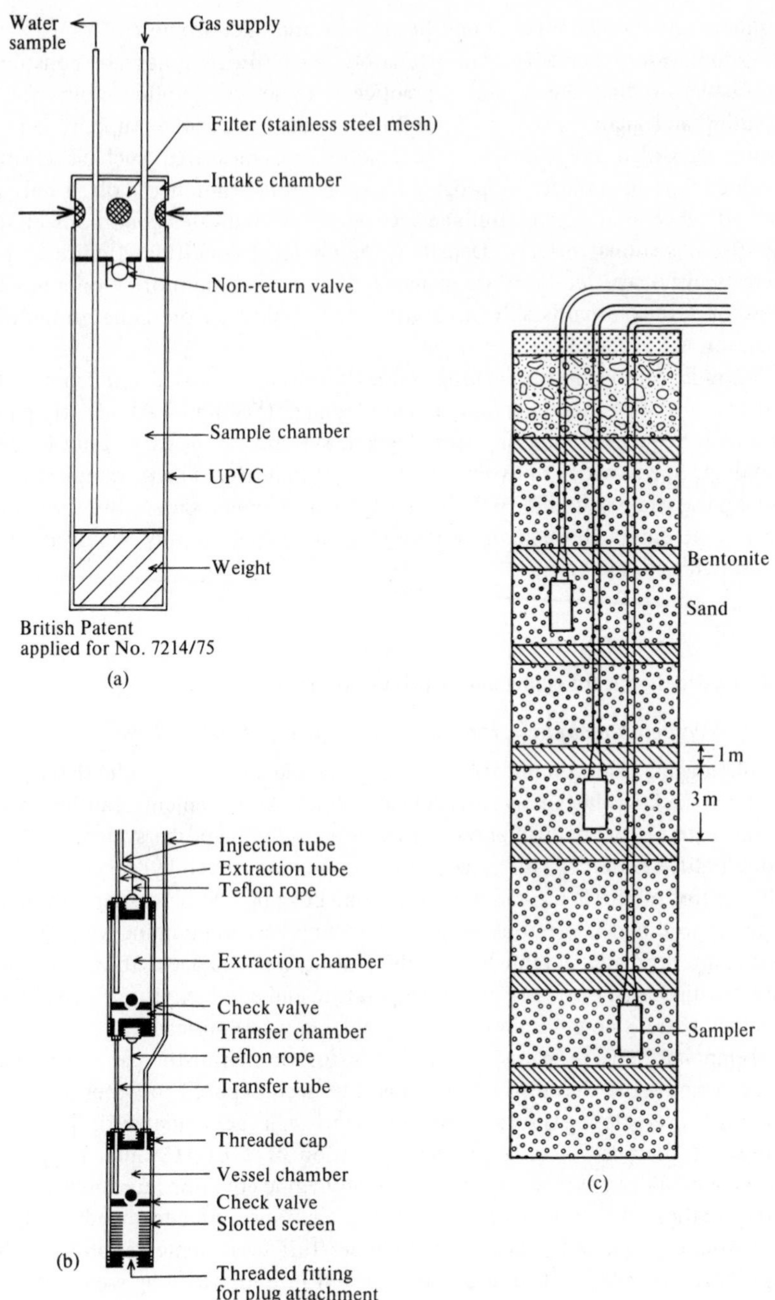


**Fig. 4.9.** *In situ* samplers: (a) gas lift sampler (Barber, Maris, and Knox 1977.); (b) gas lift sampler and transfer vessel (Morrison and Brewer 1981); (c) sampler set separated by bentonite in well.

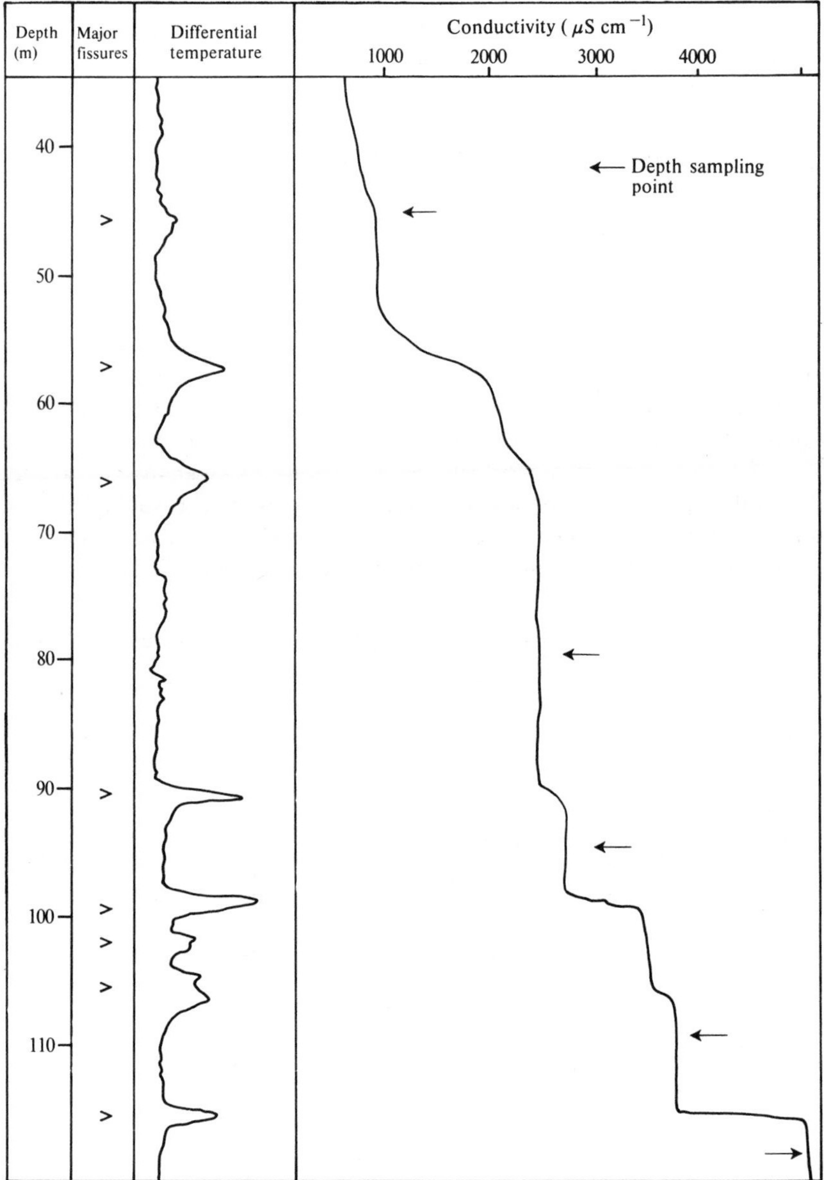


**Fig. 4.10.** Electrical conductivity and differential temperature logs from a well in a limestone aquifer with the sampling points shown.

The most efficient method of producing corrected data is by obtaining digitized output from the logging module that can then be computer processed. Differential electrical conductivity logs can be run and can prove very valuable.

An example of temperature and conductivity logging results for a limestone aquifer are shown in Fig. 4.10. Temperature differences indicate the presence of fissure zones where groundwater flow is occurring and the electrical conductivity shows marked increases at certain levels in the well. Obviously the conductivity log can be of greater importance where saline groundwaters are present in an aquifer. While interpretation of the type of log shown in Fig. 4.10 can be straightforward, careful account must be taken of well hydraulics and overall piezometric conditions in an aquifer before the measured conductivity, and indeed the chemistry from a depth sample, can be related to a specific aquifer zone. Because most aquifers are permeability stratified and many are hydraulically disturbed, differential flow and head conditions will exist to provide a conductivity profile in a well that may be sigificantly different from that in the aquifer. An example of a well conductivity profile from a sandstone aquifer is compared with a formation resistivity log, a differential temperature log, and the electrical conductivity of pore water samples in Fig. 4.11. The well fluid profile is at variance with the pore water data and the formation resistivity as a result of piezometric effects. Given that the pore water data are rarely available, con-

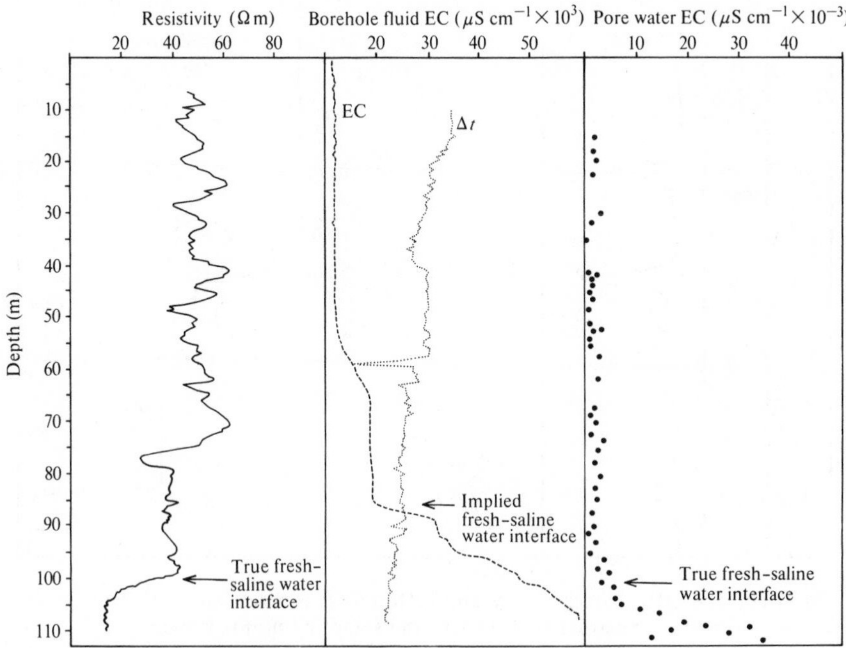


**Fig. 4.11.** Comparison of formation resistivity, borehole fluid EC, and pure water EC at a fresh–saline groundwater interface in a sandstone aquifer.

clusions drawn in this case as to the depth to the saline groundwater without a resistivity log would be totally misleading. Where conductivity logging is to be carried out to determine relationships between fresh and saline groundwaters, formational resistivity control should be established before well-fluid logs are interpreted. Time is an important factor in that, with the displacement of saline groundwater up a well, long-term diffusion from the well to the formation can mask differences. Logging should therefore be carried out during or immediately after drilling.

### *4.4.2. Formation logs and groundwater salinity interpretation*

The details of formational logging are outside of the scope of this book and the reader is referred to Scott-Keys (1968), Schlumberger (1974), Roy and Baksi (1977), and Robinson and Oliver (1981) for detailed explanations; they should, however, be considered together with hydrochemical logging techniques.

### *4.4.3. Depth sampling*

Depth sampling is an essential part of hydrochemical studies in that it allows a three-dimensional approximation of groundwater chemistry. Sampling is most advantageously carried out in combination with conductivity logging. Sample locations related to the conductivity profile are indicated in Fig. 4.10.

Various depth samplers are available, the most efficient being electrically operated units. Once the sampler is set at the required depth a capacitor is charged at the surface which when fired operates a solenoid in the sampler causing the valves to close. Sample volumes of up to 1.5 litres can be conveniently obtained.

## 4.5. Sampling for trace heavy metals

Hydrochemical studies involving heavy metals can be of value in typing groundwaters (see Chapter 6) in corrosion assessment (see Section 11.2.5), and in pollution investigations.

### *4.5.1. Storage of samples*

The very low concentration of trace heavy metals in groundwaters makes it imperative that extreme care be taken to ensure that there is no change in trace metal composition between the time of sampling and the time of analysis, as even a minor change may cause serious error in the results. The heavy metal composition of a stored sample can be affected by three processes: precipitation, adsorption, and contamination. On removal from an aquifer, the chemical environment of a groundwater sample is changed and the sample usually re-equilibrates with the exsolution of $CO_2$ and the solution of $O_2$. This results in an increase of pH and $E_H$ and a consequent reduction in the solubility of most metals. This accounts for the appearance of iron hydroxide in an initially clear sample. Even if the solubility of an element is not reduced sufficiently to cause precipitation, it may be entrained in the precipitate of another element and thus coprecipitate.

Heavy trace metals can be removed from solution by adsorption on the surface of storage bottles. Plastics and glass are supercooled liquids which have high surface energies due to the presence of distorted and broken chemical bonds, and consequently they have strong sorptive tendencies. Eicholz, Galli, and Elston (1966) show that there is no automatic advantage in using plastics in preference to glass, but that adsorption is reduced by lowering the pH. Robertson (1968*a*), in a more rigorous study of the problem, confirmed this view but recommended that polythene bottles should be used and that the sample be acidified to pH 1.5. As a specific example King, Rodriguez, and Wai (1974) showed that cadmium is strongly adsorbed by glass at pH 8, whereas it is adsorbed only very slightly by plastics; however, lowering the pH to less than 7 eliminated adsorption. Gavrishin (1968) has noted that adsorption appears to be inversely proportional to the concentration of the element concerned, and that adsorbed elements can be recovered by reducing the pH after long periods of storage. He recommends that samples should be acidified prior to analysis rather than at the time of sampling, so that the danger of leaching contaminants from the container is reduced. Robertson (1968*b*), in extensive tests, showed that glass contained relatively large amounts ($\approx$ 500 ppb) of iron, antimony, zinc, and hafnium and significant amounts ($\approx$ 80 ppb) of scandium and cobalt, whereas polyethylene generally contained low concentrations of all the elements studied except for iron. Polyethylene tends to gain adsorptive capacity with use owing to the degradation of its surface; thus bottles must be thoroughly washed in acid before use. The acidification may also cause contamination unless an analytical grade acid is used.

For sampling purposes analytical grade acid is frequently added to a sample to fix the heavy metals in solution. 10 ml of hydrochloric acid (0.5 molar) per litre of groundwater sample is usually adequate. Acidification is not, however, the most convenient means of obtaining samples so that techniques of field precipitation or extraction of heavy metals have been developed.

### *4.5.2. Field concentration methods*

The concentration technique chosen is dictated by the concentration and nature of the metals of interest in the sample and by the sensitivity and sample requirements of the final laboratory analysis to be employed. In most cases of isolation of metals, such as those of the first transition series, 0.5–5 l aliquots are sufficient for most purposes. However, where metals are expected at very low concentrations large volumes of up to several thousand litres may be treated using continuous-flow systems such as those employing large area adsorbates, e.g. manganese oxide on cellulose.

Concentration techniques which produce a solid sample, such as those involving coprecipitation, ion exchange resins, or adsorbates, are useful where analytical methods requiring solid samples, such as neutron activation or X-ray fluorescence spectrometry, are to be used. Solvent extraction methods produce liquid concentrates which are useful when methods such as atomic absorption spectrometry, plasma spectroscopy, or flame emission spectroscopy are to be used. In both cases, however, the possibilities that the concentration step may not result in 100 per cent recovery of the metals of interest must be borne in mind. In these cases careful pilot studies involving the use of spiking and recovery experiments should be performed prior to the initiation of a full analytical programme.

## 4.6. Sampling for environmental isotopes

The significance of environmental isotopes in groundwater studies is discussed in Chapter 8. Sampling in the cases of tritium and the stable isotopes is a simple process, but in the case of radiocarbon and certain of the less studied isotopes it can pose considerable problems.

### *4.6.1. Tritium sampling*

A 250 $cm^3$ bottle is adequate for groundwater sampling. The bottle should be thick plastic with an inside sealing cap and should be completely filled and sealed immediately to avoid tritium entering the sample from the atmosphere. Paper sealing washers are not satisfactory. Field concentration is not practical.

### *4.6.2. Sampling of the stable isotopes of oxygen and hydrogen*

A 20 $cm^3$ bottle is adequate for groundwater. It should be of thick plastic with an inside sealing cap and should be completely filled when the sample is taken. Paper sealing washers are not satisfactory. Contamination can be caused by molecular exchange with atmospheric water vapour. Oxygen and hydrogen isotopes are used to study evaporative processes and therefore particular care must be taken to avoid evaporation during sample collection.

### *4.6.3. Sampling for carbon isotopes*

In the existing $^{14}C$ analysis methods 2 g of carbon are normally required for positive analysis. Such an amount signifies large volumes of water by sampling standards in most groundwaters. As the $^{14}C$ is predominantly held in the dissolved carbonate species an estimate of the sample volume required can be made using the concentrations of these species as follows:

$$\text{sample volume} = \frac{334}{[HCO_3^-] + [CO_2]} \text{ litres} \tag{4.6}$$

where $[HCO_3^-]$ and $[CO_2]$ are im milliequivalents per litre. This volume is somewhat in excess of that listed elsewhere (International Atomic Energy Agency 1968) but is recommended in view of the high cost of analysis.

Sample concentration is carried out on site using a precipitation cone of volume 50 or 100 litres. The reagents required for a 50 litre sample are applied in the following concentrations and order once the sample has been introduced into the cone under non-turbulent and non-aerated conditions: 5 g $FeSO_4.7H_2O$; 0.5 litre $BaCl_2$ (saturated solution); 50 ml NaOH (carbonate-free solution containing 200 g NaOH $l^{-1}$). The sample should be stirred and then sealed from the atmosphere. Precipitation should be completed in about 30 min.

$^{13}C$ is normally analysed from the barium carbonate precipitation, but for the analysis of $^{13}C$ only from groundwater, samples can be collected in 0.5 or 1 litre bottles containing 50 ml of saturated $SrCl_2$-$NH_4OH$ solution (Friedman 1970). All dissolved inorganic carbonate is precipitated as $SrCO_3$.

In the interpretation of the carbon isotope data the $^{13}C$ value of the matrix is required. This involves control sampling of the aquifer material with the selected samples containing at least 20 mg of carbonate for reliable analyses.

## 4.7 Pore water sampling

Sampling of pore waters from cored material is becoming of increasing interest in the study of groundwater hydrochemical evolution, in saline groundwater studies, and in pollution problems. Sampling is difficult with respect to both

**Table 4.1** *A procedure for pore water sampling*

| |
|---|
| Rotary air-flush coring, various diameters, usually 100 mm<br>Mylar drilling sleeve incorporated in core barrel |
| ↓ |
| Core extruded in Mylar and taped to secure<br>Sealed, with tape, in further polythene bag |
| ↓ |
| Remove immediately to on-site freezer |
| ↓ |
| Transport and store frozen |
| ↓ |
| In laboratory, defrost, remove and discard top and bottom 4 cm and outer 1.5 cm |
| ↓ |
| Centre section broken up to a limited extent loaded into centrifuge cups and balanced to within ± 0.2 g of each other |
| ↓ |
| Centrifuged at 12 000 rev $min^{-1}$ for 30 min at 20 °C |
| ↓ |
| Extracted water combined for cups containing fractions of the same sample, filtered, and placed in the storage vessels |
| ↓ |
| Extracted water stored at 4 °C until required |
| ↓ |
| Separate sample of core dried for moisture determination |

After Wheatstone 1978

core recovery and contamination of pore water by drilling fluids. Contamination depends upon intergranular throat sizes and can be monitored in fluid drilling by the addition of a lithium chloride (LiCl) tracer to the fluid. Indications are that very significant invasion can occur in, for example, sandstones, so that fluid drilling should be avoided and air used as the flushing medium. Air-flush drilling, however, has depth limitations so that pore water sampling at greater than, say, 300 m is extremely difficult or very expensive. A flow chart for pore water sampling prcedure is given in Table 4.1.

### *4.7.1. Core sampling*

Efficient sampling can be carried out using a double-core barrel fitted with an inner Mylar drilling sleeve (Severn Trent Water Authority 1979). The Mylar drilling sleeve consists of a clear hard plastic which fits the core barrel as shown in Fig. 4.12. Face discharging bits with an oversize kerf must be used in order that the core passes smoothly into the Mylar. The core is extruded ready wrapped so that it can be immediately frozen to avoid either contamination or chemical change.

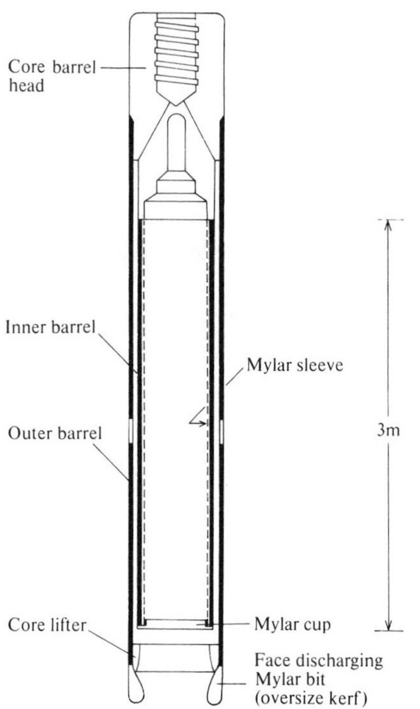


**Fig. 4.12.** Core barrel with Mylar drilling sleeve.

### 4.7.2. *Sample storage*

Unless cores are properly preserved, changes in pore water chemistry can occur before analyses are performed. Examples of chemical changes that can occur are listed in Tables 4.2 and 4.3. Total organic carbon (TOC) is included as a control parameter as it is believed that the main changes are attributable to bacterial action. To negate such changes freezing of samples immediately they are obtained is recommended. Wheatstone (1978) has shown that freezing has no adverse effect upon a wide range of determinands.

**Table 4.2** *Changes in pore water chemistry after various storage periods*

| Lithology | Extraction schedule | TOC | $Na^+$ ($mg\ l^{-1}$) | $Ca^{2+}$ ($mg\ l^{-1}$) | $Cl^-$ ($mg\ l^{-1}$) | $SO_4^{2-}$ ($mg\ l^{-1}$) |
|---|---|---|---|---|---|---|
| Chalk | a | 7.2 | 1380 | 690 | 3961 | 252 |
| | b | — | — | — | 3744 | — |
| | c | 5 | 1280 | 710 | 3778 | 461 |
| Chalk | a | 10.4 | 1280 | 620 | 3459 | 98 |
| | b | 7.6 | 1160 | 620 | 3260 | 126 |
| | c | 4.0 | 1160 | 520 | 3323 | 147 |
| Chalk | a | 6.9 | 810 | 450 | 2119 | 35 |
| | b | 5.0 | 830 | 370 | 2104 | 91 |
| | c | 2.6 | 881 | 470 | 2156 | 468 |

After Barber *et al.* 1977.
a, extraction within 2 h of recovery.
b, extraction within 3 months of recovery at 4 °C.
c, extraction within 3 months of recovery at 20 °C.

**Table 4.3** *Changes in TOC after various storage periods*

| Lithology | 1–2 days | 6 months | ΔTOC |
|---|---|---|---|
| Chalk | 10 | 16 | + 6 |
| Chalk | 13 | 7 | − 6 |
| Chalk | 19 | 10 | − 9 |
| Sandstone | 19 | 9 | −10 |
| Sandstone | 38 | 10 | −28 |
| Peat | 48 | 35 | −13 |
| Peat | 106 | 32 | −74 |
| Peat | 54 | 26 | −28 |
| Peat | 64 | 42 | −22 |
| Clay | 69 | 33 | −36 |
| Clay | 38 | 30 | − 8 |
| Clay | 29 | 28 | − 1 |
| Clay | 48 | 33 | −15 |
| Clay | 66 | 33 | −33 |

After Barber *et al.* 1977.

### *4.7.3. Pore water extraction*

Extraction from the core once it has been thawed is carried out using high speed centrifuges which are thermostatically cooled. Samples are broken and placed in a centrifuge cup as shown in Fig. 4.13 or inserted as milled plugs. Centrifuge speeds of up to 14 000 rev $min^{-1}$ are used, and as shown in Fig. 4.14 the majority of the pore water can be removed in about 20 min. Speeds, however, need not be as high as shown in Fig. 4.15.

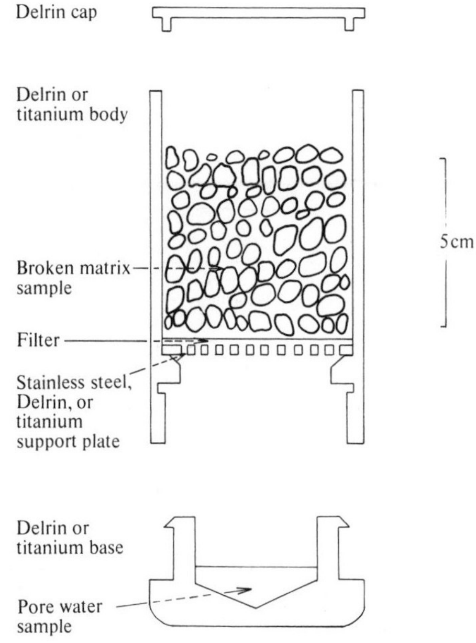


**Fig. 4.13.** Centrifuge cup for pore water sampling.

### *4.7.4. Sample volume requirements*

The amount of core material that needs to be centrifuged and the speed and time of centrifuging will be dictated by the volume of pore water sample required for analysis. This volume of pore water will depend upon the specific analysis to be undertaken and upon the method of analysis adopted. An indication of the very small amounts of pore water that can be utilized is given in Table 4.4.

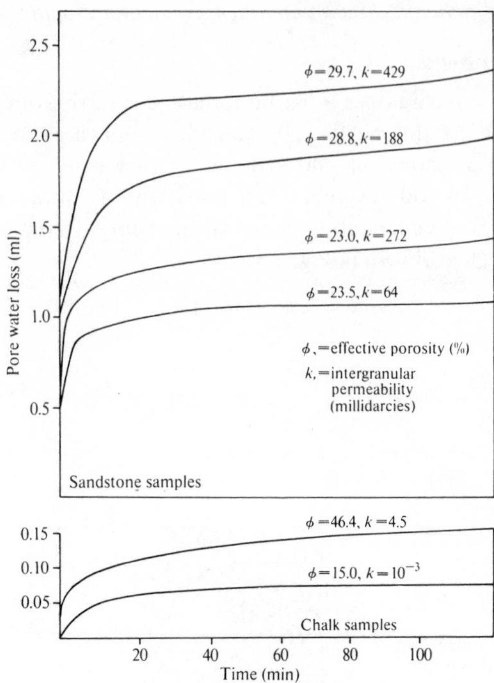


**Fig. 4.14.** Pore water loss from samples of sandstone and chalk as a function of centrifuging time: sandstone curves after Lovelock (1972) and chalk curves after Edmunds and Bath (1976).

**Table 4.4** *Volumes of pore water required for the analysis of various determinands*

| Determinand | Analytical technique | Volume (ml) |
|---|---|---|
| pH value | Combination electrode | 0.0 |
| EC | Microcell | 0.0 |
| Chloride | Autoanalyser | 0.3 |
| Ammonia | Autoanalyser | 0.3 |
| Nitrite | Autoanalyser | 0.6 |
| Nitrate | Autoanalyser | 0.4 |
| Alkalinity | Autoanalyser | 1.3 |
| Sulphate | Autoanalyser | 1.0 |
| Orthophosphate | Autoanalyser | 3.3 |
| Sodium | Flame (emission) | 2.5 |
| Potassium | Flame (emission) | 2.5 |
| Calcium | Flame (emission) | 2.5 |
| Magnesium | Flame (absorption) | 2.0 |
| Iron | Flame (absorption) | 2.0 |
| Manganese | Flame (absorption) | 2.0 |
| TOC | Tocsin (phase separation) | 6.0 |
| | Total | 26.7 |

After Wheatstone 1978.

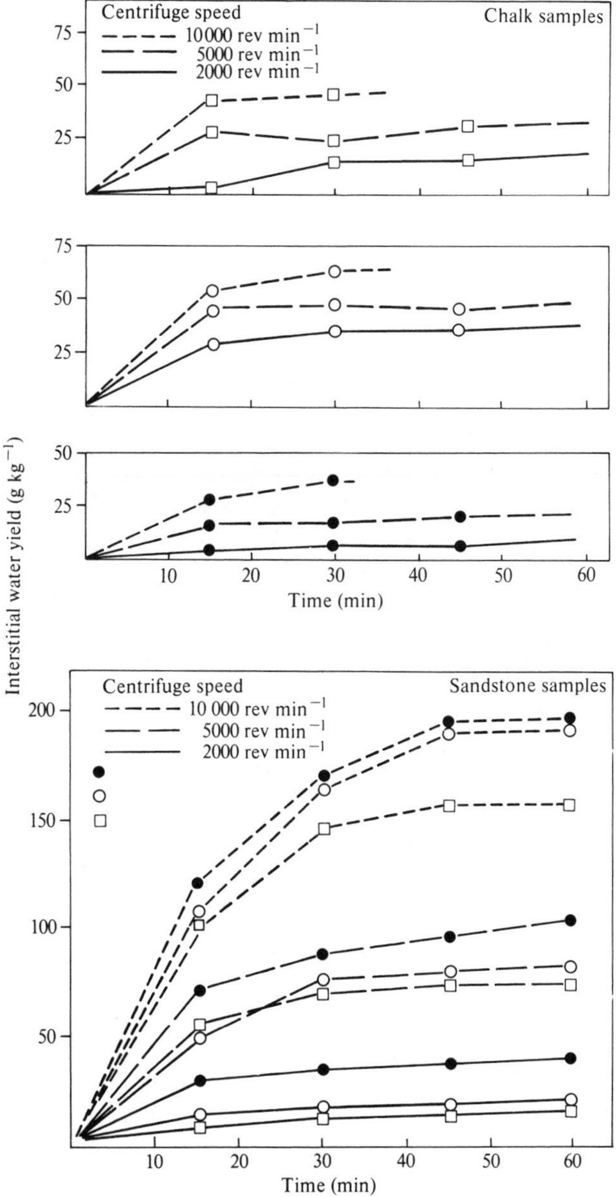


**Fig. 4.15.** Pore water yields for various centrifuge speeds. (After Barber *et al.* 1977.)

**Table 4.5** *Some recommended techniques for the analysis of major and certain minor determinands in groundwaters*

| Determinand | Analytical method | Reference | Analytical working range | Approximate detection limit | Minimum sample volume (ml) | Typical concentration in groundwaters |
|---|---|---|---|---|---|---|
| $Al^{3+}$ | Complex with 8-hydroxyquinoline, extract in field into MIBK analysis by flame AA ($N_2O$–$C_2H_2$) | 1 | 5–50 $\mu g\ l^{-1}$ | 2 $\mu g\ l^{-1}$ | 500 | 1–60 $\mu g\ l^{-1}$ |
| $Ba^{2+}$ | Flame AA ($N_2O$–$C_2H_2$) | — | 0.1–1.0 $mg\ l^{-1}$ | 0.1 $mg\ l^{-1}$ | 2 | 0.01–2.0 $mg\ l^{-1}$ |
| $HCO_3^-$ | Potentiometric titration with 0.01 M $H_2SO_4$ | 2 | 100–400 $mg\ l^{-1}$ | 10 $mg\ l^{-1}$ | 25 | 50–400 $mg\ l^{-1}$ |
| $Br^-$ | Neutron activation analysis | 3 | 0.5–100 $mg\ l^{-1}$ | 0.1 $mg\ l^{-1}$ | 1 | 0.5–100 $mg\ l^{-1}$ |
| $Ca^{2+}$ | Flame AA (air–$C_2H_2$) | — | 5–40 $mg\ l^{-1}$ | < 1 $mg\ l^{-1}$ | 2 | 5–500 $mg\ l^{-1}$ |
| $Cl^-$ | $AgNO_3$ titration using Ag–AgS ion selective electrode as monitor | — | 5–800 $mg\ l^{-1}$ | 1 $mg\ l^{-1}$ | 10 | 10–200 $mg\ l^{-1}$ |
| $Cu^{2-}$ | Flameless AA (direct) | 4 | 1–50 $\mu g\ l^{-1}$ | 1 $\mu g\ l^{-1}$ | 1 | 1–100 $\mu g\ l^{-1}$ |
| $F^-$ | Ion selective electrode[a] | — | 0.1–10 $mg\ l^{-1}$ | 0.05 $mg\ l^{-1}$ | 10 | 0.1–5 $mg\ l^{-1}$ |
| Fe(total) | Flameless AA | 5 | 2–100 $\mu g\ l^{-1}$ | 2 $\mu g\ l^{-1}$ | 1 | 10–20000 $\mu g\ l^{-1}$ |
| $Fe^{2+}$ | Spectrophotometric with 2.2′-bipyridyl[a] | 6 | 0.08–1.0 $mg\ l^{-1}$ | 0.01 $mg\ l^{-1}$ | 25 | 0.01–20 $mg\ l^{-1}$ |
| $Li^+$ | Flame AA (air–$C_2H_2$) | 7 | 20–200 $\mu g\ l^{-1}$ | 5 $\mu g\ l^{-1}$ | 2 | 2–60 $\mu g\ l^{-1}$ |

| | | | | | | |
|---|---|---|---|---|---|---|
| $Mg^{2+}$ | Flame AA (air–$C_2H_2$) | — | 0.5–6.0 mg $l^{-1}$ | 0.05 mg $l^{-1}$ | 2 | 1–200 mg $l^{-1}$ |
| Mn | Flameless AA (direct) | 8 | 1–50 $\mu$g $l^{-1}$ | 0.5 $\mu$g $l^{-1}$ | 1 | 1–50 $\mu$g $l^{-1}$ |
| $Ni^{2+}$ | Flameless AA (direct) | 4 | 10–100 $\mu$g $l^{-1}$ | 5 $\mu$g $l^{-1}$ | 1 | 1–50 $\mu$g $l^{-1}$ |
| $NO_3^-$-N | Spectrophotometric (UV)[a] | 9 | 0.1–100 mg $l^{-1}$ | 0.01 mg $l^{-1}$ | 1 | 1–100 mg $l^{-1}$ |
| $HPO_4^-$ | Spectrophotometric (molybdenum blue)[a] | 10 | 0.06–3.0 mg $l^{-1}$ | 8 $\mu$g $l^{-1}$ | 40 | 0.01–0.5 mg $l^{-1}$ |
| $K^+$ | Flame AA (air–$C_2H_2$) | 7 | 0.5–10 mg $l^{-1}$ | 0.1 mg $l^{-1}$ | 2 | 0.2–30 mg $l^{-1}$ |
| $SiO_2$ | Spectrophotometric (molybdenum blue)[a] | 11 | 0.5–4.0 mg $l^{-1}$ | 0.05 mg $l^{-1}$ | 25 | 1–10 mg $l^{-1}$ |
| $Na^+$ | Flame AA (air–$C_2H_2$) | 7 | 0.25–400 mg $l^{-1}$ | 0.1 mg $l^{-1}$ | 2 | 5–1000 mg $l^{-1}$ |
| $Sr^{2+}$ | Flame AA (air–$C_2H_2$) | 12 | 0.05–1 mg $l^{-1}$ | 0.02 mg $l^{-1}$ | 2 | 0.05–10 mg $l^{-1}$ |
| $SO_4^{2-}$ | Titration with $Ba(ClO_4)_2$ using thorin as indicator | 13 | 10–100 mg $l^{-1}$ | 10 mg $l^{-1}$ | 70 | 10–500 mg $l^{-1}$ |
| $I^-$ | Spectrophotometric catalytic oxidation (CeIV, AsIII)[a] | 14 | 0.5–30 $\mu$g $l^{-1}$ | 0.5 $\mu$g $l^{-1}$ | 5 | 1–50 $\mu$g $l^{-1}$ |
| B (total) | Spectrophotometric with Curcumin | 15 | 0.05–3 mg $l^{-1}$ | 0.02 mg $l^{-1}$ | 5 | 0.01–7 mg $l^{-1}$ |

Partly after Edmunds 1981
MIBK, methyl isobutyl ketone.
AA, atomic adsorption.
[a] Methods amenable to adoption using continuous flow or discrete sample autoanalysers (American water Works Association, American Pollution Control Federation, American Public Health Association 1975; Inland Waters Directorate 1979).

*References*: 1, Barnes 1975; 2, Barnes 1964; 3, Cawse and Pierson 1972; 4, Boyle and Edmond 1977; 5, Segar and Cantillo 1975; 6, Rainwater and Thatcher 1960; 7, Ure and Mitchell 1975; 8, McArthur 1977; 9, Miles and Espejo 1977; 10, Murphy and Riley 1962; 11, Fanning and Pilson 1973; 12, David 1975; 13, Fritz and Yamamura 1955; 14, Rodier 1975; 15, American Water Works Association, American Pollution Control Federation, American Public Health Association 1975; 16, Inland Waters Directorate 1979.

**Table 4.6** *Summary of analytical techniques for trace metals*

| Analytical method | Simultaneous multi-element capability | Easy automatic sample change | Sample preparation | Typical detection limits in raw sample | Comments | Reference[a] |
|---|---|---|---|---|---|---|
| Atomic absorption spectrometry | No | Yes | Raw sample or liquid extracts analysed | $X0.0$–$0.X$ $\mu g\ l^{-1}$ | Instrumentation is fairly inexpensive and easily used by semi-skilled personnel. Most commonly used method. | 1 |
| Plasma emission (and flourescence) spectrometry | Yes | Yes | Raw sample or liquid extracts analysed | $X0.0$–$0.X$ $\mu g\ l^{-1}$ | Expensive instrumentation but rapid multi-element analysis of large numbers of samples with high precision and accuracy | 2 |
| Spark source mass spectrometry | Yes | No | Carbon electrodes or sample filaments need to be prepared | $0.X$–$0.00X$ $\mu g\ l^{-1}$ | Very expensive instrumentation which is difficult to calibrate Slow sample preparation and often poor precision. However, very low detection limits are possible. | 3 |

| | | | | | | |
|---|---|---|---|---|---|---|
| X-ray flourescence spectrometry | Yes | No | Various liquid extracts, solid adsorbates, or co-preciptants may be analysed | *X*.0–0.*X* | Expensive instrumentation but precise rapid multi-element analysis is possible. | 4 |
| Neutron activation analysis | Yes | No | Solid extracts are preferred | *X*.0–0.0*X* $\mu g\ l^{-1}$ | Access to a nuclear reactor is required to which solid samples can usually be admitted. Subsequent long counting times (up to 10 days) are required for final analysis. | 5 |
| Polarography | Up to four or five elements | No | Little or none | *X*0.0–*X*.0 $\mu g\ l^{-1}$ | Inexpensive instrumentation ideal for the routine monitoring of trace metals in potable water. High precision is possible. | 6 |
| Anodic stripping voltammetry | Up to four or five elements | No | Little or none | *X*0.0–*X*.0 $\mu g\ l^{-1}$ | Inexpensive instrumentation ideal for the routine monitoring of trace metals in potable water. High precision is possible. | 7 |

[a] These are selected single references only. A more extensive list can be found in Mark and Mattson (1981) and Fishman, Erdmann, and Steinhiemen (1981).

*References*: 1, Prince 1974; 2, Garbarino and Taylor 1979; 3, Taylor and Taylor 1974; 4, Marsh and Lloyd 1976; 5, Lieser, Calmano, Heuss, and Neitzert 1977; 6, Abdullah and Royale 1972; 7, Ben-Bassat, Blindermann, Salomon, and Wakshal 1975.

**Table 4.7** *Summary of analytical techniques for environmental isotopes and gases*

| Isotope | Preparation | Analysis | References |
|---|---|---|---|
| Deuterium | Conversion of 10 $\mu$l water to hydrogen using zinc | $^2H/^1H$ ratio is measured by mass spectrometry. | 1 |
| Tritium | Sample is distilled to dryness to remove solids. Tritium is concentrated by electrolysing 100 ml to a volume of 10 ml. | Liquid scintillation counting or conversion to ethane and gas proportional counting. | 2, 3 |
| $^{13}C$ | Carbonate is precipitated in the field by addition of 10 ml of a solution of 400 g $SrCl_2.6H_2O\ l^{-1}$ in 0.880 $NH_3$ solution to sufficient water to contain 0.5 meq dissolved carbon. After 24 h precipitate is filtered off and washed. In the laboratory $CO_2$ gas is prepared by reaction of carbonate precipitate with 100% $H_3PO_4$. | $^{13}C/^{12}C$ ratio is measured by mass spectrometry. | 4 |
| $^{14}C$ | Carbonate is precipitated in the field by addition of 100 ml carbonate free NaOH solution (200 g NaOH $l^{-1}$) and 1 litre saturated $BaCl_2$ (500 g $BaCl_2$ $l^{-1}$) to 100 litres water. The amount of $BaCO_3$ precipitate required is at least 80 g. If the total alkalinity is less than 250 ppm then carbonate is precipitated from a second 100 litres water sample. In the laboratory the precipitate is acidified producing $CO_2$ which can be converted to benzene or methane. | Liquid scintillation counting of benzene or gas proportional counting of methane. | 5 |
| $^{18}O$ | 10 ml water is equilibrated with $CO_2$ gas permitting exchange of $^{18}O$ between water and gas. | $^{18}O/^{16}O$ ratio is measured by mass spectrometry. | 6, 7 |

| | | | |
|---|---|---|---|
| $^{234}U$, $^{238}U$, $^{230}Th$, $^{232}Th$ | 20 l sample is boiled down to 500 ml. U and Th isotopes are separated by ion exchange and electroplated onto steel planchets. | $^{230}Th$, $^{232}Th$, $^{234}U$, and $^{238}U$ activities are determined by $\alpha$ particle spectrometry. | 8 |
| $^{222}Rn$ | He or $N_2$ is bubbled through the sample to extract Rn which is then trapped on activated charcoal at $-70\ °C$. | Rn is transferred to an $\alpha$ scintillation cell and counted in a gross $\alpha$ scintillation counter. | 9 |
| $^{226}Ra$ | Acidified samples are stored for 20 days for $^{222}Rn$ in growth from $^{226}Ra$ to accumulate. The second-generation $^{222}Rn$ is extracted and analysed as in the procedure for $^{222}Rn$ above. | As for $^{222}Rn$. | 9 |
| $^{234}Ra/^{226}Ra/^{228}Ra$ activity ratios | $MnO_2$-impregnated acrylic fibres are used to scavenge Ra from 100–1000 l of sample. In the laboratory Ra is leached off the fibre and coprecipitated with $BaSO_4$. | $\gamma$ spectrometry of daughter isotopes. | 10 |
| He, Ne, Ar, Kr, Xe | 1 ml sample is vaporized and the water is removed by freezing down with solid $CO_2$.Ar, Kr, Xe are absorbed on charcoal and He and Ne are inlet to the mass spectrometer after removal of $O_2$ and $N_2$. Ar, Kr, and Xe are sequentially desorbed and analysed. | Isotope ratio mass spectrometry | 11 |
| $^{34}S$, $^{18}O$, in $SO_4^{2-}$ | 20 l of water is collected and passed through columns of Dowex 2 anion exchange resin. In the laboratory, sulphate is eluted with HCl and precipitated as $BaSO_4$. Carbon reduction of $BaSO_4$ at 1000 °C produces $CO_2$ which is analysed for $^{18}O$. $SO_4^2$ for $^{34}S$ analysis prepared by heating a mixture of $BaSO_4$ and silica glass to 1800 °C and reducing $SO_3$ evolved to $SO_2$ over hot copper. | $^{34}S/^{32}S$ and $^{18}O/^{16}O$ ratios are measured by mass spectrometry | 12, 13, 14 |

*References*: 1, Friedman 1953; 2, Cameron and Payne 1965; 3, Florkowski 1981; 4, Friedman 1970; 5, International Atomic Energy Agency 1968; 6, Epstein and Mayeda 1953; 7, O'Neil and Epstein 1966; 8, Osmond and Cowart 1976; 9, Mathieu 1977; 10, Michel, Moore, and King 1981; 11, Andrews and Lee 1979; 12, Rafter 1967; 13, Sakai and Krouse 1971; 14, Bailey and Smith 1972.

## 4.8. Laboratory measurements

Laboratory measurements and techniques are outside the scope of this book. A considerable literature covering analytical methods is available and the reader is referred to Rodier (1975), American Water Works Association (1980), and Cook and Miles (1980).

The amount of time and expense devoted to chemical analyses should be proportional to the level of interpretations required. Frequently, it is the case that analyses are performed that are too detailed and parameters are determined that are never interpreted. The choice of a particular programme should be dictated by the following factors, as far as sample and method, respectively, are concerned (Edmunds 1981):

(i) sample: collection, storage, and pre-treatment conditions; available volume; distinction between dissolved and particulate matter.

(ii) method: specificity; sensitivity; precision; accuracy and time (cost of analysis).

Analytical accuracy is the most important and neglected factor after correct sampling in hydrochemical studies. Modern rapid analytical techniques are capable of giving very precise results, although sometimes at the expense of accuracy, and this may be apparent, for example, in a monitoring programme where changes in chemistry vary with a change in analyst. The first check on accuracy is to use suitable working standards, and these must be carefully checked and reviewed frequently, otherwise deterioration either by plating out or by inadvertant evaporation may occur. The sample matrix may seriously affect the accuracy of the result and matched standards should be used where appropriate; it is also wise to use standard addition techniques at regular intervals. Similarly, it is a good idea to check results by different analytical methods if possible. Inter-laboratory comparison studies (Ellis 1976; Wilson 1978) provide the most reliable test of accuracy.

As a guide to hydrochemical analysis techniques generally in use, tables are included below for the major parameters studied in groundwaters.

### *4.8.1. Major and minor ions*

Ellis (1976) has shown that the standard of analysis for many of the major and minor ions in waters leaves much to be desired; of the major ions the greatest range in results appears to be for $SO_4^{2-}$ and $NO_3^-$. Poor results are reported for nearly all the minor ions; however, methods are rapidly improving and the techniques listed in Table 4.5 should prove reliable.

### *4.8.2. Trace metals*

The term 'trace' as used in hydrochemistry usually applies to any component present at a concentration of less than 1 mg $l^{-1}$. As such it applies to a very wide range of the total number of elements in most samples. In an increasing number of studies (i.e. pollution, epidemiological etc.) a knowledge of concentration at which trace elements are present in waters is required. The trace elements are predominantly metals, the heavy metals being important because of their toxicity. The term is used here with reference to all the metals of the first transition series (scandium to zinc) as well as mercury, aluminium, gallium, tin, lead, germanium, and some metalloids such as arsenic, selenium, and antimony.

The requirements of high sensitivity, low detection limit, high sample throughput, and multi-element capability pose severe problems for trace analysis. In most cases the problems of high sensitivity and low detection limit can be partially overcome by the use of preconcentration techniques in the laboratory or field (Section 4.5.2). Those of high throughput and multi-element capability can be overcome by the application of the numerous instrumental methods that have been developed during the last 20 years, many of which are amenable to automation. Therefore when both strategies are combined most circumstances that arise in hydrochemical studies can be catered for.

A very wide literature is available concerning the applications of the various modern instrumental methods to the analysis of trace metals in water, many of which also use preconcentration techniques. For this reason only a brief review of the methods available and some of their advantages and disadvantages can be given here and are listed in Table 4.6.

### *4.8.3. Isotopes*

A list of references related to laboratory analysis of isotopes is given in Table 4.7. A discussion of the techniques is outside the scope of this book.

## References

Abdullah, M. I. and Royale, L. G. (1972). The determination of Cu, Pb, Cd, Ni, Zn and Co in natural waters by pulse polarography. *Anal. Chim. Acta,* **58**, 283–8.

Andrews, J. N. and Lee, D. J. (1979). Inert gases in groundwater from the Bunter Sandstone of England as indicators of age and palaeoclimatic trends. *J. Hydrol.* **41**, 233–52.

American Water Works Association (1980). *Standard methods for the examination of water and waste water*. Washington, DC.

American Water Works Association, American Pollution Control Federation, American Public Health Association (1975). *Standard methods for the examination of water and waste water* (14th edn.). American Public Health Association, Washington, DC.

Bailey, S. A. and Smith, J. W. (1972). Improved method for the preparation of sulphur dioxide from barium sulphate for isotope ratio studies. *Anal. Chem.* **44**, 1542–3.

Barber, C., Maris, P. J. and Knox, K. (1977). Groundwater sampling—the extraction of interstitial water from cores or rock and sediment by high speed centrifuging. *Tech. Rep. 54*, Water Research Centre, Medmenham, England.

Barnes, I. (1964). Field measurements of alkalinity and pH. *U.S. Geol. Surv. water supply Pap. 1535-H.*

Barnes, R. B. (1975). The determination of specific forms of aluminum in natural waters. *Chem. Geol.* **15**, 177–91.

Ben-Bassat, A. H. I., Blindermann, J. M., Salomon, A., and Wakshal, E. (1975). Direct simultaneous determination of trace amounts (ppb) of zinc (II), cadmium (II), lead (II) and copper (II) in ground and spring waters using anodic stripping voltammetry: The analytical method. *Anal. Chem.* **47**, 534–7.

Boyle, E. A. and Edmond, J. M. (1977). Determination of copper, nickel and cadmium in sea water by APDC chelate coprecipitation and flameless atomic absorption spectrometry. *Anal. Chim. Acta* **91**, 189–97.

Bugg, S. F. and Lloyd, J. W. (1976). A study of fresh water lens configuration in the Cayman Islands using resistivity methods. *Q. J. Eng. Geol.* **9**, 291–302.

Cameron, S. F. and Payne, B. R. (1965). Apparatus for concentration and measurement of low tritium activities. *6th Conf. on Radiocarbon and Tritium Dating. Pullman, Washington, USAEC Conf – 650652*, pp. 454–70.

Carver, R. E. (1971). *Procedures in sedimentary petrology*. Wiley-Interscience.

Cawse, P. A. and Pierson, D. H. (1972). An analytical study of trace elements in the atmospheric environment. *Rep. R7134*, Atomic Energy Research Establishment, Harwell.

Cook, J. M. and Miles, D. L. (1980). Methods for the chemical analysis of groundwater. *Rep. 80/5*, Institute of Geological Sciences, London.

David, D. J. (1975). Magnesium, calcium, strontium and barium. In *Flame emission and atomic spectroscopy*, Vol. 3, *Elements and matrices* (eds. J. A. Dean and J. C. Rains), pp. 33–64. Marcel Dekker, New York.

Edmunds, W. M. (1981). Hydrochemical investigation. In *Case-studies in groundwater resources evaluation* (ed. J. W. Lloyd) pp. 87–112. Oxford Science Publishers, Oxford.

— and Bath, A. H. (1976). Centrifuge extraction and chemical analysis of interstitial water. *Environ. Sci. Technol.* **10**, 467–472.

Eicholz, G. G., Galli, A. N., and Elston, L. W. (1966). Problems in trace element analysis in water. *Water Resources Res.* **2**, 561–6.

Ellis, A. J. (1976). The IAGC interlaboratory water analysis comparison programme. *Geochim. cosmochim. Acta* **40**, 1359–74.

Epstein, S. and Mayeda, T. K. (1953). Variation of $^{18}O$ contents of waters from natural sources. *Geochim. cosmochim. Acta* **4**, 213–224.

Fanning, K. A. and Pilson, M. E. Q. (1973). On the spectrophotometric determination of dissolved silica in natural waters. *Anal. Chem.* **45**, 136–40.

Fishman, M. J., Erdmann, E., and Steinhiemen, T. R. 91981). Water analysis; application review. *Anal. Chem.* **53**, 182-R–214R.

Florkowski, T. (1981). Low-level tritium assay in water samples by electrolytic enrichment and liquid scintillation counting in the IAEA laboratory. *Proc. Symp. on Methods of Low-level Counting and Spectrometry, Berlin, STI/PUB/592*, pp. 335–51. International Atomic Energy Agency, Vienna.

Friedman, I. (1953). Deuterium content of natural waters. *Geochim. cosmochim. Acta* **4**, 89–103.

— (1970). Some investigations of the deposition of travertine from hot springs. The isotopic chemistry of a travertine-depositing spring. *Geochim. cosmochim. Acta*. **34**, 1303–15.

Fritz, J. S. and Yamamura, S. S. (1955). Rapid microtitration of sulphate. *Anal. Chem.* **27**, 1461–4.

Garbarino, J. R. and Taylor, H. E. (1979). An inductive-coupled plasma atomic emission spectrometric method for routine water quality testing. *Appl. Spectrosc.* **33**, 220–6.

Gavrishin, A. I. (1968) Adsorption of metals by a glass surface during storage of hydrochemical samples. *Gidrogeol. Sb*. **5**, 111–14.

Inland Waters Directorate (1979). *Analytical manual.* Inland Waters Directorate, Ottowa, Canada.

International Atomic Energy Agency (1968). Guidebook on nuclear techniques in hydrology. *Tech. Rep. Ser. 91.* International Atomic Energy Agency, Vienna.

King, W. G., Rodriguez, J. M., and Wai, C. M. (1974). Losses of trace concentrations of cadmium from aqueous solutions during storage in glass containers. *Anal. Chem.* **46**, 771-3.

Langmuir, D. (1971). The geochemistry of some carbonate groundwaters in central Pennsylvania. *Geochim. cosmochim. Acta* **35**, 1023–45.

Lieser, K. H., Calmano, W., Heuss, E., and Neitzert, W. (1977). Neutron activation as a routine method for the determination of trace elements in water. *J. Radioanal. Chem.* **37**, 717-726.

Lovelock, P. E. R. (1972). Aquifer properties of the Permo-Triassic sandstone of the United Kingdom. *Ph.D. Thesis,* University of London.

McArthur, J. M. (1977). Determination of manganese in natural water by flameless atomic absorption spectrometry. Anal. Chim. Acta **93**, 77-83.

MacDonald, Sir M. and Partners (1965). *Lower Indus Report, Physical Resources —Groundwater,* 6. Report to Water and Power Development Authority, West Pakistan.

Mark, H. B. and Mattson, J. S. (1981). *Water quality measurements. The modern analytical techniques.* Marcel Dekker, New York.

Marsh, J. M. and Lloyd, J. W. (1976). Trace element determination in waters using field concentration techniques and X-ray fluorescence. *Groundwater—Quality, Measurement, Prediction and Protection Conf., Water Research Centre, England,* pp. 298-306. Water Research Centre, Medmenham.

— and Lloyd, J. W. (1980). Details of hydrochemical variation in flowing wells. *Ground Water,* **18** (4) 366-77.

Mathieu, G. (1977). $^{222}$Rn and $^{226}$Ra technique of analysis. In *Transport and transfer rates in the waters of the continental shelf. Rep. Ey-87-S-02-2185.* U.S. Department of Energy, Washington, DC.

Michel, J., Moore, W. S., and King, P. T. (1981). $\gamma$-ray spectrometry for determination of $^{228}$Ra and $^{226}$Ra in natural waters. *Anal. Chem.* **53**, 1885-9.

Miles, D. L. and Espejo, C. (1977). Comparison between an ultraviolet procedure and the 2, 4-xylenol method for the determination of nitrate in groundwaters of low salinity. *Analyst* **102**, 104-9.

Morrison, R. D. and Brewer, P. E. (1981). Air-lift samples for zone of saturation monitoring. *Ground Water monit. Rev.* **1**, 52-5.

Murphy, J. and Riley, J. P. (1962). A modified single solution method for the determination of phosphate in natural waters. *Anal. Chim. Acta* **27**, 31-6.

O'Neil, J. R. and Epstein, S. (1966). A method for oxygen isotope analysis of milligram quantities of water and some of its applications. *J. geophys. Res.* **71**, 4955-61.

Osmond, S. K. and Cowart, J. B. (1976). The theory and uses of natural uranium isotopic variations in hydrology. *At. energy Rev.* **14**, 621-79.

Prince, W. J. (1974). *Analytical atomic absorption spectrometry*, Hyden.

Rafter, T. A. (1967). Oxygen isotopic composition of sulphates. A method for the extraction of oxygen and its quantitative conversion to $CO_2$ for isotope ratio measurements. *New Zealand. J. Sci.* **10**, 493-510.

Rainwater, F. H. and Thatcher, L. L. (1960). Methods for the collection and analysis of water samples. *Water Supply Irrigation Pap.* USGS, Washington.

Robinson, V. K. and Oliver, D. (1981). Geophysical logging of water wells.

In *Case-studies in groundwater resources evaluation* (ed. J. W. Lloyd), pp. 45–64. Oxford Science Publishers, Oxford.

Robertson, D. E. (1968*a*). The adsorption of trace elements in sea water on various container surfaces. *Anal. Chim. Acta* **42**, 533–6.

— (1968*b*). Role of contamination in trace element analysis of sea water. *Anal. Chem.* **40**, 1067–72.

Rodier, J. (1975). *Analysis of water*. Wiley, New York.

Roy, K. K. and Baksi, S. (1977). Model studies on redox logging. *J. Geophys.* **42**, 521–3.

Sakai, H. and Krouse, H. R. (1971). Elimination of memory effects in $^{18}O/^{16}O$ determinations in sulphates. *Earth planet. sci. Lett.* **11**, 369–73.

Schlumberger (1974). *Log interpretation—applications*. Schlumberger, Houston, Texas.

Scott-Keys, W. (1968). Well logging in groundwater hydrology. *Ground Water* **6**, 10–18.

Segar, D. A. and Cantillo, A. Y. (1975). Direct determination of trace metals in seawater by flameless atomic absorption spectrophotometry. *Adv. Chem. Ser.* **147**, 56–81.

Severn Trent Water Authority (1979). Chloride pollution of the Rufford Pumping Station, Central Nottinghamshire. *Rep. RP78/019*, Water Authority Research and Development Project.

Taylor, C. E. and Taylor, W. J. (1974). Multielement analysis of environmental samples by spark source mass spectrometry. *Rep. EPA-660/2-74-001*. National Environmental Research Centre, Office of Research and Development. U.S. Environmental Protection Agency, Corvallis, OR.

Ure, A. M. and Mitchell, M. C. (1975). Lithium, sodium, potassium, rubidium and cesium. In *Flame emission and atomic absorption spectrometry*, Vol. 3, *Elements and matrices* eds. J. A. Dean and T. C. Rains), pp. 1–32. Dekker, New York.

Wheatstone, K. C. (1978). The development of laboratory techniques for the extraction and analysis of pore waters from samples of borehole core. *Rep. RP78/017*, Severn Trent Water Authority, England.

Wilson, A. L. (1978). Analytical implications of harmonised monitoring schemes. *J. Inst. Water Sci.* **32**, 57–65.

Wood, W. W. (1976). Guidelines for collection and field analysis of groundwater samples for selected unstable constituents. *Techniques of Water Resources Investigations of the United States Geological Survey*, Book 1, Chapter D2. U.S. Geological Survey, Washington, DC.

# 5 Calculated parameters

## 5.1. Interconversion of units

There is no generally accepted standard unit for the reporting of hydrochemical analyses, a situation which is likely to persist because different units are useful in particular situations. Concentrations of dissolved solids can be reported as weight per unit weight of water or weight per unit volume of water; gas concentrations can also be reported in volume per unit volume.

Weight/weight units have the advantage that they are independent of volume changes caused by changes of temperature or pressure of the solution; for this reason they are favoured for thermodynamic calculations and also in oceanographic work where pressure changes are important. Weight/weight units are dimensionless and therefore are expressed as a ratio, usually multiplied by $10^6$ and referred to as parts per million. Parts per million correspond to milligrams per kilogram and to grams per tonne.

Weight/volume units are widely used, largely because it is much more convenient in the laboratory to measure the volume of liquids rather than their weight. Weight/volume units are diverse because of the different systems of units in use in different countries. Table 5.1 gives conversion factors between the units most commonly encountered. Weight per volume and weight per weight units are interrelated by

$$\text{weight per volume} = (\text{weight per weight}) \times \text{density of solution.} \qquad (5.1)$$

Most natural groundwaters have a density very close of 1 kg $l^{-1}$ (the density of seawater is 1.028 kg $l^{-1}$) and therefore in the metric system the numerical difference between milligrams per litre and parts per million can usually be neglected.

**Table 5.1** *Interconversion table for quality units*

| | | B | | | | | |
|---|---|---|---|---|---|---|---|
| | | mg $l^{-1}$ | Grains per US gallon | Grains per imperial gallon (UK degrees) | French degrees | German degrees | Short tons per acre foot |
| A | mg $l^{-1}$ | 1.00 | 0.0584 | 0.0700 | 0.100 | 0.056 | 0.00136 |
| | Grains per US gallon | 17.1 | 1.00 | 1.20 | 1.71 | 0.961 | 0.0233 |
| | Grains per imperial gallon (UK degrees) | 14.3 | 0.835 | 1.00 | 1.43 | 0.803 | 0.0195 |
| | French degrees | 10.0 | 0.584 | 0.700 | 1.00 | 0.562 | 0.0136 |
| | German degrees | 17.8 | 1.04 | 1.25 | 1.78 | 1.00 | 0.0242 |
| | Short tons per acre foot | 735 | 42.9 | 51.5 | 73.5 | 41.3 | 1.00 |

To convert A to B multiply by the value in the table.

*Example.* A brine of density 1.2 kg $l^{-1}$ at 25 °C contains 100 000 ppm Na. Express this concentration in mg $l^{-1}$. At 100 °C, 1 kg of the brine occupies 0.84 l. What is the Na concentration in mg $l^{-1}$ at this temperature?

From eqn. (5.1)

$$\text{mg l}^{-1} = \text{mg kg}^{-1} \times \text{density (in kg l}^{-1})$$
$$= 100\,000 \times 1.2.$$

Therefore

$$\text{concentration} = 120\,000 \text{ mg l}^{-1}.$$

At 100 °C, density of brine is

$$\frac{1}{0.84} \text{kg l}^{-1} = 1.19 \text{ kg l}^{-1}.$$

Thus

$$\text{concentration} = 119\,000 \text{ mg l}^{-1}.$$

This illustrates the effect of temperature on concentrations in weight per volume volume units

Gas concentrations are occasionally reported in millilitres per litre. Because of the compressibility of gases, the temperature and pressure at which the gas was measured must be quoted. Gas volumes are usually reduced to STP (standard temperature and pressure, 0 °C and 1 atm (101 325 N $m^{-2}$)) using the gas equation

$$PV = nRT \tag{5.2}$$

where $P$ is the pressure, $V$ is the volume, $T$ is the temperature in kelvins, $n$ is the number of moles present, and $R$ is the gas constant (8.3143 J $K^{-1}$ $mol^{-1}$). Equation (5.2) can also be expressed as

$$\frac{P_1 V_1}{T_1} = \frac{P_2 V_2}{T_2}. \tag{5.3}$$

Two dissolved gases are commonly analysed in groundwater: carbon dioxide and oxygen. Carbon dioxide is normally reported in milligrams per litre, but dissolved oxygen is often expressed as per cent saturation. Per cent saturation of fresh water (oxygen solubility decreases with increasing salinity) can be converted to milligrams per litre using Table 5.2.

Weights of dissolved substances may be quoted as moles or millimoles rather than as grams or milligrams. Using the definition of the mole, it is clear that

$$\text{mmol l}^{-1} = \frac{\text{mg l}^{-1}}{\text{molecular weight}}. \tag{5.4}$$

Clearly this can be done only for substances of well-defined composition. The concentration of a solution in moles per litre is called the *molarity* (M). The concentration in moles per kilogram is known as the *molality* (m). Concentrations in these units are necessary in the thermodynamic calculations discussed in Chapter 3.

**Table 5.2** *Variation of 100 per cent oxygen saturation of water with temperature*

| Temperature (°C) | Solubility (mg $l^{-1}$) | Temperature (°C) | Solubility (mg $l^{-1}$) |
|---|---|---|---|
| 0 | 14.63 | 18 | 9.46 |
| 1 | 14.23 | 19 | 9.27 |
| 2 | 13.84 | 20 | 9.08 |
| 3 | 13.46 | 21 | 8.91 |
| 4 | 13.11 | 22 | 8.74 |
| 5 | 12.77 | 23 | 8.57 |
| 6 | 12.45 | 24 | 8.42 |
| 7 | 12.13 | 25 | 8.26 |
| 8 | 11.84 | 26 | 8.12 |
| 9 | 11.55 | 27 | 7.97 |
| 10 | 11.28 | 28 | 7.84 |
| 11 | 11.02 | 29 | 7.70 |
| 12 | 10.77 | 30 | 7.57 |
| 13 | 10.53 | 31 | 7.45 |
| 14 | 10.29 | 32 | 7.33 |
| 15 | 10.07 | 33 | 7.21 |
| 16 | 9.86 | 34 | 7.09 |
| 17 | 9.65 | 35 | 6.98 |

A unit related to the mole is the *equivalent*, defined as the equivalent weight of a substance in grams. The *equivalent weight* of a substance is (molecular weight of ion)/(size of ionic charge) or (molecular weight of substance)/(oxidation number change in reaction). The equivalent weight of a substance therefore depends on the reaction under consideration. The equivalent weight of a substance is useful in investigating the proportions in which substances react; this aspect of chemistry is called *stoichiometry*. In groundwater hydrochemistry, the equivalent is usually used to understand the interrelationships of dissolved ions. For example, in an ion exchange reaction calcium ions in solution may be replaced by sodium ions, but throughout this reaction the total concentration of calcium and sodium ions in equivalents per litre will remain constant since one equivalent of calcium is replaced by one equivalent of sodium.

*Example.* An outcrop groundwater contains 150 mg $l^{-1}$ calcium and 10 mg $l^{-1}$ sodium. Express these concentrations in mmol $l^{-1}$ and meq $l^{-1}$. If all the calcium were replaced by sodium during subsequent ion exchange, what would be the resulting sodium concentration?

The molecular weight of the $Ca^{2+}$ ion is 40; therefore its equivalent weight is $40/2 = 20$.

$$\begin{aligned}\text{Concentration of calcium ion} &= 150/40\\ &= 3.75 \text{ mmol l}^{-1}\\ &= 7.5 \text{ meq l}^{-1}.\end{aligned}$$

The molecular weight of the $Na^{+}$ ion is 23; therefore its equivalent weight is $23/1 = 23$.

$$\begin{aligned}\text{Concentration of sodium ion} &= 10/23\\ &= 0.43 \text{ mmol l}^{-1}\\ &= 0.43 \text{ meq l}^{-1}.\end{aligned}$$

N.B. It has been assumed that both calcium and sodium are present as their ions. This is effectively true for all waters.

During ion exchange, 1 eq of calcium is replaced by 1 eq of sodium. Therefore the new sodium concentration after ion exchange is

$$8.5 + 0.43 = 7.93 \text{ meq l}^{-1}$$
$$= 7.93 \times 23 \text{ mg l}^{-1}$$
$$= 182.4 \text{ mg l}^{-1}.$$

N.B. 1 mol of calcium is replaced by 2 mol of sodium; 40 g of calcium is replaced by 46 g of sodium.

## 5.2. Expression in terms of substances

Analyses are frequently reported in terms of substances that differ from the species analysed. This may be done when the species analysed is uncertain, e.g. dissolved boron, or when the total concentration of an element is of more interest than the species present, e.g. nitrogen. Analyses from older literature which predate widespread acceptance of the ionic concept may also be quoted as (sometimes hypothetical) compounds. The basic approach in all these calculations is to work in terms of moles or equivalents of the substance of interest.

Analyses of nitrogen compounds ($NH_3$, $NH_4^+$, $NO_2^-$, $NO_3^-$) are often reported in terms of nitrogen (N, not $N_2$), expressions such as 'nitrate N' and '$NH_3$ N' being used. Interconversion of these species is simple since each molecule contains only one nitrogen atom and therefore a mole of each contains the same quantity of nitrogen. Hence

$$\underset{14.01\text{ g}}{N} \equiv \underset{17.03\text{ g}}{NH_3} \equiv \underset{18.04\text{ g}}{NH_4^+} \equiv \underset{46.01\text{ g}}{NO_2^-} \equiv \underset{62.00\text{ g}}{NO_3^-}.$$

Total oxidized nitrogen (TON) is the sum of $NO_2$ N and $NO_3$ N. Organically bound nitrogen present in amino acids etc. may be reported as albuminoid nitrogen. This cannot be expressed in other terms because the exact nature of the nitrogen-containing species is not known.

Analyses for orthophosphate are quoted as phosphate or phosphate P. This is done because orthophosphoric acid, the common form of dissolved phosphorus in natural waters, is tribasic, producing the ions $H_2PO_4^-$, $HPO_4^{2-}$, and $PO_4^{3-}$ in proportions solely depending on pH. The analysis for 'phosphate' detects all three forms, the exact distribution being calculated from pH as in the carbonate system (see Section 3.4). It is understood that the term 'phosphate' includes mono and dihydrogen orthophosphate as well as orthophosphate *sensu strictu*, which is only significant at very high pH. Hence, as for nitrogen,

$$\underset{30.97\text{ g}}{P} \equiv \underset{96.99\text{ g}}{H_2PO_4^-} \equiv \underset{95.98\text{ g}}{HPO_4^{2-}} \equiv \underset{94.97\text{g}}{PO_4^{3-}}.$$

Rock analyses are reported as per cent oxides and there was an early tendency to report water analyses in a similar fashion (usually in parts per hundred thousand). Conversion is simplified because the formula is almost invariably quoted, although the compounds may be unfamiliar. They certainly do not exist in aqueous solution. The conversions are summarized in Table 5.3.

**Table 5.3** *Conversion factors for oxides (in weight units)*

| Oxide | Multiply by | To convert to |
|---|---|---|
| $CaO$ | 0.7147 | $Ca^{2+}$ |
| $MgO$ | 0.6030 | $Mg^{2+}$ |
| $Na_2O$ | 0.7419 | $Na^+$ |
| $K_2O$ | 0.8302 | $K^+$ |
| $SO_3$ | 1.1998 | $SO_4^{2-}$ |
| $CO_2$ | 1.3864 | $HCO_3^-$ |
| $N_2O_5$ | 0.5741 | $NO_3^-$ |

Historical analyses are frequently reported as a hypothetical mixture of salts, which, if dissolved in water, would produce the observed chemistry. Whilst this approach may shed some light on the hydrochemical processes involved, the result is entirely artificial and may be very misleading. An analysis in these terms must be broken down into individual ions which are then summed.

*Example.* Express the analysis given in terms of individual ions: calcium carbonate, 263 mg $l^{-1}$; magnesium carbonate, 22 mg $l^{-1}$; calcium sulphate, 62 mg $l^{-1}$; sodium chloride, 47 mg $l^{-1}$; sodium sulphate, 19 mg $l^{-1}$.

Each salt is broken down into its individual ions by calculating the number of moles of substance present, e.g.

$$263 \text{ mg l}^{-1} \text{ calcium carbonate } (CaCO_3) = \frac{263}{100} \text{ mmol l}^{-1}.$$

1 mol of $CaCO_3$ contains 1 mol of Ca (40 g).
Therefore 263 mg $l^{-1}$ $CaCO_3$ contains

$$\frac{263}{100} \times 40 = 105.2 \text{ mg l}^{-1} \text{ Ca}.$$

Proceeding similarly for the other salts (N.B. 1 mol $Na_2SO_4$ = 2 mol Na + 1 mol $SO_4$), the result is as follows.

| | | $Ca^{2+}$ | $Mg^{2+}$ | $Na^+$ | $Cl^-$ | $SO_4^{2-}$ | $CO_3^{2-}$ |
|---|---|---|---|---|---|---|---|
| $CaCO_3$ | 263 | 105.2 | | | | | 157.8 |
| $MgCO_3$ | 22 | | 6.3 | | | | 15.7 |
| $CaSO_4$ | 62 | 18.2 | | | | 43.8 | |
| $NaCl$ | 47 | | | 18.5 | 28.5 | | |
| $Na_2SO_4$ | 19 | | | 6.2 | | 12.8 | |
| Totals | | 123.4 | 6.3 | 24.7 | 28.5 | 56.6 | 173.5 |

It is unrealistic to express the dissolved carbon as $CO_3^{2-}$ since the species present (at pH 7) is $HCO_3^-$. Since $[CO_3^{2-}]$ is equivalent to $2[HCO_3^-]$

$$173.5 \text{ mg l}^{-1}\ CO_3^{2-} = 352.8 \text{ mg l}^{-1}\ HCO_3^-.$$

The final result (in mg $l^{-1}$) is $Ca^{2+}$, 123.4; $Mg^{2+}$, 6.3; $Na^+$, 24.7; $Cl^-$, 28.5; $SO_4^{2-}$, 56.6; $HCO_3^-$, 352.8.

By following these general principles it should be possible to use an analysis, no matter how strange the units or the reported species. All that is needed are the relevant atomic weights and the concept of chemical equivalence.

## 5.3. Alkalinity and acidity

These terms are used to describe measurable parameters of the carbonate system, as discussed in Section 3.4. From them it is usually desirable to calculate the concentrations of carbonate, bicarbonate, and carbonic acid $H_2CO_3^*$. Occasionally it may be necessary to reverse this procedure.

Alkalinity is the quantity of strong acid required to bring a water to a specified carbonate system end point. Alkalinity may be expressed either in milliequivalents per litre of acid used or in terms of the equivalent amount of calcium carbonate (see Section 5.4). Phenolphthalein alkalinity (which includes caustic alkalinity) is determined at the $CO_3^{2-}$ end point and total alkalinity (which includes p-alkalinity) is determined at the $HCO_3^-$ end point. In natural waters the carbonate system is not necessarily the only acid–base system present; others may be $H_3PO_4$, $HBO_3$, organic acids, and suspended metal hydroxides. Organic acids are negligible in most groundwaters and metal hydroxides are usually removed by filtration before analysis. Phosphate and boron can be measured and a correction applied to the alkalinity measurement to calculate the carbonate-system-only alkalinities using the information given in Fig. 5.1.

For normal purposes, the carbonate parameters can be calculated easily from the alkalinities from the following equations, using the approximation that activity coefficients are unity:

$$[CO_3^{2-}] = [\text{p-alk}] - 10^{\log K_w + \text{pH}} \tag{5.5}$$

if p-alkalinity is present;

$$\log [CO_3^{2-}] = \log [HCO_3^-] + \log K_2 + \text{pH} \tag{5.6}$$

if p-alkalinity is absent;

$$[HCO_3^-] \approx [\text{total alk}] - 2\ [\text{p-alk}] \tag{5.7}$$

if [p-alk] $<\frac{1}{2}$ [total alk] (otherwise use eqn (5.6));

$$\log [H_2CO_3^*] = \log [HCO_3^-] - \log K_1 - \text{pH}; \tag{5.8}$$

$$[\text{total alk}] = [HCO_3^-] + 2\ [CO_3^{2-}] + 10^{\log K_w + \text{pH}} \tag{5.9}$$

In the above equations, alkalinities are in *equivalents*, and other quantities are in *moles* (N.B. not millimoles).

Values of the constants are given in Tables 3.3 and 3.4. If exact values are needed, activity coefficients will have to be evaluated and then computer calculations are necessary (see Section 3.4). Accurate values, which are needed only for saturation calculations, are given in Section 5.7.

Acidity is the quantity of strong alkali required to bring a water to a specified

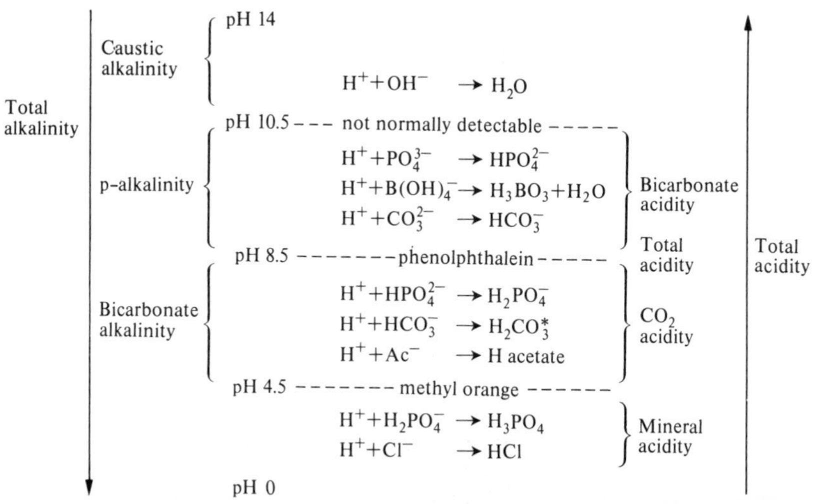


**Fig. 5.1.** Reactions during alkalinity and acidity titrations.

carbonate end point, and is complementary to alkalinity. Acidity is usually reported in milliequivalents per litre. Mineral acidity is determined at the $H_2CO_3$ end point (methyl orange) and $CO_2$ acidity is determined by the additional quantity of acid added to the $HCO_3^-$ end point. The reactions that take place during the acidity titration are summarized in Fig. 5.1. For approximate purposes

$$[H_2CO_3^*] \approx [CO_2 \text{ acidity}]. \tag{5.10}$$

Other species are calculated using eqns (5.6) and (5.8).

*Example.* Evaluate the carbonate and bicarbonate concentrations in the following waters. What would their $CO_2$ acidities be?
(1) pH 11, p-alk = 3 meq $l^{-1}$, and total alkalinity = 5.214 meq $l^{-1}$.
(2) pH 7.1 and total alkalinity = 5 meq $l^{-1}$.

(1) From the log composition diagrams (Figs 3.3 and 3.4) it can be seen that the dominant species in this water are $CO_3^{2-}$ and $OH^-$. Using eqn (5.5) (N.B. quantities are equivalents not milliequivalents)

$$\begin{aligned}[CO_3^{2-}] &= 3 \times 10^{-3} - 10^{-14+11} \\ &= 2 \times 10^{-3} \text{ mol l}^{-1} \\ &= 120 \text{ mg l}^{-1}\end{aligned}$$

Because [p-alk] > $\frac{1}{2}$ [total alk], eqn (5.6) is used to calculate $[HCO_3^-]$. On rearranging we obtain

$$\log [HCO_3^-] = \log [CO_3^{2-}] - \log K_2 - pH$$

and on substituting

$$\log [HCO_3^-] = \log \left\{ 2 \times [10^{-3}] \right\} + 10.331 - 11 \quad \text{at } 25\ ^\circ C$$
$$= -3.368$$

Therefore

$$[HCO_3^-] = 4.29 \times 10^{-4}\ \text{mol}\ l^{-1}$$
$$= 26.14\ \text{mg}\ l^{-1}.$$

Since the pH of the water is greater than 8.5, the $CO_2$ acidity is not measurable. The actual amount of $H_2CO_3^*$ can be deduced from eqn (5.8):

$$\log [H_2CO_3^*] = -3.368 + 6.352 - 11 \quad \text{at } 25^\circ C$$
$$[H_2CO_3^*] = 9.64 \times 10^{-9}\ \text{mol}\ l^{-1}$$
$$= 5.98 \times 10^{-4}\ \text{mg}\ l^{-1}.$$

(2) From the log composition diagrams given in Figs 3.5 and 3.8 it can be seen that the dominant species are $HCO_3^-$ and $H_2CO_3^*$.

Since the pH of the water is below 8.5, p-alkalinity is not measurable. From eqn (5.7)

$$[HCO_3^-] = 5\ \text{mmol}\ l^{-1}$$
$$= 305\ \text{mg}\ l^{-1}.$$

From eqn (5.6)

$$\log [CO_3^{2-}] = \log [5 \times 10^{-3}] - 10.331 + 7.1 \quad \text{at } 25^\circ C$$
$$= -5.532.$$

Therefore

$$[CO_3^{2-}] = 2.937 \times 10^{-6}\ \text{mol}\ l^{-1}$$
$$= 0.176\ \text{mg}\ l^{-1}.$$

From eqn (5.5), the contribution to the total alkalinity from $CO_3^{2-}$ is $5.87 \times 10^{-3}$ meq $l^{-1}$ and can safely be neglected in comparison with the total alkalinity. Thus the approximation of eqn (5.7) is valid. From eqn (5.8)

$$\log [H_2CO_3^*] = \log [5 \times 10^{-3}] + 6.352 - 7.1$$
$$= -3.049$$
$$[H_2CO_3^*] = 8.932 \times 10^{-4}\ \text{mol}\ l^{-1}$$
$$= 55.38\ \text{mg}\ l^{-1}.$$

From eqn (5.10)

$$CO_2 \text{ acidity} = 0.8932\ \text{meq}\ l^{-1}.$$

## 5.4. Hardness

*Hardness* is a property of water which causes difficulty of lathering with soap. Many hard waters also cause scaling in boilers, kettles, etc. These properties of water are of considerable economic significance. Both these undesirable properties of water are caused by the presence of divalent cations in the water, particularly $Ca^{2+}$ and $Mg^{2+}$, but also $Fe^{2+}$ and $Sr^{2+}$.

Soap consists of a soluble mixture of sodium and potassium salts of long-chain carboxylic acids such as stearic acid ($C_{17}H_{35}CO_2H$). These salts react with polyvalent cations to produce insoluble salts that precipitate as a sticky scum, thus inhibiting the cleansing action of the soap until sufficient has been added to precipitate all the polyvalent metal ions. Scale formation when a water is heated results from the volatilization of carbon dioxide, thus driving the equilibrium

$$2HCO_3^- \rightleftharpoons CO_3^{2-} + H_2O + CO_2 \qquad (5.11)$$

to the right. All divalent ions common in natural waters have insoluble carbonates, in contrast with the monovalent ions which have soluble carbonates, and therefore the divalent ions are precipitated by the rising $CO_3^{2-}$ ion concentration as $CO_2$ is lost.

The precipitation of insoluble carbonates during boiling reduces the divalent cation concentration of the water and thereby partially softens it. The part of the hardness removable in this way is called *temporary*, the remainder being *permanent*. Temporary hardness is sometimes referred to as carbonate hardness, and permanent hardness is sometimes called non-carbonate hardness. Temporary hardness arises from the dissolution of carbonates such as calcite in the presence of carbon dioxide, essentially the reverse of the precipitation process on boiling. Permanent hardness arises from the dissolution of salts other than carbonates, e.g. gypsum ($CaSO_4.2H_2O$). In reality the chemistry of waters is more complex than this and it is not in general possible to ascribe parts of the hardness to different minerals.

Hardness is conventionally expressed in terms of calcium carbonate, either in milligrams per litre or as degrees of hardness (Table 5.1). A more rational unit would be milliequivalents per litre. 1 mol (100.088 g) of calcium carbonate ($CaCO_3$) produces 2 equivalents of divalent calcium ion; therefore a hardness of 1 meq $l^{-1}$ equals 50.044 mg $l^{-1}$ $CaCO_3$. Alkalinity is also often expressed as calcium carbonate.

Total hardness can be measured directly, or it can be calculated from the analysis by summing the equivalents of the divalent ions and expressing in terms of $CaCO_3$. The process of carbonate precipitation by boiling is exemplified by

$$Ca^{2+} + 2HCO_3^- \rightarrow CaCO_3 \downarrow + H_2O + CO_2 \uparrow \qquad (5.12)$$

From this equation it can be deduced that the temporary hardness is given by the lesser of the divalent cation concentration and the bicarbonate concentration, both expressed in equivalents. In hard waters the bicarbonate concentration is equal to the alkalinity (the carbonate concentration must be very small if the water contains significant dissolved calcium or magnesium); thus temporary hardness is the lesser of alkalinity and total hardness. The difference between total hardness and alkalinity, if positive, is the permanent hardness.

*Example.* Evaluate the temporary and permanent hardness of these waters:

(1) Ca, 150 mg $l^{-1}$; Mg, 10 mg $l^{-1}$; alkalinity, 300 mg $l^{-1}$ $CaCO_3$.

(2) Ca, 50 mg $l^{-1}$; Mg, 5 mg $l^{-1}$; Na, 200 mg $l^{-1}$; alkalinity, 5 meq $l^{-1}$.

(1) Convert cations to milliequivalents per litre:

$$\text{equivalent weight of } Ca^{2+} = 40/2$$
$$\text{calcium concentration} = 150/20 \text{ meq } l^{-1} = 7.5 \text{ meq } l^{-1}$$
$$\text{equivalent weight of } Mg^{2+} = 24/2$$
$$\text{magnesium concentration} = 10/12 \text{ meq } l^{-1} = 0.83 \text{ meq } l^{-1};$$
$$\text{total divalent cation concentration} = 7.5 + 0.83 = 8.33 \text{ meq } l^{-1} = 8.33 \times 50 \text{ mg } l^{-1}\ CaCO_3 = 416.5 \text{ mg } l^{-1}\ CaCO_3;$$
$$\text{total hardness} = 416.5 \text{ mg } l^{-1}\ CaCO_3.$$

Alkalinity is less than total hardness; therefore

$$\text{temporary hardness} = \text{alkalinity} = 300 \text{ mg } l^{-1}\ CaCO_3;$$
$$\text{permanent hardness} = \text{total hardness} - \text{temporary hardness} = 116.5 \text{ mg } l^{-1}\ CaCO_3.$$

(2)

$$\text{Calcium concentration} = 50/20 \text{ meq } l^{-1} = 2.5 \text{ meq } l^{-1}$$
$$\text{magnesium concentration} = 5/12 \text{ meq } l^{-1} = 0.42 \text{ mg } l^{-1};$$
$$\text{total divalent cation concentration} = 2.5 + 0.42 = 2.92 \text{ meq } l^{-1};$$
$$\text{total hardness} = 146 \text{ mg } l^{-1}\ CaCO_3.$$

Total hardness is less than alkalinity (compare in meq $l^{-1}$); therefore

$$\text{temporary hardness} = \text{total hardness} = 146 \text{ mg } l^{-1}\ CaCO_3;$$
$$\text{permanent hardness} = 0.$$

N.B. Sodium forms a monovalent ion only, and therefore is irrelevant to the hardness calculation

## 5.5. Analysis checks

### *5.5.1. General*

An analysis should first be checked for unreasonable values. What is reasonable often depends on experience and will also depend on the aquifer being sampled. Very low alkalinity in a water from a carbonate aquifer may be considered suspicious, for example, but might not be unusual from a fractured granite; however, high calcium from the latter aquifer might be suspect. Guidance may be obtained by intercomparison if a batch of analyses is available.

Some values reported might be chemically impossible, for example p-alkalinity exceeding total alkalinity! The authors received a batch of analyses from the Middle East which regularly reported carbonate values of the same order of magnitude as the bicarbonate values, despite pH values near 7. Mistakes of these kinds occur through copying errors and are usually easily spotted. If an analysis passes this inspection, numerical checks are available.

### *5.5.2. Ion balance*

Water samples are are electrically neutral and therefore the total charges on the cations and anions reported in the analysis should be equal. The total positive and negative charges are obtained by summing the equivalents of cations and anions respectively. The ion balance error is normally expressed by the difference as a percentage of the sum.

*Example.* Calculate the ion balance in error for the following analysis: $Ca^{2+}$, 99 mg $l^{-1}$; $Mg^{2+}$, 13 mg $l^{-1}$; $Na^+$, 39 mg $l^{-1}$; $K^+$, 12.5 mg $l^{-1}$; $Cl^-$, 72 mg $l^{-1}$; $SO_4^{2-}$, 34 mg $l^{-1}$; alkalinity, 306 mg $l^{-1}$ $CaCO_3$; $NO_3$ N, 3.2 mg $l^{-1}$; pH 7.2.

At near neutral pH, the contribution to the ionic balance from $H^+$ and $OH^-$ can be ignored. The sum of equivalents of cations is

$$\begin{aligned} &Ca^{2+} + Mg^{2+} + Na^{+} + K^{+} \\ &(99/20) + (13/12) + (39/23) + (12.5/39) \\ &= 8.049 \text{ meq } l^{-1}. \end{aligned}$$

The sum of equivalents of anions is

$$\begin{aligned} &\{HCO_3^- + CO_3^{2-}\} + Cl^- + SO_4^{2-} + NO_3^- \\ &= (306/50) + (72/35.5) + (34/48) + (3.2/14) \\ &= 9.085 \text{ meq } l^{-1}. \end{aligned}$$

Note that there is no need to convert alkalinity into $HCO_3^-$ and $CO_3^{2-}$; the number of equivalents of charge can be calculated directly using the equivalent weight of $CaCO_3$, and similarly for nitrate nitrogen.

$$\text{Ion balance error} = \frac{\Sigma\text{cations} - \Sigma\text{anions}}{\Sigma\text{cations} + \Sigma\text{anions}} \times 100 \text{ per cent}$$

$$= -6.046 \text{ per cent}$$

The method has certain drawbacks. It is always possible for two errors to cancel. The method is only valid if all major ions are analysed. If there is an unanalysed ion there will be an ion balance error; conversely if one ion is calculated by difference, as sodium or sulphate often were, the ion balance error will be zero. If the analysis is dominated by a few ions, for example saline waters dominated by $Na^+$ and $Cl^-$, large errors in the other ions will produce only small errors in the ion balance.

The ion balance error of a modern analysis should be less than 5 per cent and certainly less than 10 per cent. An error outside this range suggests an analytical error or an unanalysed ion.

### *5.5.3. Total dissolved solids*

A rough check of an analysis can be made by comparing the total analysed solids with the measured total dissolved solids. When calculating the total dissolved solids, allowance must be made for the loss of carbon dioxide that occurs during evaporation to dryness according to eqns (5.11) and (5.12). 2 mol of bicarbonate (122 g) become 1 mol of carbonate (60 g) in the dry residue. Agreement between calculated and measured dissolved solids should be within 20 per cent; the large margin of error arises because total dissolved solids can be difficult to determine accurately, particularly on high sulphate and high chloride waters.

## 5.6. Completing partial analyses

A complete analysis is advantageous for hydrochemical studies but many water analyses, particularly historical ones, do not report all the major ions. However, it is sometimes possible to deduce additional parameters from the data given by using the relationships developed elsewhere in this chapter.

In most groundwaters, total hardness is caused by the presence of calcium and magnesium. Assuming this to be so, if any two of these parameters are given, the third can be calculated by difference. Moreover, experience with complete analyses from the area being studied may show that either magnesium or calcium are negligible, or that their ratio is fairly constant and therefore approximate calcium and magnesium values can be estimated from the total hardness alone.

If a water displays permanent hardness, its alkalinity equals its temporary hardness. If carbonate and bicarbonate are reported in unreasonable proportions it is possible that the alkalinity from which they were calculated is correct, in which case correct values can be obtained by calculating the alkalinity from the data given, and then recalculating $HCO_3^-$ and $CO_3^{2-}$.

A single missing ion is readily calculated by ionic balance. The drawback with this procedure is that the net error in all the other analysed ions is accumulated in the ion calculated by difference; therefore the alternative procedures above should be used first.

*Example*. Complete, if possible, the following analyses:

(1) Permanent hardness, 75 mg $l^{-1}$ $CaCO_3$; temporary hardness, 345 mg $l^{-1}$ $CaCO_3$; $Mg^{2+}$, 6 mg $l^{-1}$; NaCl, 35 mg $l^{-1}$; pH 7.1.

(2) $Ca^{2+}$, 93 mg $l^{-1}$; $Mg^{2+}$, 7 mg $l^{-1}$; $Na^+$, 18 mg $l^{-1}$; $K^+$, 2.4 mg $l^{-1}$; $HCO_3^-$, 183 mg $l^{-1}$; $CO_3^{2-}$, 45 mg $l^{-1}$; $Cl^-$, 150 mg $l^{-1}$; total dissolved solids, 320 mg $l^{-1}$; pH 6.7.

(1) The analysis is lacking values for $Ca^{2+}$, $K^{+}$, $HCO_3^-$ and $CO_3^{2-}$, and $SO_4^{2-}$. Ca can be obtained from the hardness values:

$$\begin{aligned}\text{total hardness} &= 345 + 75\\ &= 420 \text{ mg l}^{-1} \text{ CaCO}_3\\ &= 8.4 \text{ meq l}^{-1}\\ \text{magnesium} &= 6/12\\ &= 0.5 \text{ meq l}^{-1}.\end{aligned}$$

If it is assumed that $Ca^{2+}$ and $Mg^{2+}$ are the only hardness-producing ions

$$\begin{aligned}Ca^{2+} &= 8.5 - 0.5\\ &= 7.9 \text{ meq l}^{-1}\\ &= 7.9 \times 20 \text{ mg l}^{-1}\\ Ca^{2+} &= 158 \text{ mg l}^{-1}.\end{aligned}$$

Permanent hardness is present; therefore

$$\begin{aligned}\text{alkalinity} &= \text{temporary hardness}\\ &= 345 \text{ mg l}^{-1} \text{ CaCO}_3\\ \text{at pH 7 } [HCO_3^-] &= \text{alkalinity}\\ &= (345/50) \times 61 \text{ mg l}^{-1}\\ &= 421 \text{ mg l}^{-1} \text{ HCO}_3^-.\end{aligned}$$

From eqn (5.6)

$$\log [CO_3^{2-}] = \log [HCO_3^-] + \log K_2 + \text{pH}$$

and on substitution

$$\begin{aligned}\log [CO_3^{2-}] &= -2.16 - 10.5 + 7.1 \text{ at } 10\,^{\circ}\text{C}\\ [CO_3^{2-}] &= 0.17 \text{ mg l}^{-1}.\end{aligned}$$

Strictly it is possible to prodeed no further as two ions are still unknown: $K^+$ and $SO_4^{2-}$. If the reasonable assumption is made that $K^+$ is negligible, $SO_4^{2-}$ can be found by ionic balance.

First, the concentrations of $Na^+$ and $Cl^-$ must be determined:

$$\begin{aligned}35 \text{ mg l}^{-1} \text{ NaCl} &= 35/(23 + 35.5)\\ &= 0.598 \text{ mmol l}^{-1} \text{ NaCl}\\ &= 0.598 \text{ mmol l}^{-1} \text{ Na}^+ + 0.598 \text{ mmol l}^{-1} \text{ Cl}^-\\ Na^+ &= 0.598 \times 23\\ &= 13.8 \text{ mg l}^{-1}\\ Cl^- &= 0.598 \times 35.5\\ &= 21.2 \text{ mg l}^{-1}.\end{aligned}$$

The sum of equivalents of cations is

$$\begin{aligned}\text{Ca} + \text{Mg} + \text{Na} &= 7.9 + 0.5 + 0.598\\ &= 8.998 \text{ meq l}^{-1}\end{aligned}$$

The sum of equivalents of anions is

$$\{HCO_3^- + CO_3^{2-}\} + Cl^- = 6.9 + 0.598$$
$$= 7.498 \text{ meq l}^{-1}.$$

The difference is ascribed to $SO_4^{2-}$:

$$SO_4^{2-} = 8.998 - 7.498$$
$$= 1.5 \text{ meq l}^{-1}$$
$$= 1.5 \times (96/2) \text{ mg l}^{-1}$$
$$= 72 \text{ mg l}^{-1}.$$

Thus the completed analysis is

| | | | | |
|---|---|---|---|---|
| $Ca^{2+}$ | 158 mg $l^{-1}$ | $Cl^-$ | 21.2 mg $l^{-1}$ | pH 7.1 |
| $Mg^{2+}$ | 6.0 mg $l^{-1}$ | $SO_4^{2-}$ | 72.0 mg $l^{-1}$ | |
| $Na^+$ | 13.8 mg $l^{-1}$ | $HCO_3^-$ | 421 mg $l^{-1}$ | |
| $K^+$ | assumed zero | $CO_3^{2-}$ | 0.17 mg $l^{-1}$ | |

(2) A first glance shows an impossible combination of $HCO_3^-$ and $CO_3^{2-}$ at pH 6.7. First the alkalinity is calculated:

$$\text{Alk} = [HCO_3^-] + 2[CO_3^{2-}]$$
$$= (183/61) + 2\,(45/60)$$
$$= 4.5 \text{ meq l}^{-1}$$
$$= 225 \text{ mg l}^{-1}\ CaCO_3.$$

At pH 6.7

$$[HCO_3^-] = \text{Alk}$$
$$= 4.5 \times 61$$
$$= 274.5 \text{ mg l}^{-1}\ HCO_3^-.$$

From eqn (5.6)

$$\log\,[CO_3^{2-}] = \log\,[HCO_3^-] + \log K_2 + \text{pH}$$
$$= -2.35 - 10.5 + 6.7 \text{ at } 10\,^{\circ}\text{C}$$

Therefore

$$[CO_3^{2-}] = 0.044 \text{ mg l}^{-1}.$$

Sulphate is the only major ion not explicitly stated and can therefore be determined from ion balance. The sum of equivalents of cations is

$$Ca + Mg + Na + K = 4.65 + 0.58 + 0.78 + 0.06$$
$$= 6.07 \text{ meq l}^{-1}.$$

The sum of equivalents of anions is

$$\{HCO_3^- + CO_3^{2-}\} + Cl = 4.5 + 4.23$$
$$= 8.73 \text{ meq l}^{-1}.$$

It should be noted that the anionic charge is already greater than the cationic charge and therefore a positive sulphate content cannot be calculated. The ion balance error as the analysis stands is −18 per cent, which suggests error. The total dissolved solids are checked as follows:

$$\begin{aligned}\text{total dissolved solids} &= \text{Ca} + \text{Mg} + \text{Na} + \text{K} + \text{Cl} + (\text{HCO}_3 \times 30/61)\\ &= 93 + 7 + 18 + 2.4 + 150 + 135\\ &= 405.4\ \text{mg}\ \text{l}^{-1}.\end{aligned}$$

This is 26 per cent greater than the observed value, tending to confirm that the analysis contains an error. On inspection, the wide disparity between the sodium and chloride concentrations is unusual and a repeat analysis would start with these two ions.

It is concluded that this analysis cannot be completed by the determination of $SO_4^{2-}$ because the analysis contains other errors.

## 5.7. Saturation indices

In the study of the chemistry of groundwaters, it is often necessary to express the extent to which a water has reached chemical equilibrium with the minerals of the aquifer matrix. In carbonate-containing aquifers in particular, the ability of circulating groundwater to dissolve the aquifer matrix and therefy to increase the permeability of the aquifer is most important. It is therefore necessary to express whether or not the water is saturated with aquifer carbonate. Groundwaters saturated with dissolved substances are prone to depositing some of their solute load on abstraction, causing incrustation and clogging of pipework.

Various methods of representing the degree of saturation of a water with respect to a mineral have been used, the most useful of which are *saturation percentage*, defined by

$$\text{saturation percentage} = \frac{\text{IAP}}{K_s} \times 100 \text{ per cent} \tag{5.13}$$

and *saturation index* defined by

$$\text{saturation index} = \log\left(\frac{\text{IAP}}{K_s}\right). \tag{5.14}$$

IAP is the ionic activity product of the appropriate ions and $K_s$ is the solubility product of the mineral. The saturation index is the form most commonly used in groundwater chemistry. It is easy to show that a water is in equilibrium with a mineral when the saturation index is zero, is undersaturated when the index is negative, and is oversaturated when the index is positive. Obviously this approach is only valid for minerals that dissolve reversibly and whose chemical constitution is sufficiently well defined for $K_s$ to be meaningful. Unfortunately these provisos exclude many important minerals including most silicates such as quartz and the clay minerals which are important in many aquifers. The use of the saturation index is best restricted to sparingly soluble simple salts, of which calcite and dolomite are the most common. Gypsum is also suitable for saturation index treatment, as are some minor ion minerals like fluorite ($CaF_2$) and celestine ($SrSO_4$). Highly soluble minerals like halite (NaCl) cannot be described satisfactorily by the thermodynamics of solutions developed in Chapter 3 and therefore the saturation index technique cannot be used.

The basic procedure for all saturation indices is the same: evaluate the activities of all the ions under the temperature and ionic strength conditions prevailing, calculate the equilibrium constant at the required temperature from available data, and then use eqn (5.14). The calculations for calcite and dolomite, which are the minerals in which the hydrochemist is most often interested, are quite extensive because the required activities are not directly measurable. Therefore most of the rest of this section will be taken up by the development of equations and associated tables to simplify the calculation of calcite and dolomite saturation.

The calcite and dolomite saturation indices $SI_C$ and $SI_D$ are defined by

$$SI_C = \log\left\{\frac{(Ca^{2+})(CO_3^{2-})}{K_C}\right\} \tag{5.15}$$

$$SI_D = \log\left\{\frac{(Ca^{2+})(Mg^{2+})(CO_3^{2-})^2}{K_D}\right\}. \tag{5.16}$$

Relevant parameters available from water analyses are $[Ca^{2+}]$, $[Mg^{2+}]$, [Alk], and pH. The ionic strength of the water, which is necessary to calculate activity coefficients, can be calculated from a full analysis if this is available, or estimated from the conductivity of the water.

For calcite, eqn (5.15) can be rewritten

$$SI_C = \log(Ca^{2+}) + \log(CO_3^{2-}) - \log K_C \tag{5.17}$$

$(CO_3^{2-})$ must be evaluated from [Alk]. The carbonate alkalinity is defined as

$$[Alk'] = [Alk] - 10^{K_w + pH} \tag{5.18}$$

From eqn (5.9)

$$[Alk'] = [HCO_3^-] + 2[CO_3^{2-}] \tag{5.19}$$

$$K_2 = \frac{(H^+)(CO_3^{2-})}{(HCO_3^-)} = \frac{(H^+)[CO_3^{2-}]\gamma_{CO_3^{2-}}}{[HCO_3^-]\gamma_{HCO_3^-}}. \tag{5.20}$$

Solving these equations gives

$$[CO_3^{2-}] = [Alk'] \frac{K_2\gamma_{HCO_3^-}}{(H^+)\gamma_{CO_3^{2-}} + 2K\gamma_{HCO_3^-}} \tag{5.21}$$

whence

$$\log(CO_3^{2-}) = \log[Alk'] + \log K_2 + \log\gamma_{HCO_3^-} + \log\gamma_{CO_3^{2-}} - \log\{(H^+)\gamma_{CO_3^{2-}} + 2K_2\gamma_{HCO_3^-}\}.$$

Thus the full expression is

$$SI_C = \log [Ca^{2+}] + \log [Alk'] - \log \{(H^+)\gamma_{CO_3^{2-}} + 2K_2\gamma_{HCO_3^-}\} + A + B \quad (5.22)$$

where

$$A = \log \gamma_{Ca^{2+}} + \log \gamma_{HCO_3^-} + \log \gamma_{CO_3^{2-}} \quad (5.23)$$

$$B = \log K_2 - \log K_C. \quad (5.24)$$

Values of $A$ and $B$ are given in Tables 5.4 and 5.5. Similarly, the dolomite saturation index is given by

$$SI_D = \log [Ca^{2+}] + \log [Mg^{2+}] + 2 \log [Alk'] - 2 \log \{(H^+)\gamma_{CO_3^{2-}} + 2K_2\gamma_{HCO_3^-}\} + C + D \quad (5.25)$$

where

$$C = \log \gamma_{Ca^{2+}} + \log \gamma_{Mg^{2+}} + 2 \log \gamma_{HCO_3^-} + 2 \log \gamma_{CO_3^{2-}} \quad (5.26)$$

$$D = 2 \log K_2 - \log K_D. \quad (5.27)$$

**Table 5.4** *Values of A for calcite saturation index calculation*

| Ionic strength (mol l$^{-1}$) | 0 °C | 10 °C | 20 °C | 30 °C | 40°C |
|---|---|---|---|---|---|
| $0.1 \times 10^{-5}$ | −0.0044 | −0.0045 | −0.0046 | −0.0046 | −0.0047 |
| $0.2 \times 10^{-5}$ | −0.0062 | −0.0063 | −0.0064 | −0.0065 | −0.0067 |
| $0.5 \times 10^{-5}$ | −0.0099 | −0.0100 | −0.0102 | −0.0103 | −0.0105 |
| $0.1 \times 10^{-4}$ | −0.0139 | −0.0141 | −0.0143 | −0.0146 | −0.0148 |
| $0.2 \times 10^{-4}$ | −0.0196 | −0.0199 | −0.0202 | −0.0206 | −0.0209 |
| $0.5 \times 10^{-4}$ | −0.0309 | −0.0314 | −0.0318 | −0.0324 | −0.0330 |
| $0.1 \times 10^{-3}$ | −0.0435 | −0.0441 | −0.0448 | −0.0456 | −0.0464 |
| $0.2 \times 10^{-3}$ | −0.0611 | −0.0620 | −0.0629 | −0.0640 | −0.0651 |
| $0.5 \times 10^{-3}$ | −0.0953 | −0.0966 | −0.0981 | −0.0998 | −0.1015 |
| $0.1 \times 10^{-2}$ | −0.1326 | −0.1345 | −0.1366 | −0.1389 | −0.1414 |
| $0.2 \times 10^{-2}$ | −0.1836 | −0.1862 | −0.1891 | −0.1922 | −0.1956 |
| $0.5 \times 10^{-2}$ | −0.2786 | −0.2825 | −0.2868 | −0.2915 | −0.2966 |
| $0.1 \times 10^{-1}$ | −0.3767 | −0.3819 | −0.3877 | −0.3940 | −0.4008 |
| $0.2 \times 10^{-1}$ | −0.5015 | −0.5083 | −0.5159 | −0.5242 | −0.5332 |
| $0.5 \times 10^{-1}$ | −0.7092 | −0.7187 | −0.7292 | −0.7406 | −0.7531 |
| 0.1 | −0.8945 | −0.9062 | −0.9191 | −0.9333 | −0.9488 |
| 0.2 | −1.0927 | −1.1067 | −1.1222 | −1.1393 | −1.1579 |
| 0.5 | −1.3408 | −1.3578 | −1.3766 | −1.3974 | −1.4201 |
| 1 | −1.4772 | −1.4961 | −1.5172 | −1.5405 | −1.5659 |
| 2 | −1.5126 | −1.5332 | −1.5562 | −1.5816 | −1.6094 |

**Table 5.5** *Values of B, D, and F for index calculations*

| Temperature (°C) | B | D | F |
|---|---|---|---|
| 0 | −2.2751 | −4.8076 | −7.6916 |
| 5 | −2.2026 | −4.5443 | −7.7121 |
| 10 | −2.1298 | −4.2987 | −7.7350 |
| 15 | −2.0566 | −4.0698 | −7.7602 |
| 20 | −1.9831 | −3.8569 | −7.7875 |
| 25 | −1.9092 | −3.6591 | −7.8168 |
| 30 | −1.8351 | −3.4757 | −7.8481 |
| 35 | −1.7607 | −3.3060 | −7.8811 |
| 40 | −1.6860 | −3.1492 | −7.9160 |

Values of $C$ and $D$ are given in Table 5.6 and Table 5.5 respectively. In both eqns (5.22) and (5.25) the $(H^+)$ term simplifies at pH < 9 so that

$$\log \{(H^+)\gamma_{CO_3^{2-}} + 2\,K_2\,\gamma_{HCO_3^-}\} \approx -\text{pH} + \log \gamma_{CO_3^{2-}} \qquad (5.28)$$

Values of $\log \gamma_{CO_3^{2-}}$ are given in Table 5.7. The calculation of saturation indices is straightforward using these relations.

**Table 5.6** *Values of C for dolomite saturation index calculation*

| Ionic strength mol l$^{-1}$ | 0 °C | 10 °C | 20 °C | 30 °C | 40 °C |
|---|---|---|---|---|---|
| $0.1 \times 10^{-5}$ | −0.0088 | −0.0090 | −0.0091 | −0.0093 | −0.0094 |
| $0.2 \times 10^{-5}$ | −0.0125 | −0.0127 | −0.0129 | −0.0131 | −0.0133 |
| $0.5 \times 10^{-5}$ | −0.0197 | −0.0200 | −0.0203 | −0.0207 | −0.0210 |
| $0.1 \times 10^{-4}$ | −0.0278 | −0.0282 | −0.0287 | −0.0292 | −0.0297 |
| $0.2 \times 10^{-4}$ | −0.0393 | −0.0398 | −0.0405 | −0.0411 | −0.0419 |
| $0.5 \times 10^{-4}$ | −0.0618 | −0.0627 | −0.0637 | −0.0648 | −0.0659 |
| $0.1 \times 10^{-3}$ | −0.0870 | −0.0882 | −0.0896 | −0.0911 | −0.0927 |
| $0.2 \times 10^{-3}$ | −0.1221 | −0.1329 | −0.1258 | −0.1279 | −0.1302 |
| $0.5 \times 10^{-3}$ | −0.1903 | −0.1931 | −0.1961 | −0.1993 | −0.2029 |
| $0.1 \times 10^{-2}$ | −0.2650 | −0.2688 | −0.2729 | −0.2775 | −0.2824 |
| $0.2 \times 10^{-2}$ | −0.3666 | −0.3718 | −0.3775 | −0.3838 | −0.3906 |
| $0.5 \times 10^{-2}$ | −0.5557 | −0.5635 | −0.5721 | −0.5815 | −0.5917 |
| $0.1 \times 10^{-1}$ | −0.7507 | −0.7611 | −0.7726 | −0.7852 | −0.7988 |
| $0.2 \times 10^{-1}$ | −0.9981 | −1.0117 | −1.0268 | −1.0432 | −1.0611 |
| $0.5 \times 10^{-1}$ | −1.4084 | −1.4271 | −1.4479 | −1.4707 | −1.4954 |
| 0.1 | −1.7720 | −1.7951 | −1.8208 | −1.8490 | −1.8796 |
| 0.2 | −2.1579 | −2.1854 | −2.2162 | −2.2500 | −2.2868 |
| 0.5 | −2.6313 | −2.6647 | −2.7019 | −2.7430 | −2.7877 |
| 1 | −2.8758 | −2.9130 | −2.9545 | −3.0004 | −3.0505 |
| 2 | −2.9003 | −2.9407 | −2.9859 | −3.0359 | −3.0906 |

**Table 5.7** *Values of log* $\gamma_{CO_3^{2-}}$

| Ionic strength $mol\,l^{-1}$ | 0 °C | 10 °C | 20 °C | 30°C | 40 °C |
|---|---|---|---|---|---|
| $0.1 \times 10^{-5}$ | −0.0020 | −0.0020 | −0.0020 | −0.0021 | −0.0021 |
| $0.2 \times 10^{-5}$ | −0.0028 | −0.0028 | −0.0029 | −0.0029 | −0.0030 |
| $0.5 \times 10^{-5}$ | −0.0044 | −0.0044 | −0.0045 | −0.0046 | −0.0047 |
| $0.1 \times 10^{-4}$ | −0.0062 | −0.0063 | −0.0064 | −0.0065 | −0.0066 |
| $0.2 \times 10^{-4}$ | −0.0087 | −0.0089 | −0.0090 | −0.0091 | −0.0093 |
| $0.5 \times 10^{-4}$ | −0.0137 | −0.0139 | −0.0142 | −0.0144 | −0.0146 |
| $0.1 \times 10^{-3}$ | −0.0193 | −0.0196 | −0.0199 | −0.0202 | −0.0206 |
| $0.2 \times 10^{-3}$ | −0.0271 | −0.0275 | −0.0280 | −0.0284 | −0.0289 |
| $0.5 \times 10^{-3}$ | −0.0423 | −0.0429 | −0.0436 | −0.0443 | −0.0451 |
| $0.1 \times 10^{-2}$ | −0.0589 | −0.0598 | −0.0607 | −0.0617 | −0.0628 |
| $0.2 \times 10^{-2}$ | −0.0816 | −0.0827 | −0.0840 | −0.0854 | −0.0869 |
| $0.5 \times 10^{-2}$ | −0.1237 | −0.1255 | −0.1274 | −0.1294 | −0.1317 |
| $0.1 \times 10^{-1}$ | −0.1673 | −0.1696 | −0.1722 | −0.1750 | −0.1780 |
| $0.2 \times 10^{-1}$ | −0.2229 | −0.2259 | −0.2292 | −0.2329 | −0.2368 |
| $0.5 \times 10^{-1}$ | −0.3159 | −0.3200 | −0.3246 | −0.3297 | −0.3352 |
| 0.1 | −0.4001 | −0.4052 | −0.4108 | −0.1417 | −0.4238 |
| 0.2 | −0.4929 | −0.4990 | −0.4048 | −0.5133 | −0.5214 |
| 0.5 | −0.6208 | −0.6282 | −0.6364 | −0.6454 | −0.6553 |
| 1 | −0.7142 | −0.7223 | −0.7315 | −0.7416 | −0.7526 |
| 2 | −0.7992 | −0.8080 | −0.8180 | −0.8290 | −0.8410 |

In the discussion above, ion pairing has been neglected. However, in many groundwaters significant ion pairing does take place, producing an overestimate in the saturation indices in excess of 0.1 units. A full correction for ion pairing requires the use of a large computer but the partial correction described below will reduce errors considerably. The ion pairs which have the greatest effect on the saturation index in groundwaters are listed with their dissociation constants in Table 5.8. From this table it can be seen that calcium and magnesium ion pairs have similar dissociation constants for the same anion. The carbonate complexes are the strongest, but carbonate concentrations are low in groundwaters with $pH < 9$ and groundwaters with $pH > 9$ are usually low in calcium and magnesium. Sulphate is probably the most important complexing ion in groundwaters since the complex is fairly strong and high sulphate concentrations are possible.

The effect of the $CaSO_4^0$ ion pair is demonstrated below; equations for other ion pairs are of similar form. The total calcium (i.e. the analytical value) is given by

$$[Ca_{tot}] = [Ca^{2+}] + [CaSO_4^0]. \tag{5.29}$$

**Table 5.8** *Ion pair dissociation constants at 25 °C*

| Reaction | log $K$ |
|---|---|
| $CaHCO_3^+ \rightleftharpoons Ca^{2+} + HCO_3^-$ | −0.85 |
| $CaSO_4^0 \rightleftharpoons Ca^{2+} + SO_4^{2-}$ | −2.31 |
| $CaCO_3^0 \rightleftharpoons Ca^{2+} + CO_3^{2-}$ | −3.07 |
| $MgHCO_3^+ \rightleftharpoons Mg^{2+} + HCO_3^-$ | −1.04 |
| $MgSO_4^0 \rightleftharpoons Mg^{2+} + SO_4^{2-}$ | −2.24 |
| $MgCO_3^0 \rightleftharpoons Mg^{2+} + CO_3^{2-}$ | −2.92 |

However,

$$(CaSO_4^0) = \frac{(Ca^{2+})(SO_4^{2-})}{K} \tag{5.30}$$

or

$$(CaSO_4^0) = \frac{[Ca^{2+}]\,[SO_4^{2-}]\,\gamma_{Ca^{2+}}\gamma_{SO_4^{2-}}}{K\gamma_{CaSO_4^0}}. \tag{5.31}$$

Substitution in eqn (5.29) gives

$$[Ca^{2+}] = [Ca_{tot}] \Bigg/ \left(1 + \frac{[SO_4^{2-}]\,\gamma_{Ca^{2+}}\gamma_{SO_4^{2-}}}{K\gamma_{CaSO_4^0}}\right). \tag{5.32}$$

Approximate activity coefficients for use in this equation can be obtained from Figure 3.1. If large numbers of saturation indices are to be calculated on analyses of varied ionic background, a computer program using a full aqueous model becomes highly desirable for both speed and accuracy (Section 3.4).

The equilibrium carbon dioxide partial pressure, $pP_{CO_2}$, is of related interest to the carbonate saturation indices. The derivation is similar to that of the calcite saturation index and gives

$$pP_{CO_2} = 2\text{ pH} - \log\,[Alk'] + \log\,\{(H^+)\gamma_{CO_3^{2-}} + 2K_2\gamma_{HCO_3^-}\} - E + F \tag{5.33}$$

where

$$E = \log \gamma_{HCO_3^-} + \log \gamma_{CO_3^{2-}} \tag{5.34}$$

$$F = \log K_H + \log K_1. \tag{5.35}$$

Values of $E$ and $F$ are given in Table 5.9 and Table 5.5 respectively. Note that the approximation used to obtain eqn (5.28) also relevant here.

It is interesting to note that in the range where the approximation of eqn (5.28) is valid the indices $SI_C$, $SI_D$, and $pP_{CO_2}$ are linearly dependent on pH. This has led to the concept of saturation pH, defined as the pH at which a water is exactly saturated with a specified mineral. It is easy to show that for calcite the saturation pH or $pH_s$, is given by

**Table 5.9** *Values of E for* p$P_{CO_2}$ *calculation*

| Ionic strength (mol $l^{-1}$) | 0 °C | 10 °C | 20 °C | 30 °C | 40°C |
|---|---|---|---|---|---|
| $0.1 \times 10^{-5}$ | −0.0025 | −0.0025 | −0.0025 | −0.0026 | −0.0026 |
| $0.2 \times 10^{-5}$ | −0.0035 | −0.0035 | −0.0036 | −0.0036 | −0.0037 |
| $0.5 \times 10^{-5}$ | −0.0055 | −0.0056 | −0.0056 | −0.0057 | −0.0058 |
| $0.1 \times 10^{-4}$ | −0.0077 | −0.0078 | −0.0080 | −0.0081 | −0.0082 |
| $0.2 \times 10^{-4}$ | −0.0109 | −0.0111 | −0.0112 | −0.0114 | −0.0116 |
| $0.5 \times 10^{-4}$ | −0.0172 | −0.0174 | −0.0177 | −0.0180 | −0.0183 |
| $0.1 \times 10^{-3}$ | −0.0242 | −0.0245 | −0.0249 | −0.0253 | −0.0258 |
| $0.2 \times 10^{-3}$ | −0.0339 | −0.0344 | −0.0350 | −0.0355 | −0.0362 |
| $0.5 \times 10^{-3}$ | −0.0529 | −0.0537 | −0.0545 | −0.0544 | −0.0564 |
| $0.1 \times 10^{-2}$ | −0.0737 | −0.0747 | −0.0759 | −0.0771 | −0.0785 |
| $0.2 \times 10^{-2}$ | −0.1019 | −0.1034 | −0.1050 | −0.1067 | −0.1086 |
| $0.5 \times 10^{-2}$ | −0.1547 | −0.1568 | −0.1592 | −0.1618 | −0.1646 |
| $0.1 \times 10^{-1}$ | −0.2092 | −0.2120 | −0.2152 | −0.2187 | −0.2225 |
| $0.2 \times 10^{-1}$ | −0.2786 | −0.2823 | −0.2865 | −0.2911 | −0.2960 |
| $0.5 \times 10^{-1}$ | −0.3949 | −0.4001 | −0.4058 | −0.4121 | −0.4189 |
| 0.1 | −0.5001 | −0.5065 | −0.5136 | −0.5213 | −0.5298 |
| 0.2 | −0.6162 | −0.6238 | −0.6323 | −0.6416 | −0.6518 |
| 0.5 | −0.7760 | −0.7852 | −0.7954 | −0.8067 | −0.8191 |
| 1 | −0.8927 | −0.9029 | −0.9144 | −0.9270 | −0.9408 |
| 2 | −0.9989 | −1.0100 | −1.0225 | −1.0362 | −1.0512 |

$$pH_s = pH - SI_C. \tag{5.36}$$

For dolomite, the expression is

$$pH_s = pH - SI_D/2. \tag{5.37}$$

Clearly the concept of saturation pH is valid only for minerals for which there is a clear dependence of saturation index on pH. The technique can also be used to calculate changes in saturation index on charges of carbon dioxide partial pressure. For example, many carbonate groundwaters were formed in contact with carbon-dioxide-rich soil gas and therefore have equilibrium partial pressures above the atmospheric value. In contact with the atmosphere, such waters lose carbon dioxide and become supersaturated with calcite, resulting in incrustation.

*Example.* A water with a pH of 7.1 and $SI_C = -0.05$, p$P_{CO_2} = 2.55$ is allowed to equilibrate with the atmosphere. What will be the pH and $SI_C$ values?

The p$P_{CO_2}$ value of the atmosphere is 3.52. Use of eqn (5.33) shows that the pH must be raised by $3.52 - 2.55 = 0.97$ units to raise p$P_{CO_2}$ to the required value. Thus the new pH is 8.07. The new $SI_C$ is $+ 0.92$; thus this water will tend to precipitate calcite.

It is not possible to calculate the loss of calcite directly; instead recourse must be made to models such as MIX2 (Plummer, Parkhurst, and Kosiur 1975).

The error in calcite saturation index calculation as a result of the accumulation of analytical errors, the most significant of which is pH, has been estimated as ±0.1 unit (Langmuir 1971). The error in the dolomite index is larger, about ±0.2, because of the 2 × pH term. In addition there is the effect of errors in the solubility products $K_C$ and $K_D$. Picknett (1972) has shown that the value of log $K_C$ may be altered up to be 0.1 units by the presence of magnesium ions in solution. The value of $K_D$ is rather poorly known because of experimental difficulties (Langmuir 1971). Thus a value of $SI_C$ within ± 0.2 units of zero may indicate equilibrium, the error being larger for $SI_D$.

## References

Langmuir, D. (1971). The geochemistry of some carbonate groundwaters in central Pennsylvania. *Geochim. cosmochim. Acta* **35**, 1023–45.

Picknett, R. G. (1972). The pH of calcite solutions with and without $MgCO_3$ present and the implications concerning rejuvenated aggressiveness. *Trans. cave res. Group G.B.* **14**, 141–150.

Plummer, L. N., Parkhurst, D., and Kosiur, D. R. (1975). MIX2, a computer program for modelling chemical reactions in natural water. *U.S. geol. Surv., Rep.* **61–75**.

# 6 The representation of hydrochemical data as a basis for interpretation

## 6.1. Introduction

As can be seen from the discussions in the previous chapters, well-head measurements, laboratory analyses, and calculations provide a plethora of hydrochemical parameters on which interpretation can be based. The multitude of chemical data frequently amassed in groundwater studies can, however, pose problems, as without a methodical approach to the analysis of the data the hydrochemical processes occurring, the influence of hydrogeological controls, etc. can be confused and the data may appear contradictory.

In this chapter the representation of hydrochemical data is discussed with respect to understanding hydrochemical distributions, the basis for assessing the relationship between differing hydrochemical water types, and some of the processes involved in the chemical evolution of groundwaters.

It is stressed that although a sensible general approach to the analysis of data can be adopted and normally general hydrochemical processes can be recognized in any hydrogeological situation, each situation should be considered as being unique so that only some of the techniques described below will be applicable in each individual study, and applicability may vary from study to study.

In the procedure of representing data for interpretation it is recommended that a twofold approach be adopted which should consist firstly of maps of distribution of pertinent parameters and secondly of relevant parameter relationship diagrams. It is emphasized that the two types of representation should be used concurrently. Further, while representation is essential for interpretation it is also important for displaying data coherently and for communicating the complexity of interpretation.

## 6.2. Distribution maps

The most frequently used distribution maps are those depicting total dissolved solids or electrical conductivity. Such maps provide useful preliminary information about a system together with water quality indications.

In Fig. 6.1 a total dissolved solids distribution is compared with a groundwater flow net for the Lower Palaeozoic Disi Sandstone aquifer in Southern Jordan. The groundwater flow in the area is partly influenced by a vertical dyke and the control is reflected by the hydrochemistry. Similarly, the geological structure in the Lincolnshire Limestone (Fig. 6.2) is believed to control the

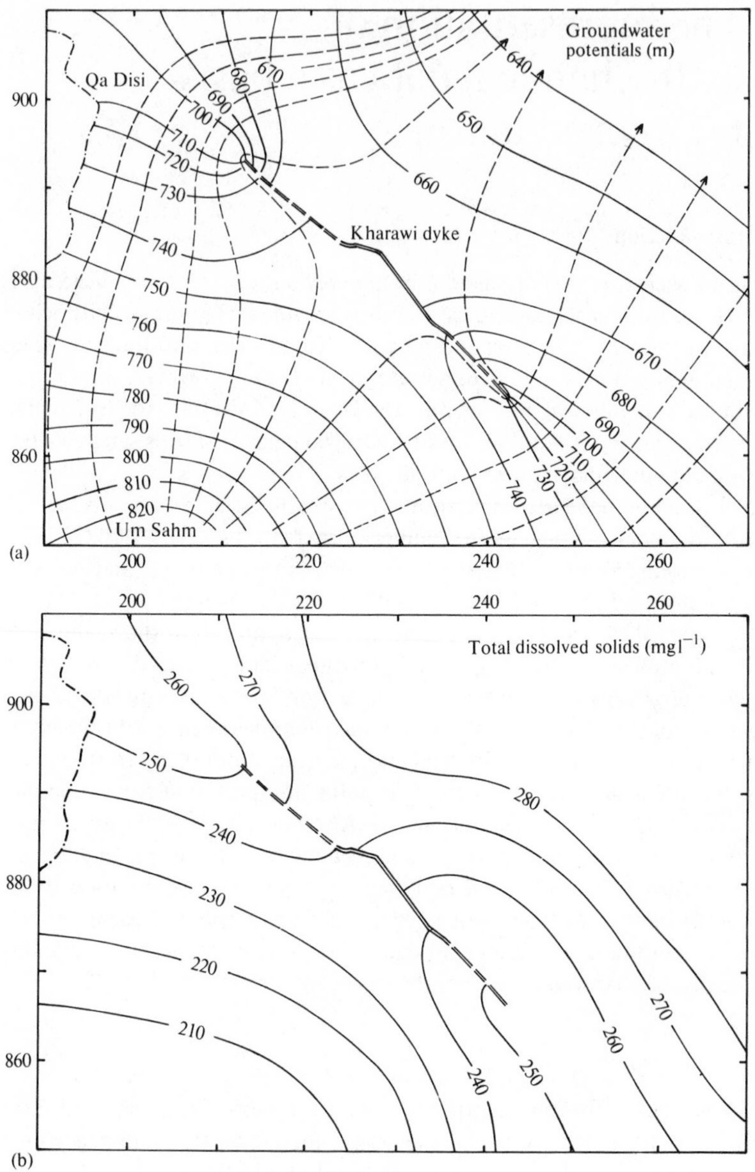


**Fig. 6.1.** Relationship between a flow net (map (a)) and total dissolved solids distribution (map (b)) for the southern desert of Jordan. (After Lloyd 1969.)

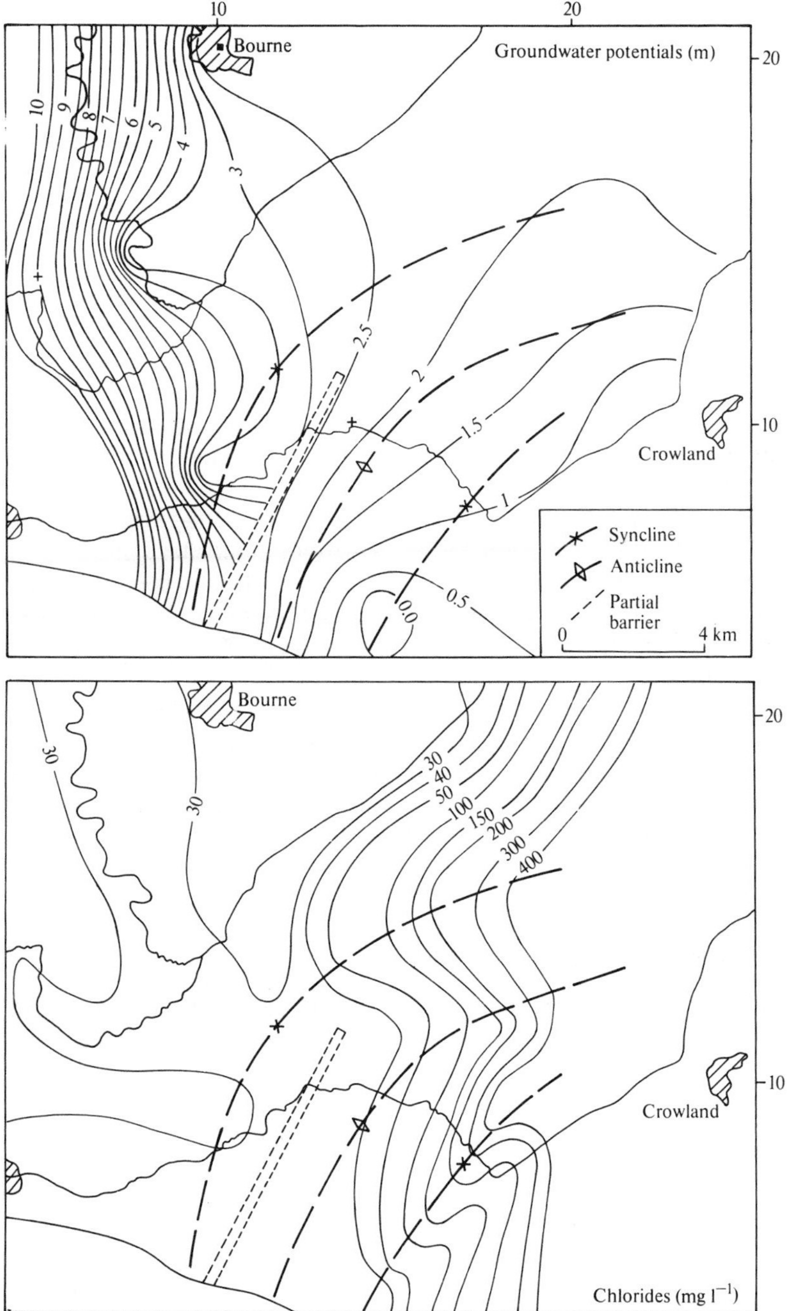


**Fig. 6.2.** Groundwater potential distribution for part of the Lincolnshire Limestone aquifer and chloride distribution showing influences of geological structural control.

hydrochemistry (Marsh 1977). Flow is shown conforming to a partial barrier boundary attributed to a low permeability zone on an anticlinal flank with chloride concentrations reflecting the boundary in a subdued manner. Low flow conditions along the eastern syncline are clearly reflected by the chlorides. Total dissolved solids maps, however, have limitations in that they are bulk parameters and not unique. For distribution maps it is therefore preferable to use individual ions, ion ratios, or certain calculated parameters.

In hydrochemical groundwater evolution the chloride ion tends to be the most conservative in that it is readily removed from matrix materials but rarely precipitated under dilute solution conditions. Chloride concentrations therefore normally increase down the hydraulic gradient and with groundwater flow experience and residence. The normal chloride concentration increase is only disturbed where pollution or dilution occurs so that chloride is an excellent indicator of groundwater flow direction and preferential permeability conditions.

In constructing distribution maps it is essential that they relate to the hydrogeological controls and reflect the piezometry of the aquifer system. One of the main values of distribution maps is that they provide an independent check on piezometry and can reveal hydrogeological conditions which are not necessarily apparent from the piezometry. This is particularly so where piezometry is highly disturbed by abstraction.

Maps of hydrochemical ratios have long been favoured by French workers and can provide useful information concerning such features as ion exchange and the onset of brackish water influences. An example of ratio maps and the associated total dissolved solids is given in Fig. 6.3 for the Mornag alluvial aquifer in Tunisia.

## 6.3. Hydrochemical sections

While distribution maps indicate rates of change of chemical concentration regionally and can be thus interpreted, the overall trends are probably better represented and understood using hydrochemical sections on which a number of parameters can be shown. Such diagrams depict associated changes in parameter values and aid in interpreting the controlling hydrochemical processes.

Hydrochemical sections can be constructed along specific groundwater flow lines, but normally are constructed in the direction of declining hydraulic head from either a recharge mound or the start of a confining cover using an associated set of regional data. Where complex groundwater flow occurs total dissolved solids can be used as the abscissa.

On Fig. 6.4 a confined aquifer hydrochemical section is shown for the Lincolnshire Limestone in which a classical redox barrier with the associated hydrochemical changes along the hydraulic groundwater gradient are depicted. Changes in alkalinity and the various hardness parameters can also be depicted

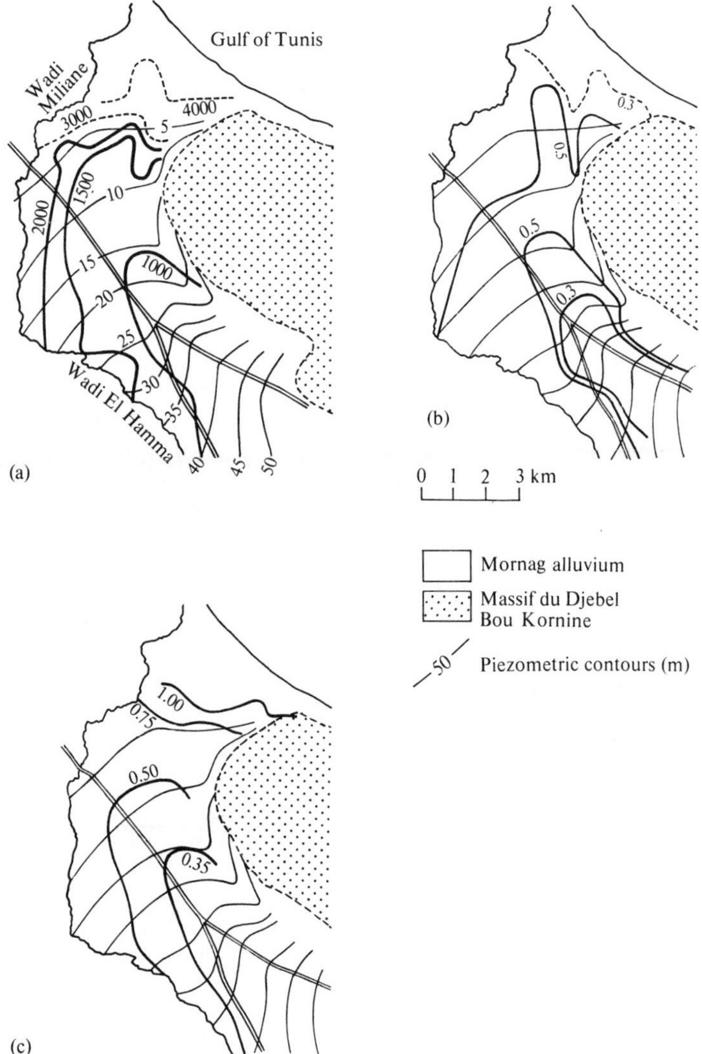


**Fig. 6.3.** Hydrochemical ratio distributions from the Mornag alluvium aquifer, Tunisia: (a) map of total dissolved solids (mg $l^{-1}$); (b) map of $SO_4^{2-}/Cl^-$ ratios; (c) map of $Mg^{2+}/Ca^{2+}$ ratios. N.B. Location and piezometric contours are shown in all three maps. (After Schoeller 1959.)

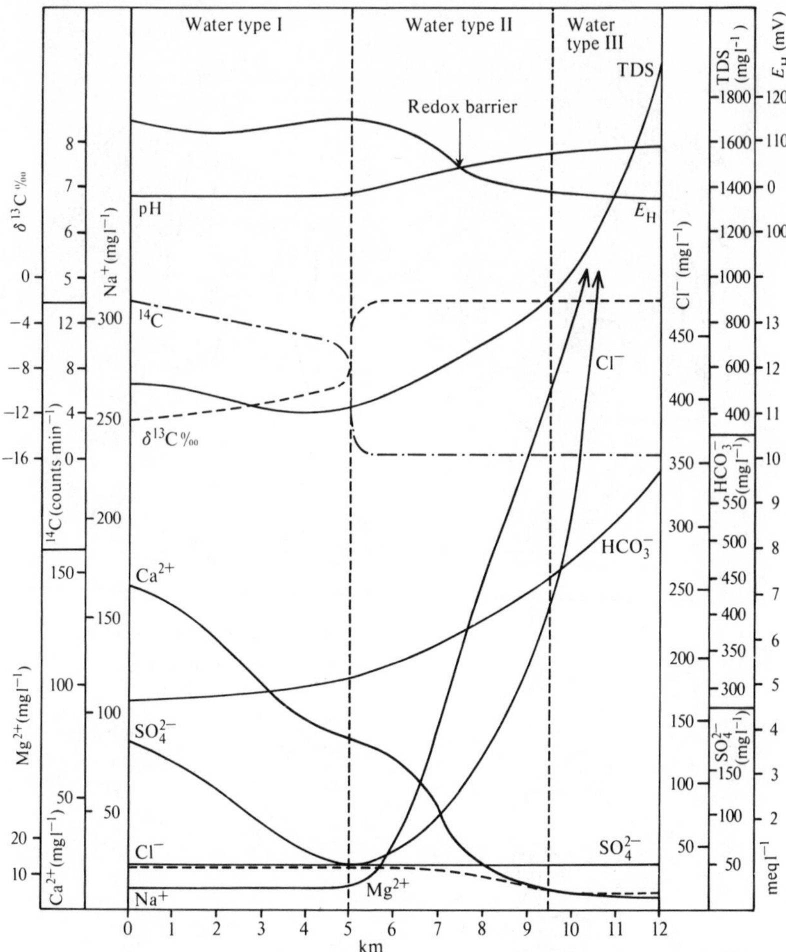


**Fig. 6.4.** Hydrochemical section for part of the central Lincolnshire Limestone. (After Lawrence, Lloyd, and Marsh 1976.)

to advantage by hydrochemical sections as shown in Fig. 6.5 for part of the London Basin chalk aquifer.

## 6.4. Hydrochemical diagrams

Graphical representations of hydrochemical data pose considerable problems. Dimension limitations severely restrict illustration to all but the most simple forms, while graphical representation of statistically derived hydrochemical functions, although attractive, is frequently difficult to comprehend. Hydrochemical diagrams need to be as comprehensive as possible and in addition

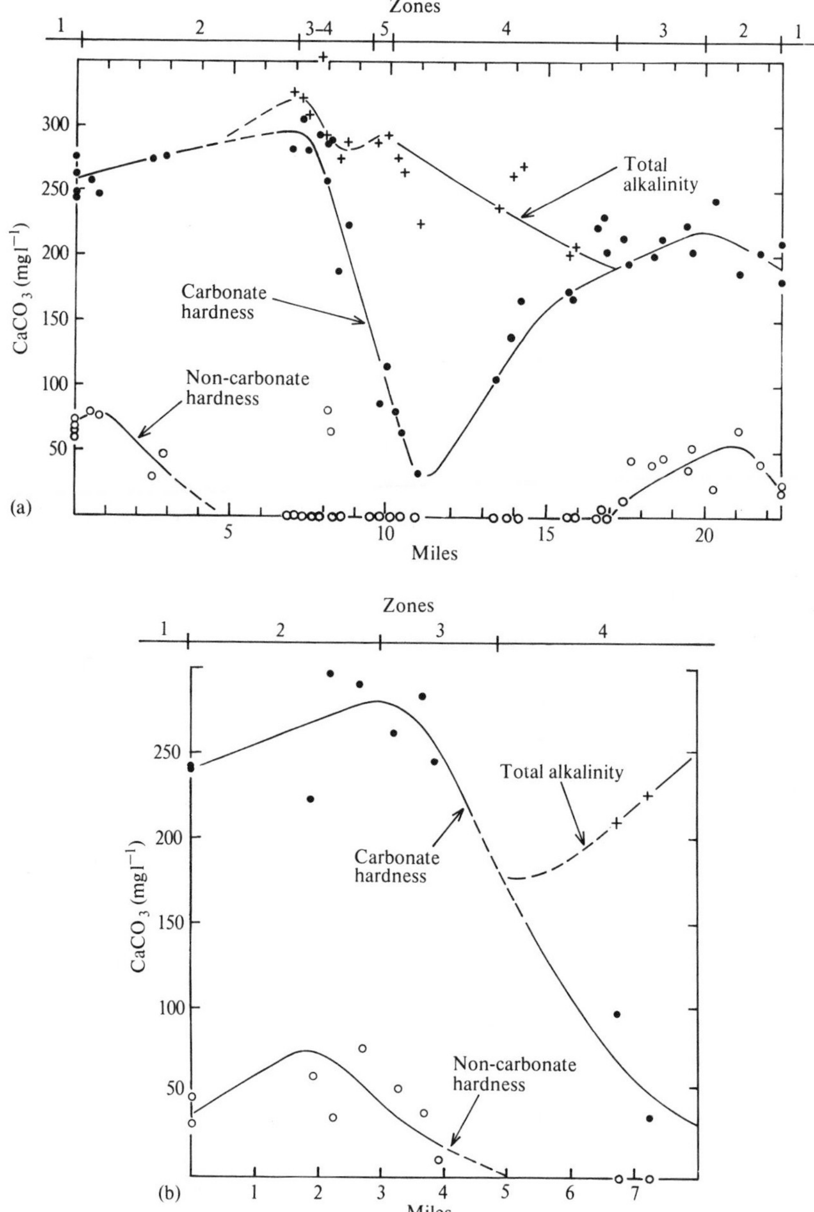


**Fig. 6.5.** Hydrochemical sections of alkalinity and hardness in the chalk of (a) the London Basin and (b) North Kent: +, total alkalinity; •, carbonate hardness; ○, non-carbonate hardness. (After Ineson and Dowing 1963.)

should facilitate interpretation of evolutionary trends and hydrochemical processes. Few diagrams really possess these features. In using hydrochemical diagrams it is re-emphasized that their real value is realized when they are interpreted in conjunction with distribution maps.

### *6.4.1. Vertical bar graphs*

A vertical bar graph of groundwater analyses is shown in Fig. 6.6(a). Each analysis appears as a vertical bar with a height proportional to the total concentration of anions or cations, normally expressed in milliequivalents per litre. Individual ion concentrations are depicted proportionally in a consistent pattern. While the major ions are shown on Fig. 6.6, clearly any parameter permutation can be used. As shown in Fig. 6.6(b) an analysis can be demonstrated in terms of the percentage of total milliequivalents per litre.

### *6.4.2. Vector diagrams*

A vector diagram representing major ion chemistry for several groundwater analyses is shown in Fig. 6.7. The lengths of the six vectors represent ionic concentrations in milliequivalents per litre. Such diagrams appear to have little merit.

### *6.4.3. Circular and radial diagrams*

Circular diagrams of hydrochemical analyses (Fig. 6.8) are constructed with their radii proportional to the total ionic concentrations. Sectors within a circle represent fractions of the different ions expressed as milliequivalents per litre. A radial diagram is shown in Fig. 6.9. On the diagram the ions are scaled and the intercepts are joined to represent the water. As with vector diagrams only one analysis is represented on circular and radial diagrams so they are of limited value.

### *6.4.4. Pattern diagrams*

Pattern diagrams were first devised by Stiff (1951) to represent analyses in distinctive graphical shapes by plotting ions in milliequivalents per litre on on parallel axes as illustrated in Fig. 6.10. While pattern diagrams can show differences between individual chemical analyses, they suffer from the same limitations as discussed for the previous diagrams.

### *6.4.5. Semilogarithmic diagrams*

Semilogarithmic diagrams were developed by Schoeller (1962) to represent major ion analyses in milliequivalents per litre and to demonstrate different hydrochemical water types on the same diagram. A semilogarithmic diagram is shown in Fig. 6.11 with two water types clearly depicted. Because of the line work, unfortunately only a few analyses can usually be illustrated. The diagram has one advantage over the more universally used trilinear diagrams discussed below in that actual parameter concentrations are given. The diagrams can be

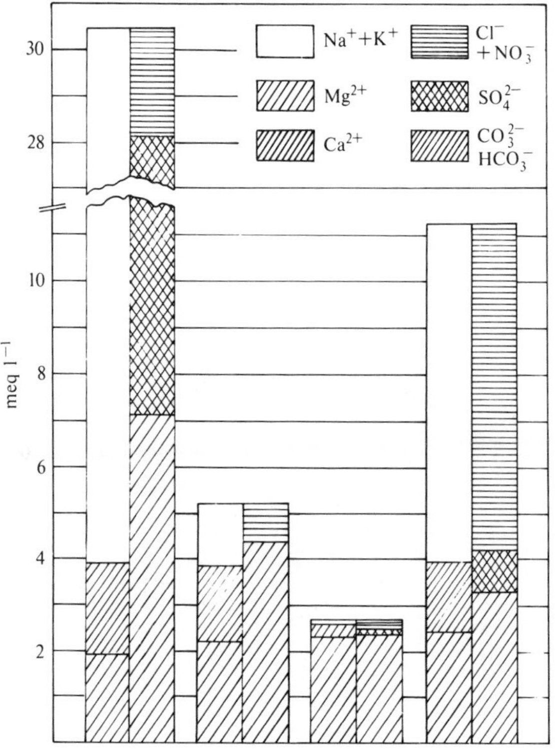


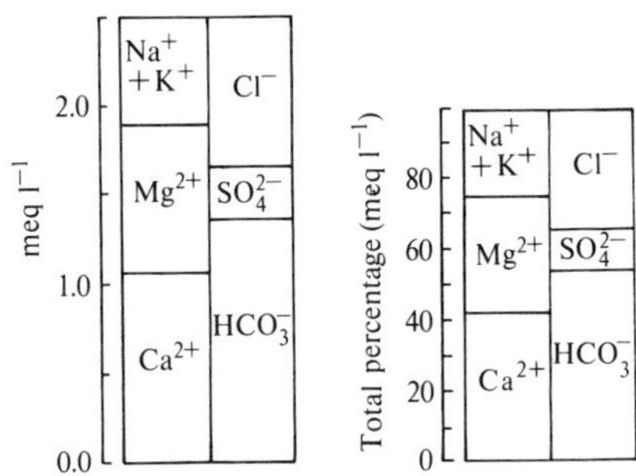


**Fig. 6.6.** Major ion chemistry represented by bar charts.

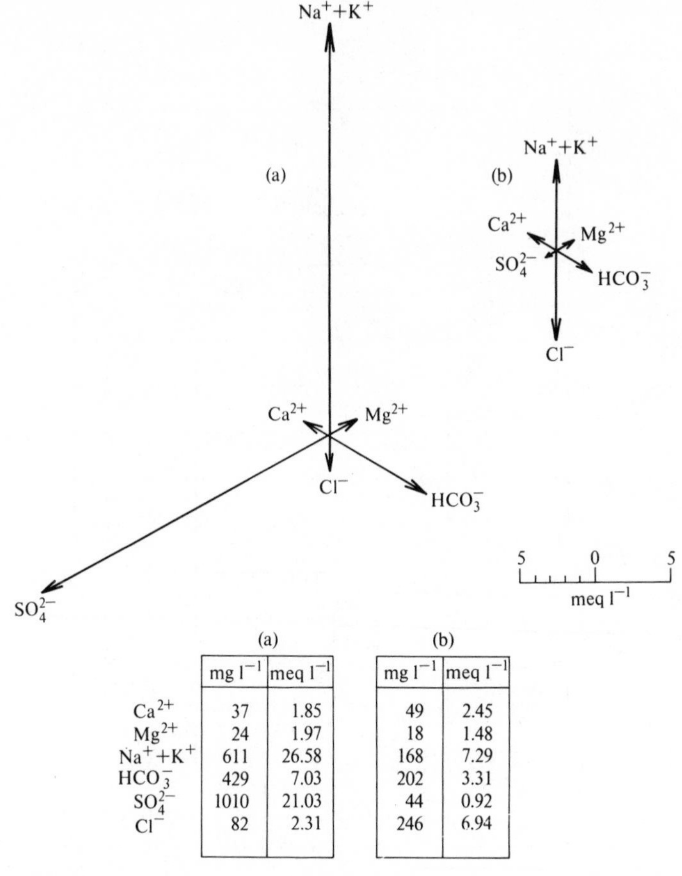


| | (a) mg $l^{-1}$ | (a) meq $l^{-1}$ | (b) mg $l^{-1}$ | (b) meq $l^{-1}$ |
|---|---|---|---|---|
| $Ca^{2+}$ | 37 | 1.85 | 49 | 2.45 |
| $Mg^{2+}$ | 24 | 1.97 | 18 | 1.48 |
| $Na^+ + K^+$ | 611 | 26.58 | 168 | 7.29 |
| $HCO_3^-$ | 429 | 7.03 | 202 | 3.31 |
| $SO_4^{2-}$ | 1010 | 21.03 | 44 | 0.92 |
| $Cl^-$ | 82 | 2.31 | 246 | 6.94 |

**Fig. 6.7.** Vector diagram for major ions. (After Maucha 1949.)

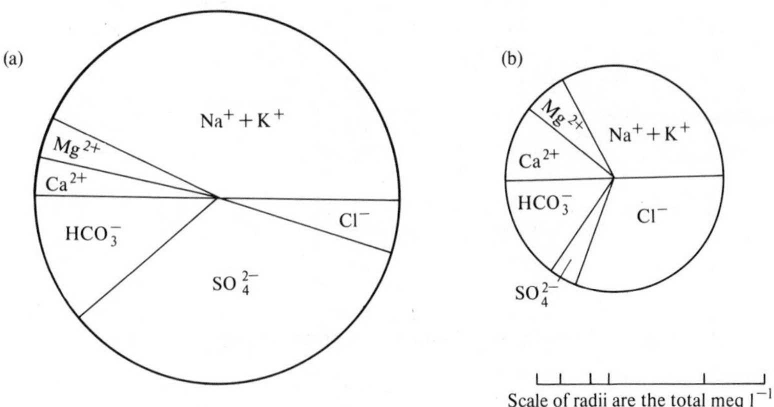


**Fig. 6.8.** Circular diagram for major ions (analyses shown on Fig. 6.7). (After Hem 1970.)

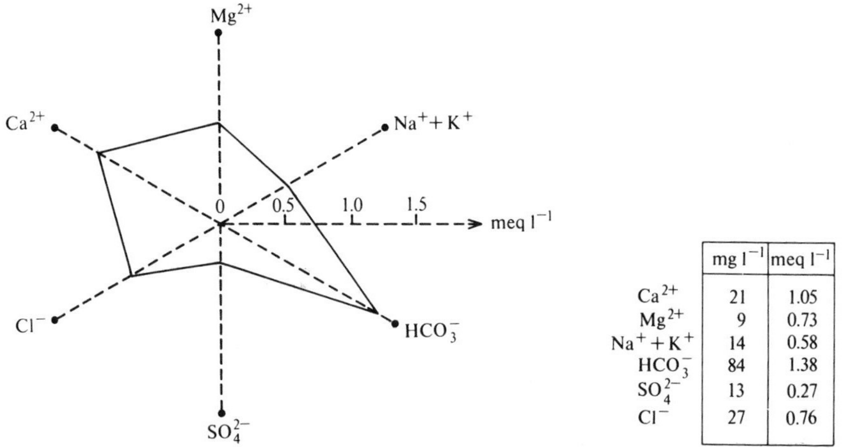


| | mg $l^{-1}$ | meq $l^{-1}$ |
|---|---|---|
| $Ca^{2+}$ | 21 | 1.05 |
| $Mg^{2+}$ | 9 | 0.73 |
| $Na^{+} + K^{+}$ | 14 | 0.58 |
| $HCO_3^{-}$ | 84 | 1.38 |
| $SO_4^{2-}$ | 13 | 0.27 |
| $Cl^{-}$ | 27 | 0.76 |

**Fig. 6.9.** Radial diagram for major ions.

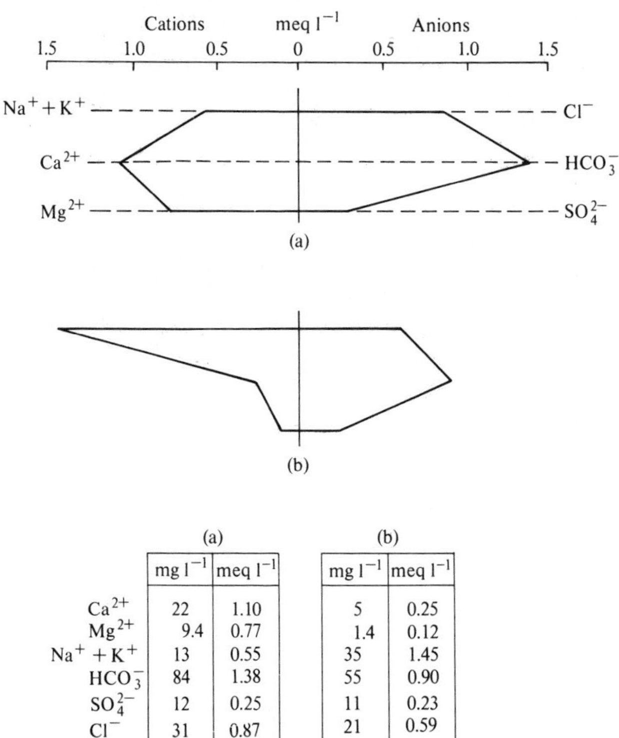


| | (a) mg $l^{-1}$ | (a) meq $l^{-1}$ | (b) mg $l^{-1}$ | (b) meq $l^{-1}$ |
|---|---|---|---|---|
| $Ca^{2+}$ | 22 | 1.10 | 5 | 0.25 |
| $Mg^{2+}$ | 9.4 | 0.77 | 1.4 | 0.12 |
| $Na^{+} + K^{+}$ | 13 | 0.55 | 35 | 1.45 |
| $HCO_3^{-}$ | 84 | 1.38 | 55 | 0.90 |
| $SO_4^{2-}$ | 12 | 0.25 | 11 | 0.23 |
| $Cl^{-}$ | 31 | 0.87 | 21 | 0.59 |

**Fig. 6.10.** Pattern diagram for major ions. (After Stiff 1951.)

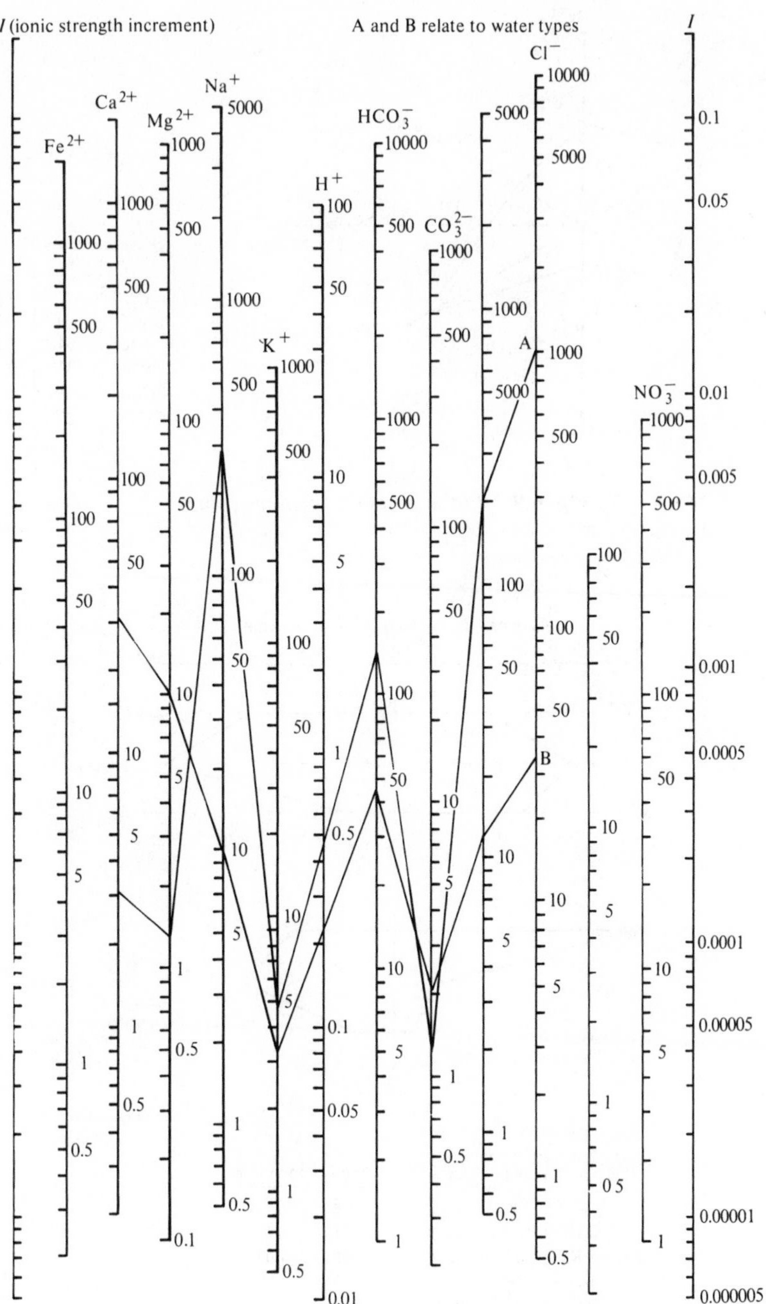


**Fig. 6.11.** Semilogarithmic diagram for major ions (mg $l^{-1}$). (After Schoeller 1962.)

adapted to determine the degree of calcite saturation as demonstrated by Schoeller (1962), although this has little advantage today in view of computational facilities.

Semilogarithmic diagrams can be used in a simplified form to represent any parameters, for example a heavy metal suite.

### *6.4.6. X–Y plots*

*X–Y* plots are the most simple and obvious initial approach to the interpretation of hydrochemical data. With modern computer graphical facilities such plots can rapidly be compiled for a large number of *X–Y* relationships which will allow a preliminary assessment of those parameters that may assist in interpretation. Plots of single ion relationships and those ion ratios that may prove meaningful can easily be established. Ratios considered include $Ca^{2+}$: $Mg^{2+}$, $Ca^{2+}$: $Na^+$, $Na^+$: $Ca^{2+} + Mg^{2+}$, $HCO_3^-$: $Cl^-$, $SO_4^{2-}$: $Cl^-$, $Cl^-$-$Na^+$: $Cl^-$, etc.

For ion *X–Y* plots the stoichiometric balance between ions, if pertinent, can be represented for pairs of ions to examine possible process effects. For example the stoichiometry shown on Fig. 6.12 for $Ca^{2+}/(HCO_3^- + SO_4^{2-})$ and $Na^+/Cl^-$ indicates a ratio close to 1.0, showing that for the particular groundwaters being studied neither ion exchange nor sulphate reduction, for example, are complicating the carbonate chemistry and simple solution is the main process.

### *6.4.7. Dilution diagrams*

A variation of the *X–Y* plot is the dilution diagram in which the groundwaters in a system are compared with respect to two end-point waters. Such diagrams are usually plotted semilogarithmically or logarithmically. The dilution line represents a percentage mixing between the two end-point waters so that waters plotting away from the line indicate preferential enrichment or depletion by one ion with respect to the other which may infer a certain chemical process control.

A dilution diagram for a set of saline groundwaters in the Lincolnshire Chalk is shown in Fig. 6.13; the end-point waters are a low chloride water and the local modern seawater. The diagram shows an enrichment in potassium with respect to chloride for many of the groundwaters, which is attributed to ion exchange occurring in the aquifer.

### *6.4.8. Mixing diagrams*

While dilution diagrams represent a theoretical mixing between two waters, their main function is to indicate deviations from the theoretical condition. Where waters fall on a dilution line a possible mixing percentage can be calculated; similarly, as shown in Fig. 6.14 mixing percentages between three groundwaters can be calculated. Mixing based on two parameters, however, can lead to fallacious conclusions so that the process is better represented using the whole suite of major ions as depicted in Fig. 6.15. In this figure intercepts on the straight-line relationships between the end-point waters give the percentage of true

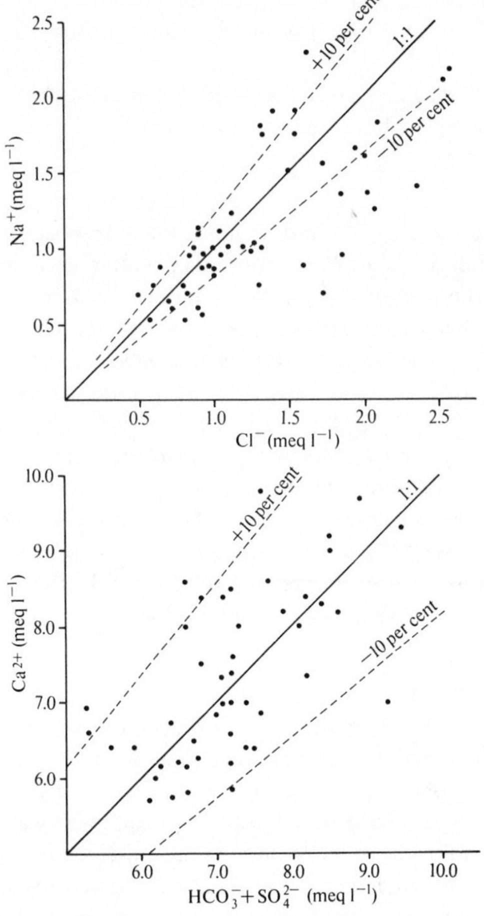


**Fig. 6.12.** *X–Y* plot of major ions showing simple dissolution.

mixing that has occurred. Deviations from a straight-line relationship indicate a complicating process occurring in addition to mixing.

Where actual mixing is found to occur between three waters multiparameter confirmation is difficult graphically although it has been attempted by McKinnell (1958).

### *6.4.9. Trilinear diagrams*

The use of trilinear diagrams to represent groundwaters was first attempted by Hill (1940) and refined by Piper (1944). The Piper diagram has become universally used and is shown in Fig. 6.16. Major ions are plotted in the two base triangles of the diagram as cation and anion percentages of milliequivalents per litre. Total cations and total anions are each considered as 100 per cent. The

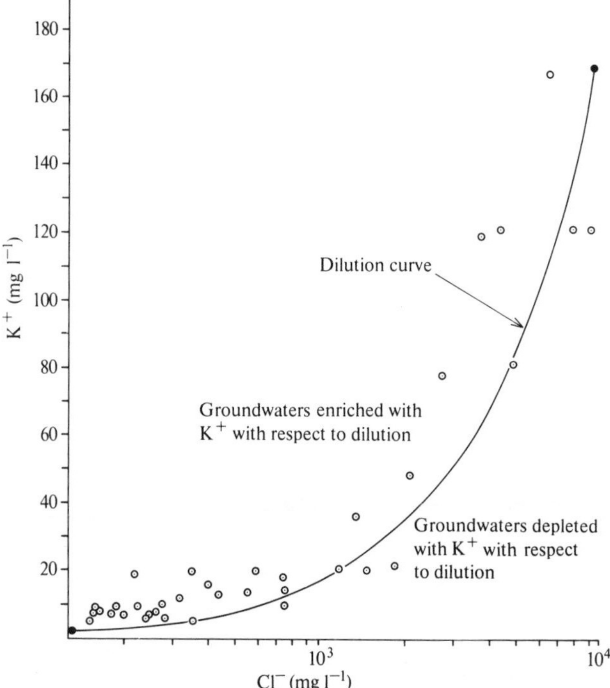


**Fig. 6.13.** Dilution diagram for a groundwater data set relating $Cl^-$ and $K^+$: •, end-point waters.

respective cation and anion locations for an analysis are projected into the rectangle which represents the total ion relationships, as shown on Fig. 6.16.

The Piper diagram allows comparisons to be made between a large number of analyses but has the drawback that all trilinear diagrams have in not portraying actual ion concentrations. The distribution of ions within the main field rectangle is unsystematic in hydrochemical process terms so that the diagram lacks a certain logic. Further, the large amount of line work is a hindrance.

As shown on Fig. 6.16 total dissolved solids can also be represented on a Piper diagram by circles drawn at analysis locations in the rectangle. Circle radii as shown are normally proportional to logarithmic values of total dissolved solids.

The diagram can be used to calculate resultant mixing between two groundwaters if the groundwaters plot on a straight line in each of the three fields. To demonstrate conclusively that simple mixing is occurring Piper (1944) recommends that the graphical conclusion is substantiated by calculations.

An alternative diagram to that of Piper has been devised by Durov (1948) and is shown in Fig. 6.17. As with the former diagram it is normally based upon percentage major ion milliequivalent values, but in this case the cations and

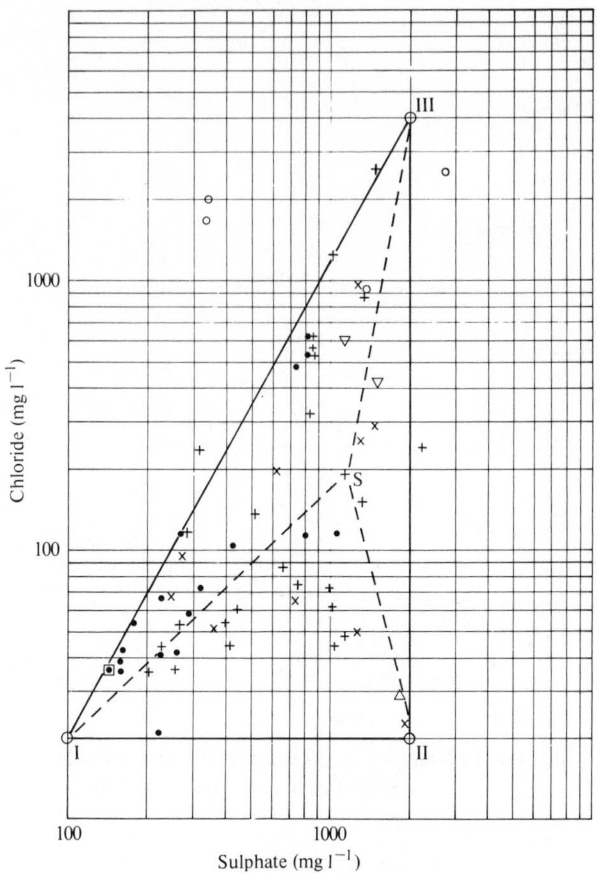


**Fig. 6.14.** Set of groundwaters representing mixing of the three end-point waters I, II, and III. The mixed water S is composed of 47.5%, I, 48.3% II, and 4.2% III.

anions together total 100 per cent. The cation and anion values are plotted in the appropriate triangle and projected into the square main field as shown on Fig. 6.17.

An expanded version of the Durov diagram is shown in Fig. 6.18. In this diagram, developed by Burdon and Mazloum (1958) and Lloyd (1965), the cation and anion triangles are recognized and are separated along the 25 per cent axes so that the main field is conveniently divided. On Fig. 6.17 the extension of the diagram to include a seventh parameter is shown. While total dissolved solids are shown in this case, clearly any parameter can be used.

The expanded Durov diagram has the distinct advantage over the Piper diagram in that it provides a better display of hydrochemical types and some processes, and in practical terms has less line work in the main field. Although,

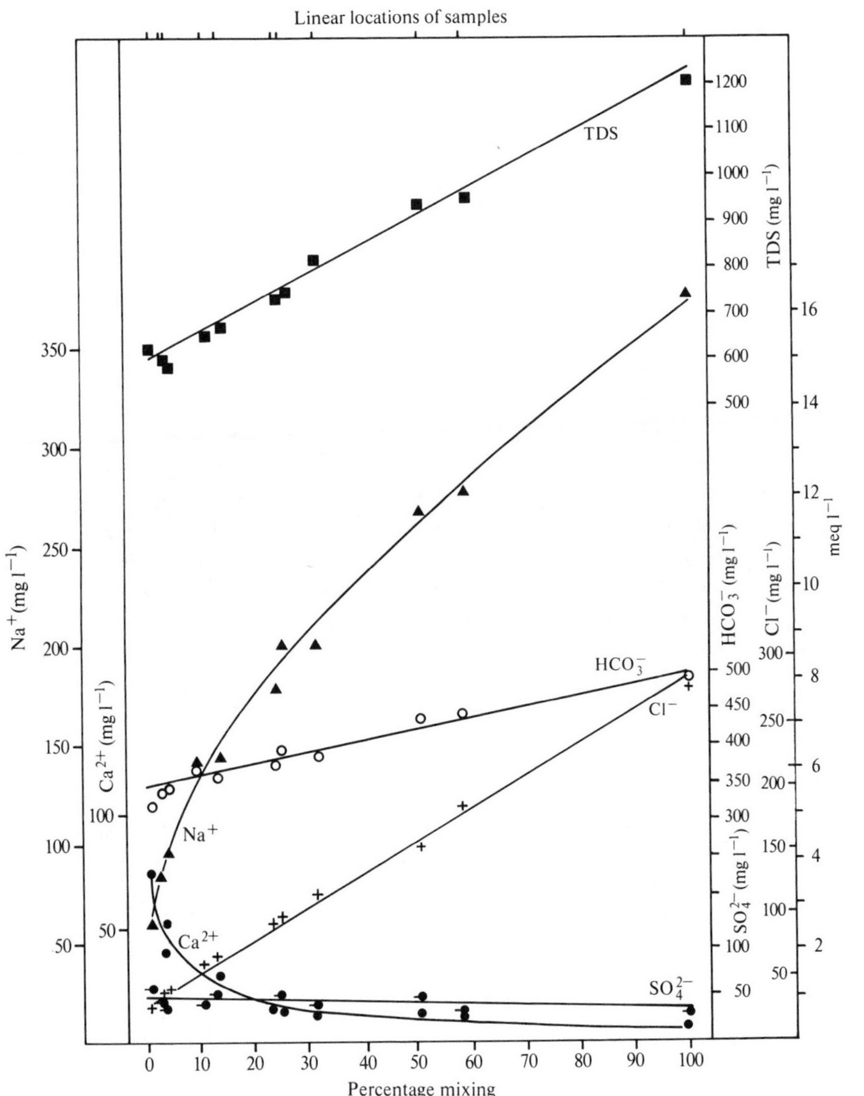


**Fig. 6.15.** Mixing diagram showing ion exchange. (After Lawrence *et al.* 1976.)

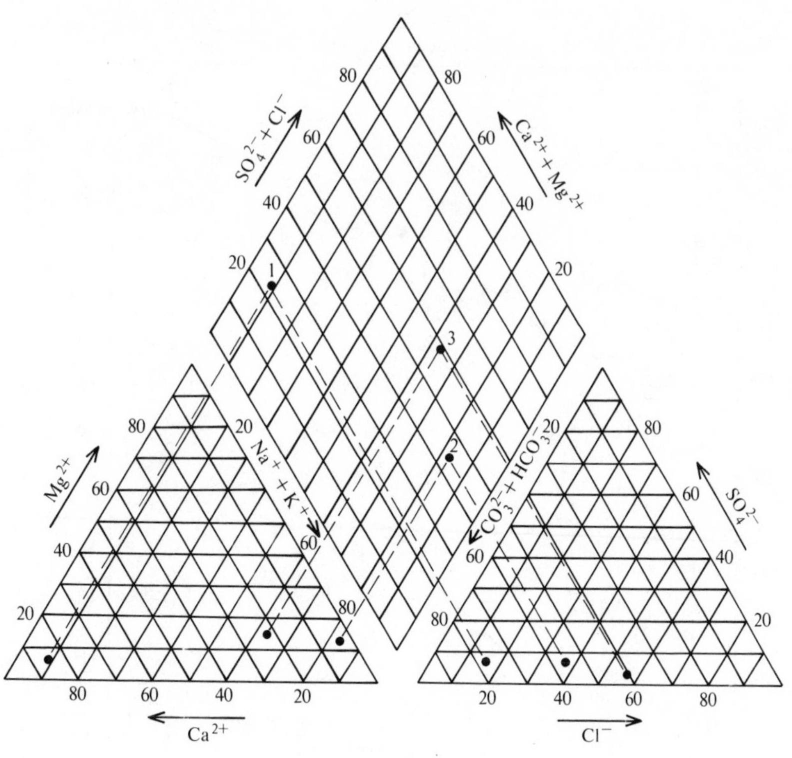


| | 1 | | | 2 | | | 3 | | |
|---|---|---|---|---|---|---|---|---|---|
| | mg $l^{-1}$ | meq $l^{-1}$ | Per cent | mg $l^{-1}$ | meq $l^{-1}$ | Per cent | mg $l^{-1}$ | meq $l^{-1}$ | Per cent |
| $Ca^{2+}$ | 152 | 7.58 | 85.5 | 8 | 0.40 | 4.0 | 85 | 4.24 | 22.3 |
| $Mg^{2+}$ | 5.8 | 0.48 | 5.4 | 15.5 | 1.28 | 12.8 | 35 | 2.88 | 15.2 |
| $Na^+$ | 17 | 0.74 | 8.3 | 186 | 8.09 | 81.4 | 270 | 11.74 | 61.8 |
| $K^+$ | 2.8 | 0.07 | 0.8 | 7 | 0.18 | 1.8 | 6 | 0.15 | 2.8 |
| $HCO_3^-$ | 404 | 6.61 | 77.7 | 326 | 5.34 | 55.8 | 479 | 7.84 | 40.6 |
| $SO_4^{2-}$ | 29 | 0.60 | 7.1 | 37 | 0.77 | 8.0 | 29 | 0.60 | 3.1 |
| $Cl^-$ | 46 | 1.30 | 15.2 | 123 | 3.47 | 36.2 | 286 | 10.89 | 56.3 |

**Fig. 6.16.** Trilinear diagram of major ions. (After Piper 1944.)

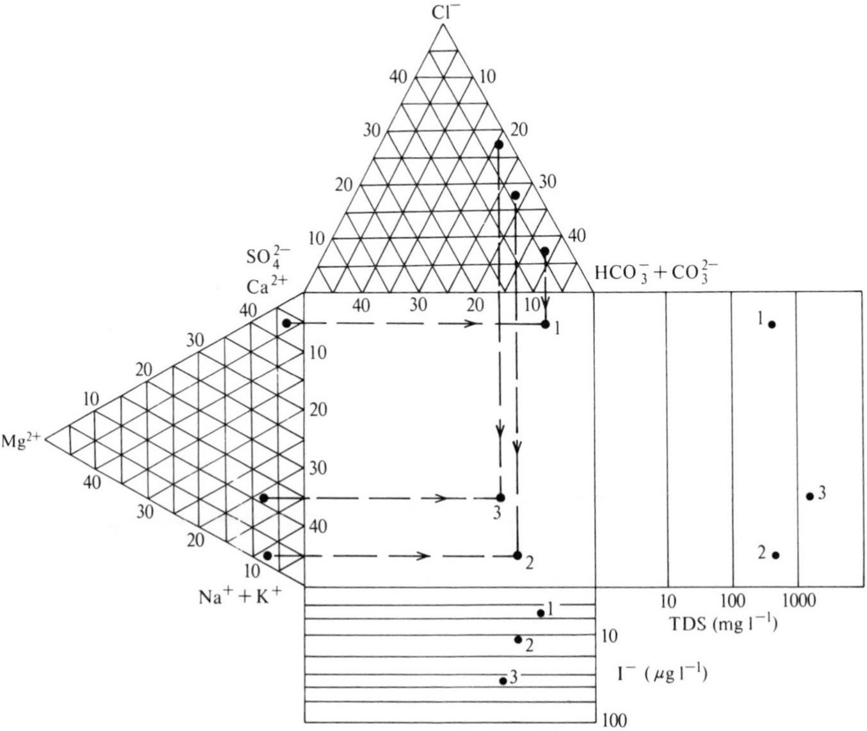

Analyses as in Fig. 6.16 with percentages divided by 2 and

| | $I^-$ | TDS |
|---|---|---|
| 1 | 2.7 | 660 |
| 2 | 10.4 | 710 |
| 3 | 3.4 | 1190 |

**Fig. 6.17.** The original type of Durov diagram.

as is shown on Fig. 6.18, waters with a 25 per cent concentration of a certain ion can theoretically plot ambiguously, in practice the ambiguity has little relevance and can be solved by the association of the problem water with neighbouring waters.

The significance of the nine fields on the expanded Durov diagram can be discussed with respect to Fig. 6.19 as follows:

1. $HCO_3^-$ and $Ca^{2+}$ dominant, frequently indicates recharging waters in limestone, sandstone, and many other aquifers.

2. $HCO_3^-$ dominant and eight $Mg^{2+}$ dominant or cations indiscriminant, with $Mg^{2+}$ dominant or $Ca^{2+}$ and $Mg^{2+}$ important, indicates waters often associated with dolomites; where $Ca^{2+}$ and $Na^+$ are important partial ion exchange may be indicated.

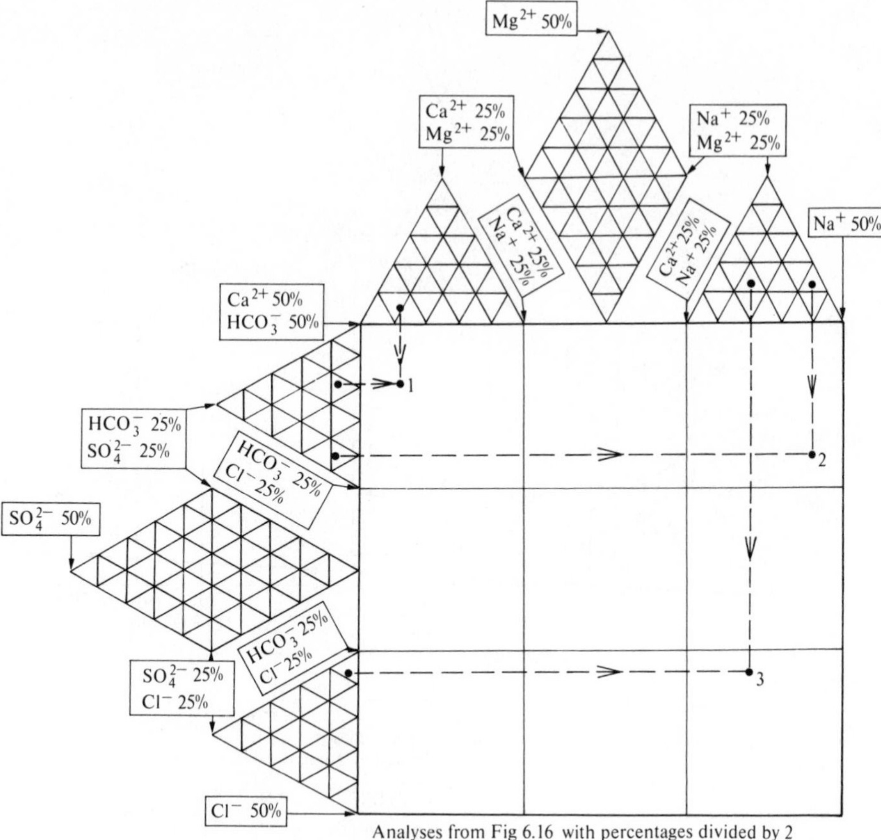


**Fig. 6.18.** Expanded Durov diagram demonstrating major ions.

3. $HCO_3^-$ and $Na^+$ dominant, normally indicates ion-exchanged waters although the generation of $CO_2$ at depth can produce $HCO_3^-$ where $Na^+$ is dominant under certain circumstances (Winograd and Farlekas 1974).

4. $SO_4^{2-}$ dominant or anions indiscriminant and $Ca^{2+}$ dominant, $Ca^{2+}$ and $SO_4^{2-}$ dominant frequently indicates a recharge water in lavas and gypsiferous deposits; otherwise a mixed water or a water exhibiting simple dissolution may be indicated (see Fig. 6.20).

5. No dominant anion or cation, indicates waters exhibiting simple dissolution or mixing.

6. $SO_4^{2-}$ dominant or anions indiscriminant and Na dominant, is a water type not frequently encountered and indicates probable mixing influences.

7. $Cl^-$ and $Ca^{2+}$ dominant, is infrequently encountered unless cement pollution is present in a well; otherwise the waters may result from reverse ion exchange of $Na^+ - Cl^-$ waters.

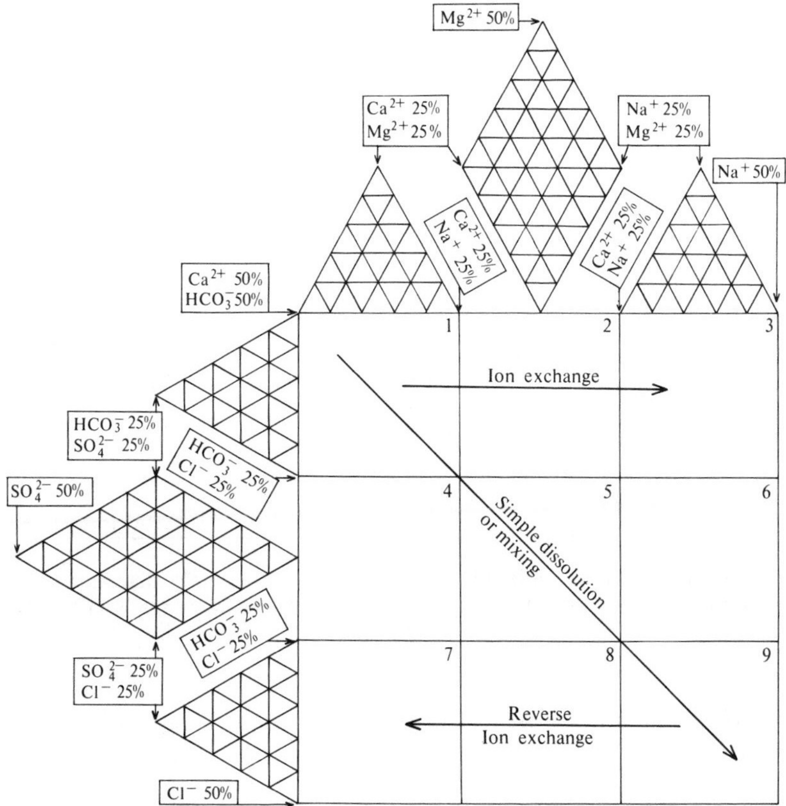


**Fig. 6.19.** Expanded Durov diagram with subdivisions and processes demonstrated.

8. $Cl^-$ dominant and no dominant cation indicates that the groundwaters may be related to reverse ion exchange of $Na^+$-$Cl^-$ waters.
9. $Cl^-$ and $Na^+$ dominant frequently indicates end-point waters. The Durov diagram does not permit much distinction between $Na^+$-$Cl^-$ waters.

The arrows in Fig. 6.19 indicate possible process paths such as ion exchange or dissolution. Care, however, has to be exercised as demonstrated in Fig. 6.20 in which a simple mixing between a recharge water and brackish water is shown to fall in five of the plotting fields.

By sensible interpretation of the expanded Durov diagram groundwaters can be classified or typed as shown on Fig. 6.21 using simple groupings. The same method can also be applied to parts of the Durov diagram if distinctive waters are present (Fig. 6.22).

Although the Durov diagram has been extensively applied to major ion

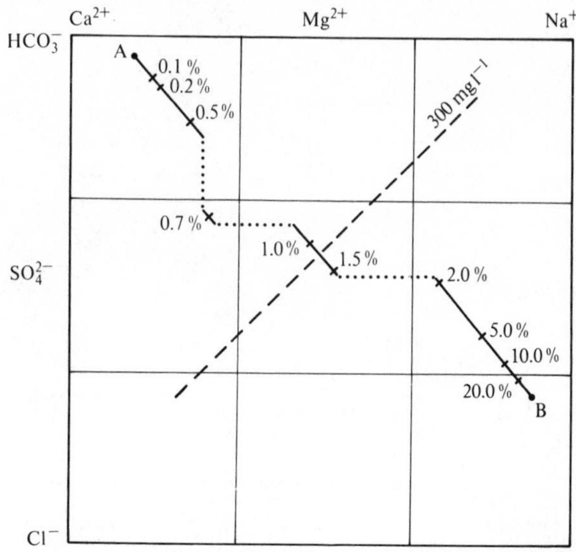


**Fig. 6.20.** Groundwater mixing demonstrated on the Durov diagram: curve A, fresh groundwater; curve B, brackish groundwater. (Percentage is mix between A and B.)

chemistry it can also be used to represent minor ions for groundwater typing purposes as shown on Fig. 6.23.

### *6.4.10. Statistical diagrams and techniques*

Statistical diagrams are not extensively used in groundwater chemistry because the complexity of hydrochemical processes does not readily permit statistics to be realistically applied.

Hem (1970) discusses the use of cumulative percentage composition diagrams. Such diagrams have similar representation characteristics to Schoeller diagrams but are of very limited value.

Probability diagrams provide more information and can be used particularly in the early stages of interpretation to distinguish between waters. They are frequently favoured in isotope hydrochemistry and as shown in Fig. 6.24 changes in probability trend are considered to reflect differing $^{18}O$ groundwaters from Southern Arabia.

Multivariate analyses, although attempted, have been difficult to understand in relation to hydrochemical studies. Dalton and Upchurch (1978) have used principal component analysis, but generally the method provides little hydrochemical insight while Ashley and Lloyd (1978) have reviewed cluster and factor methods and applications, and found them to be of marginal value.

Superficially cluster and factor analyses would appear to offer more powerful methods of interpreting hydrochemical data than the conventional ones.

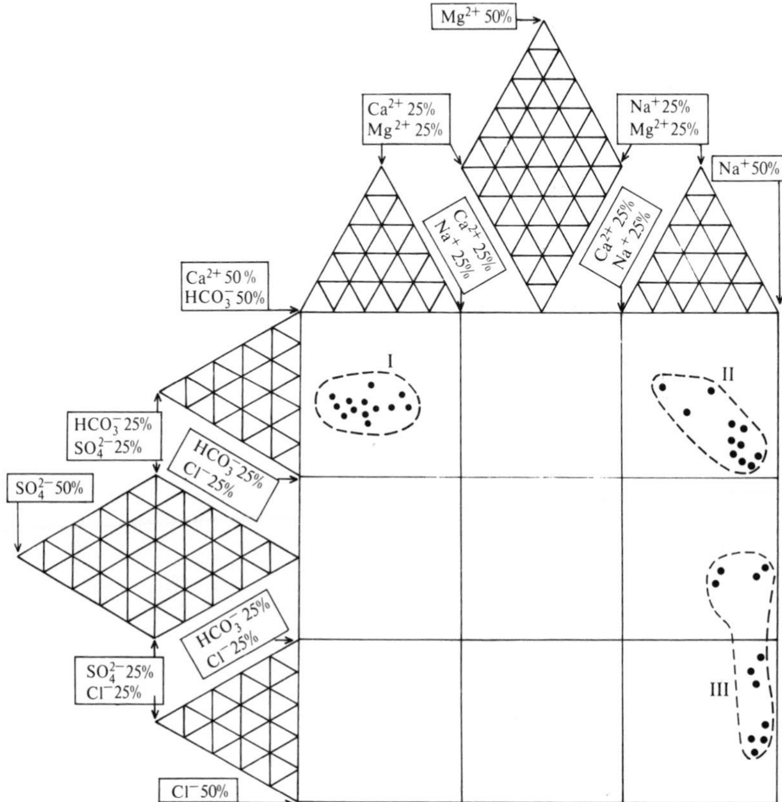


**Fig. 6.21.** Water types grouped using a Durov diagram: I, fresh recharge water; II, ion exchanged water; III, old brackish water.

Ashley and Lloyd (1978), however, have found that the statistical and conventional approaches usually provide the same results. Unfortunately cluster and factor analyses require considerable data handling with data normalization and can only be effective with good computer facilities. A cluster analysis output for groundwaters in the Gamogara area in Southern Africa is shown in Fig. 6.25 (Smith 1980); five groups of groundwaters are identified at or below a significance of 1.5 (Davis 1973) and are represented as a distribution on Fig. 6.26. While the distribution proved hydrogeologically significant in the study, the same grouping was more easily obtained using trilinear grouping methods. Further, many of the groundwaters could not be grouped using the dendrogram.

The approach of mapping statistical groupings has also been followed in factor analysis. The statistical method is outside of the scope of this book (see Davis 1973); however, the method resolves a set of multivariate factors in which

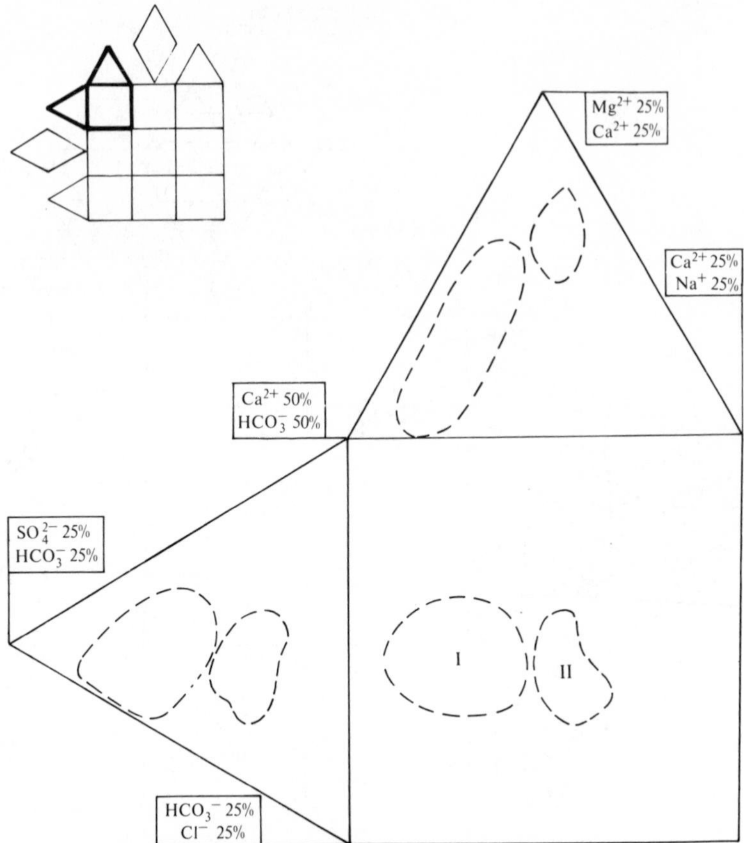


**Fig. 6.22.** Water-type grouping in a sector of a Durov diagram: I, $Ca^{2+}$-dominant recharge: II, $Mg^{2+}$-dominant recharge.

certain hydrochemical parameters may prove significant. The factors themselves, though statistically significant, have no hydrochemical significance, so if plotted on a distribution map their significance relates to the dominant hydrochemical parameter weighting the factor. A factor distribution from groundwaters in the alluvial Santiago Basin in Chile is shown in Fig. 6.27. The factor has a high silica weighting with the 'extreme' values related to:

(i) where the Mapocho River enters the basin after draining part of a granite intrusion further east;

(ii) down gradient of basement high areas where groundwater in the alluvium is moving slowly;

(iii) where runoff to the basin occurs from granitic intrusion along the northwestern and western margins of the basin.

The value of the factor analysis in this case was to draw attention to the significance of silica which otherwise may have been overlooked.

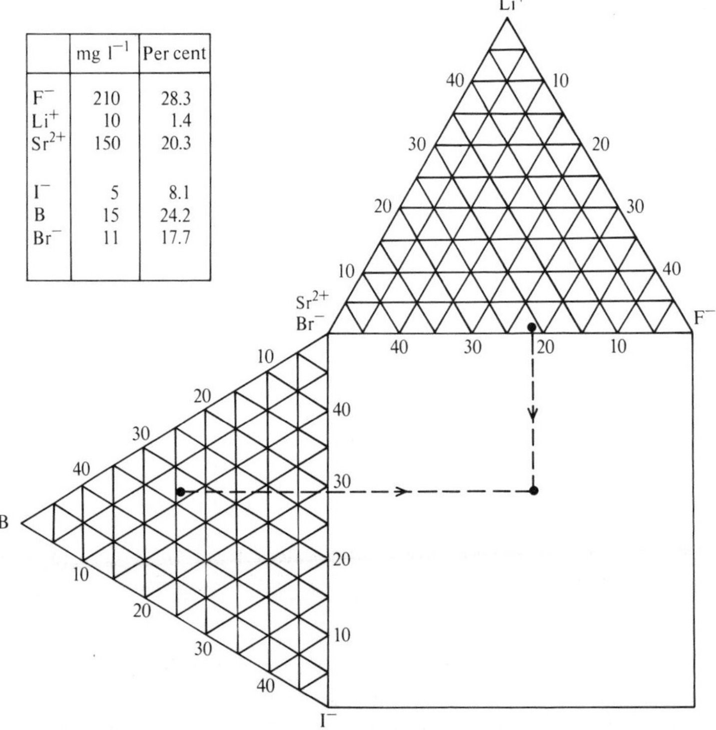

| | mg l$^{-1}$ | Per cent |
|---|---|---|
| $F^-$ | 210 | 28.3 |
| $Li^+$ | 10 | 1.4 |
| $Sr^{2+}$ | 150 | 20.3 |
| $I^-$ | 5 | 8.1 |
| B | 15 | 24.2 |
| $Br^-$ | 11 | 17.7 |



**Fig. 6.23.** Minor ion chemistry represented by a Durov diagram.

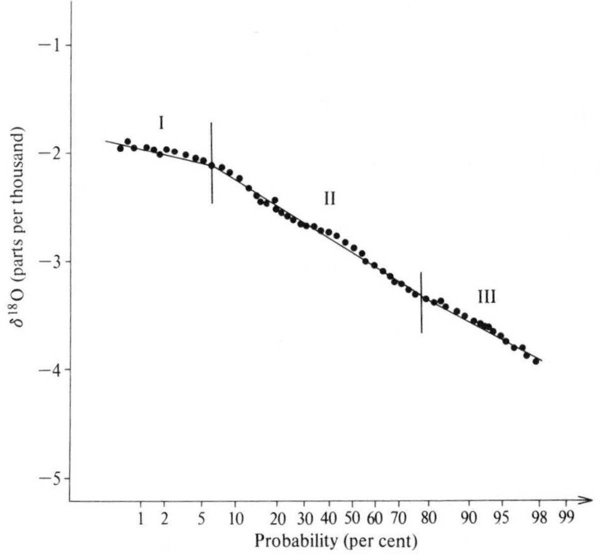


**Fig. 6.24.** Probability diagram used for grouping groundwaters: I–III, possible $^{18}O$ groupings based upon probability.

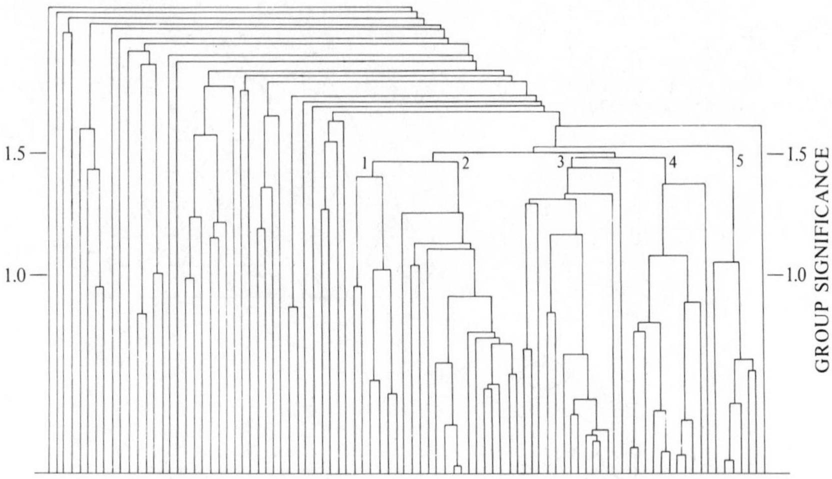


**Fig. 6.25.** Cluster dendrogram for groundwaters from Gamogara in Southern Africa: 1–5, groups at a significance of 1.5. (After Smith 1980.)

Trend surface analyses are frequently used for determining parameter distribution in geology. An example of their application to groundwater chemistry is shown on Fig. 6.28 for the Murray Basin of Australia. Such analyses can provide an initial guide for interpretation but tend to produce unrealistic hydrochemical patterns owing to data distribution bias. One useful aspect of trend surface analysis is that differences between surfaces of different degrees can indicate areas of hydrochemical anomaly.

## 6.5. Conclusions

To represent groundwater chemistry a simple but effective approach which can easily be computerized can be adopted as follows:

(i) analyse data to provide initial *X–Y* plots which should allow recognition of variations in data relationships and indications of chemical processes and anomalies;

(ii) prepare distribution maps and hydrochemical sections of pertinent parameters;

(iii) in conjunction with (ii) distinguish water types or groupings using preferably a Durov-type plot;

(iv) represent any particular hydrochemical relationship on the most appropriate diagram (e.g. a dilution diagram).

While hydrochemical representation appears simple, it is very important in helping to interpret hydrochemical conditions and time spent in sensibly attaining a good representation can be very worthwhile.

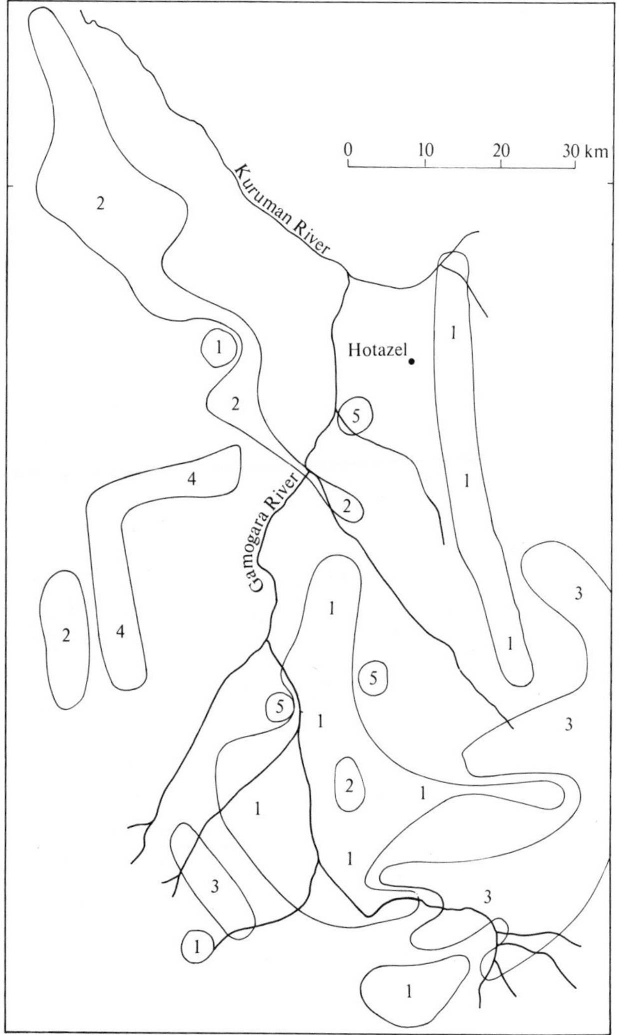


**Fig. 6.26.** Distribution of cluster dendrogram groups from Fig. 6.25. 1–5 relates to groups at a significance of 1.5 as shown in Fig. 6.25. Groundwaters lying outside these areas cannot be grouped on the dendrogram.

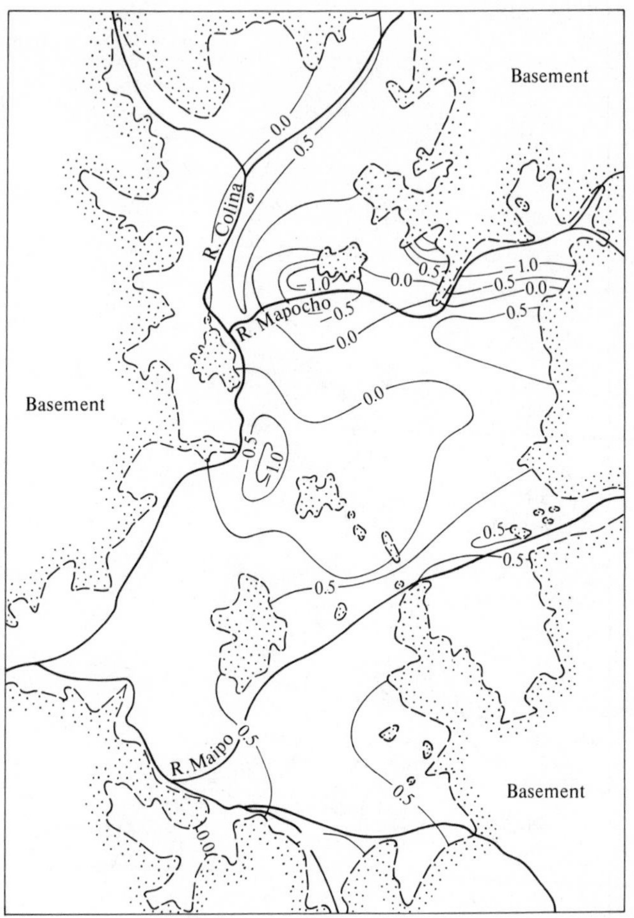


**Fig. 6.27.** Factor distribution for groundwaters in the alluvial aquifer of the Santiago Basin, Chile.

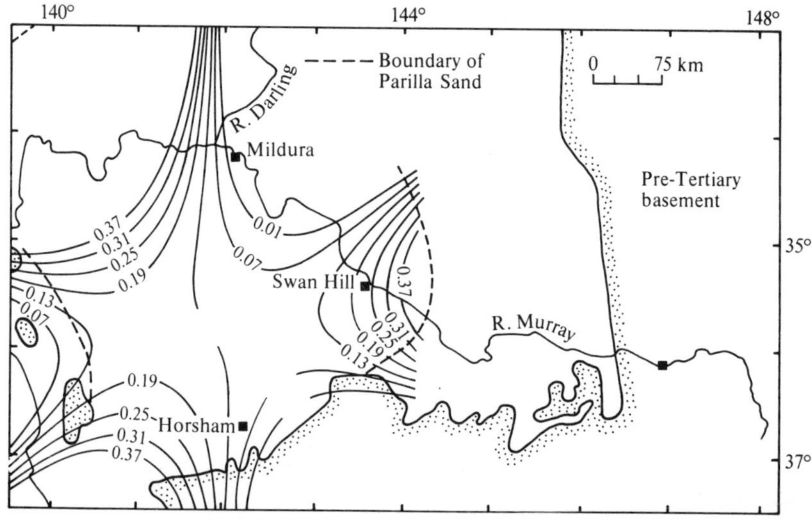


**Fig. 6.28.** Third degree surface map of $HCO_3^-/Cl^-$ ratios of groundwater associated with the Parilla Sand, River Murray Basin, Australia. The percentage fit is 25.89%. (After Lawrence 1975.)

## References

Ashley, R. P. and Lloyd, J. W. (1978). An examination of the use of factor analysis and cluster analysis in groundwater chemistry interpretation. *J. Hydrol.* **41**, 329–44.

Burdon, D. J. and Mazloum, S. (1958). Some chemical types of groundwater from Syria. *UNESCO Symp. Teheran*, pp. 73–90. Unesco, Paris.

Dalton, M. G. and Upchurch, S. G. (1978). Interpretation of hydrochemical facies by factor analysis. *Ground Water* **16**, 228–33.

Davis, J. C. (1973). *Statistics and data analysis in geology*. Wiley, New York.

Durov, S. A. (1948). Natural waters and graphic representation of their composition. *Dokl. Akad. Nauk SSSR* **59**, 87–90.

Hem, J. D. (1970). Study and interpretation of the chemical characteristics of natural water. *U.S. geol. Surv. water supply Pap. 1473.*

Hill, R. A. (1940). Geochemical patterns in the Coachella Valley, California. *Trans. Am. geophys. Union* **21**, 46–9.

Ineson, J. and Downing, R. A. (1963). Changes in the chemistry of ground waters of the Chalk passing beneath argillaceous strata. *Bull. geol. Surv. G.B.* **20**, 519–41.

Lawrence, A. R., Lloyd, J. W., and Marsh, J. M. (1976). Hydrochemistry and groundwater mixing in part of the Lincolnshire Limestone aquifer, England. *Ground Water* **14**, 12–20.

Lawrence, C. R. (1975). Geology, hydrodynamics and hydrochemistry of the southern Murray Basin. *Geol. Surv. Victoria, Mem. 30.*

Lloyd, J. W. (1965). The hydrochemistry of the aquifers of north-eastern Jordan. *J. Hydrol.* **3**, 319–30.

— (1969). The hydrogeology of the southern desert of Jordan. *UNDP/FAO. Pub. tech. Rep. 1. Special Fund 212.*

Marsh, J. M. (1977). Groundwater chemistry and its relation to flow in the southern Lincolnshire Limestone. *Univ. Birmingham Rep. Dept. Geol. Sci.*

Maucha, R. (1949). The graphical symbolisation of the chemical computation of natural waters. *Hydrol. Közölny* **13**, 117–18.

McKinnell, J. C. (1958), Indentification of mixtures of waters from chemical water analysis. *J. petrol. Technol. Tech. Note* **2016**, 79–82.

Piper, A. M. (1944). A graphic procedure in the geochemical interpretation of water analyses. *Trans. Am. geophys. Union*, 25, 914–23.

Schoeller, H. (1959). Arid zone hydrology—recent developments. *Arid Zone Research XII* UNESCO, Paris.

— (1962). *Les eaux souterraines.* Massio, Paris.

Smith, P. (1980). A re-assessment of the hydrogeological data from the Gamogora catchment, South Africa. *M. Sc. Project Rep., Dept. Geol. Sci. University of Birmingham.*

Stiff, H. A. (1951). The interpretation of chemical water analysis by means of patterns. *J. petrol. technol.* **3**, 15–17.

Winograd, I. J. and Farlekas, G. M. (1974). Problems in $^{14}C$ dating of waters from aquifers of deltaic origin. In *Isotope hydrology*, pp. 69–93. International Atomic Energy Agency, Vienna.

# 7 Saline groundwaters

## 7.1. Introduction

Brackish and saline groundwaters frequently occur in hydraulic contact with fresh groundwaters and can cause very considerable constraints upon the exploitation of the fresh water resources. In understanding the response of a saline groundwater body to fresh water abstraction a knowledge of the total system including head controls, flows, etc. is necessary. In part the response will be dictated by the disposition of the saline water which in turn may be dictated by its origin. The determination of the origin is usually largely hydrochemical. The attainment of a good understanding of a saline groundwater body will clearly be of benefit in the management of adjacent fresh water resources although it is frequently a difficult proposition. The three-dimensional distribution and chemistry of saline groundwaters is normally not understood because of the natural reluctance to invest in drilling in such areas. Unfortunately also many wells inadvertently drilled into saline groundwaters are often back-filled because of the lack of yield potential. Surface geophysics may alleviate the problem partially, and certainly borehole geophysics (Section 4.4) is invaluable when holes are available; however, hydrochemical interpretations are usually based upon limited sample numbers so that analyses need to be as comprehensive as possible.

Traditionally groundwaters have been classified upon their TDS content which has been applied particularly to the non-fresh groundwaters as given in Table 7.1. Such a classification, however, has little relevance so that in this account the term 'saline' will be adopted for those waters with TDS in excess of about 1000 mg $l^{-1}$. Saline groundwaters associated with exploitable fresh waters usually have salinities less than 100 000 mg $l^{-1}$ so that brines will not be considered in any detail.

**Table 7.1** *Groundwater classification using total dissolved solids*

| Class | TDS (mg $l^{-1}$) |
|---|---|
| Fresh | 0–1000 |
| Brackish | 1000–10 000 |
| Saline | 10 000–100 000 |
| Brine | >100 000 |

## 7.2. Occurrence and origin of groundwater salinity

As a result of chemical and biochemical interactions between waters and the material through or over which they flow, and to a lesser extent because of contributions from the atmosphere, the waters acquire salinity proportional to their flow experience. The salinities may be acquired within the ground as groundwater chemistry evolves or maybe introduced during aquifer matrix deposition or subsequent groundwater or surface water movement. Because of the various ways in which salinity is imparted to a groundwater certain chemical signatures can be recognized which may also indicate the origin of the salinity. Most saline waters are dominantly sodium chloride in type so that the major ions are not always of significance in interpretation and reliance is placed on other parameters.

### *7.2.1. 'Modern' hydrogeological influences*

*Cyclic salting.* Cyclic salting is a term applied to salt input to the ground from precipitation. Where recharge flows are significant salts in the recharge are flushed into the aquifer without concentration. Only in small islands or coastal aquifers are recharge waters likely to be high in salts under such conditions, because of sea spray effects. In areas of low or intermittant recharge, however, cyclic salting and the ensuing salt concentration in the unsaturated zone may lead to sigificant salination of groundwaters when recharge pulses occur. Important cyclic salting effects are recorded in a number of arid zone countries including the United States and Australia.

In Australia, where groundwater salinity is a major problem, the salinity is derived from a number of sources one of which is thought to be cyclic salting under the low intermittent recharge conditions. In Western Australia, Hingston and Gailitis (1976) have demonstrated the significance of salt spray effects on dry fall-out and Martin and Harris (1982) have shown the influence of the fall-out in imparting a chemistry to groundwaters. In Table 7.2 the groundwater ratios are comparable with seawater with the exception of that including $Ca^{2+}$ which is derived from another source in the groundwater.

**Table 7.2** *Ratios of cations to chloride ions from the Perth area, Western Australia*

| | $Na^+/Cl^-$ | $K^+/Cl^-$ | $Ca^{2+}/Cl^-$ | $Mg^{2+}/Cl^-$ |
|---|---|---|---|---|
| Seawater | 0.556 | 0.020 | 0.021 | 0.067 |
| Rainfall | 0.548 | 0.032 | 0.075 | 0.071 |
| Dry fall-out | 0.366 | 0.085 | 0.392 | 0.085 |
| Groundwater | 0.560 | 0.050 | 0.450 | 0.093 |

Jenkin (1981) considered that salt build-up in the soil profile for Victoria, which eventually effects aquifer recharge, is a function of weathering intensity and is therefore chiefly related to past climatic history on a geological time scale. Van Dijk (1969) has postulated a similar source for areas of New South Wales, while Evans (1982) has demonstrated that rock types alone cannot explain the recharge water chemistry, particularly the chloride concentrations. The influence of cyclic salting in Australia is far from clear and is subject to much research; however, it would appear that modern fall-out is not a dominant factor regionally, but that cyclic salting has occurred under relatively low recharge regimes for a considerable time period providing one important salinity component to the groundwater. High groundwater salinity under low recharge conditions is further discussed in Section 7.2.3.

*Modern seawater intrusion*. As the sea provides the hydrochemical sink in the hydrological cycle, seawaters have a high salinity. In estuaries and adjacent to coastlines modern seawater intrusion frequently occurs into aquifers either under natural flow controls or because of flows induced by abstraction. The recognition of modern seawater intrusion normally should not pose difficulties; however, it may be associated with saline groundwaters of other origins or may have been modified by residence in the aquifer. In any case a reasonably comprehensive hydrochemical interpretation is worthwhile as it may shed light upon the mechanisms controlling intrusion or the rates of flow.

In Table 7.3 an average seawater composition is given. The composition is important in that it shows the suite of parameters in addition to the major ions that may be recognized in significant concentrations (greater than say 0.01 $mg\ l^{-1}$). Seawater composition varies around the globe under the influence of, for example, closeness to an ice cap or limited circulation in an area such as the Arabian Gulf. In intrusion studies therefore the local seawater should be analysed.

Changes in modern seawater composition upon its entry into an aquifer can be caused by the loss or acquisition of certain parameters in addition to the mixing with fresh water. Simple mixing can be identified and quantified as discussed in Section 6.4 numerous workers, however, relate preferential changes most frequently in $Ca^{2+}$, $Na^+$, $SO_4^{2-}$, and $HCO_3^-$. Ion exchange and dissolution are most commonly cited as the reasons for the changes (Burdon and Dounas 1963; Arad, Kafri, and Fleisher 1975; Desai, Gupta, Shah, and Sharma 1979).

Reverse ion exchange in which $Na^+$ concentrations reduce preferentially for $Ca^{2+}$ is a frequent feature of waters on the fringe of a saline intrusion, as shown in Fig. 7.1, and can readily be distinguished as shown in Fig. 6.19. While it is difficult to ascertain rates of exchange the phenomenon does tend to indicate fairly slow movement of the saline body.

Dissolution of carbonate material with increases in $Ca^{2+}$ and $HCO_3^-$ can occur when saline and fresh waters mix as undersaturation with respect to carbonate minerals occurs irrespective of the fact that the two mixing waters are supersaturated. The ion activity coefficients for calculations of saturation indices in the high ionic strength saline waters are given by eqn (3.16) and Table 3.2.

Minor ions such as iodide, strontium, and fluoride may also appear in greater concentrations in the intrusive waters than in the adjacent seawaters. These increases can be attributed to enrichment as the seawater passes through muds in an estuary or at the coast before entering the aquifer. An iodide example can be

**Table 7.3.** *Average sea water composition*

| Constituent | Concentration (mg $l^{-1}$) | Principal forms in which constituent occurs |
|---|---|---|
| Cl | 19,000 | $Cl^-$ |
| Na | 10,500 | $Na^+$ |
| $SO_4$ | 2,700 | $SO_4^{2-}$ |
| Mg | 1,350 | $Mg^{2+}$, $MgSO_4$(aq) |
| Ca | 400 | $Ca^{2+}$, $CaSO_4$(aq) |
| K | 380 | $K^+$ |
| $HCO_3$ | 142 | $HCO_3^-$, $H_2CO_3$(aq), $CO_3^{2-}$ |
| Br | 65 | $Br^-$ |
| Sr | 8 | $Sr^{2+}$, $SrSO_4$(aq) |
| $SiO_2$ | 6.4 | $H_4SiO_4$(aq), $H_3SiO_4^-$ |
| B | 4.6 | $H_3BO_3$(aq), $H_2BO_3^-$ |
| F | 1.3 | $F^-$ |
| N | 0.5 | $NO_3^-$, $NO_2^-$, $NH_4^+$ |
| Li | 0.17 | $Li^+$ |
| Rb | 0.12 | $Rb^+$ |
| P | 0.07 | $HPO_4^{2-}$, $H_2PO_4^-$, $PO_4^{3-}$, $H_3PO_4$(aq) |
| I | 0.06 | $IO_3^-$, $I^-$ |
| Ba | 0.03 | $Ba^{2+}$, $BaSO_4$(aq) |
| Al | 0.01 | |
| Fe | 0.01 | $Fe(OH)_3$ |
| Mo | 0.01 | $MoO_4^{2-}$ |
| Zn | 0.01 | $Zn^{2+}$, $ZnSO_4$(aq) |
| Se | 0.004 | $SeO_4^{2-}$ |
| As | 0.003 | $HAsO_4^{2-}$, $H_2AsO_4^-$, $H_3AsO_4$(aq), $H_3AsO_3$(aq) |
| Cu | 0.003 | $Cu^{2+}$, $CuSO_4$(aq) |
| Sn | 0.003 | |

| | | |
|---|---|---|
| U | 0.003 | $UO_2(CO_3)_3^{4-}$ |
| Mn | 0.002 | $Mn^{2+}$, $MnSO_4(aq)$ |
| Ni | 0.002 | $Ni^{2+}$, $NiSO_4(aq)$ |
| V | 0.002 | $VO_2(OH)_3^{2-}$ |
| Ti | 0.001 | |
| Co | 0.0005 | $CO^{2+}$, $CoSO_4(aq)$ |
| Cs | 0.0005 | $Cs^+$ |
| Sb | 0.0005 | |
| Ce | 0.0004 | |
| Ag | 0.0003 | $AgCl_2^-$, $AgCl_3^{2-}$ |
| La | 0.0003 | |
| Y | 0.0003 | |
| Cd | 0.0001 | $Cd^{2+}$, $CdSO_4(aq)$ |
| W | 0.0001 | $WO_4^{2-}$ |
| Ge | 0.00007 | $Ge(OH)_4(aq)$, $H_3GeO_4^-$ |
| Cr | 0.00005 | |
| Th | 0.00005 | |
| Sc | 0.00004 | |
| Ga | 0.00003 | |
| Hg | 0.00003 | $HgCl_3^-$, $HgCl_4^{2-}$ |
| Pb | 0.00003 | $Pb^{2+}$, $PbSO_4(aq)$ |
| Bi | 0.00002 | |
| Nb | 0.00001 | |
| Au | 0.000004 | $AuCl_4^-$ |
| Be | 0.0000006 | |
| Pa | $2 \times 10^{-9}$ | |
| Ra | $1 \times 10^{-10}$ | $Ra^{2+}$, $RaSO_4(aq)$ |

After Goldberg 1963.

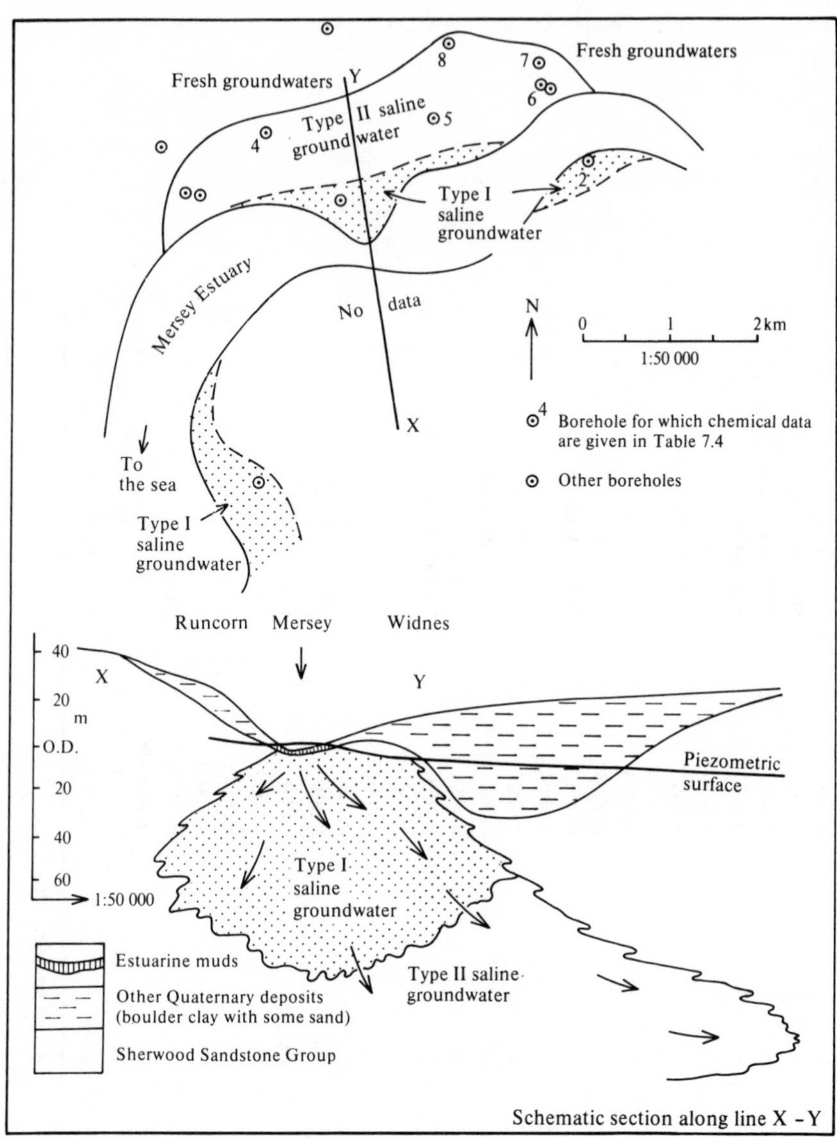


**Fig. 7.1.** Saline groundwaters in the Widnes area of the Lower Mersey Valley sandstone aquifer, England.

cited from the Mersey Estuary in western England (Lloyd, Howard, Pacey, and Tellam 1982). From the data given in Table 7.4 two saline groundwater types can be recognized in the Sherwood Sandstone, Type I waters (wells 1-3) which have an $Na^+/Cl^-$ ratio of the order of 0.51-0.55, close to that of the local seawater (0.55) and are characterized by significant $NH_4^+$ concentrations. Type II waters (wells 4-8) have much lower $Na^+/Cl^-$ ratios and no $NH_4^+$ content. Type I waters are related to modern seawater intrusion as shown on Fig. 7.1, while Type II waters are slightly older estuary reverse ion-exchanged waters. If the $I^-/Cl^-$ ratios are considered, marked iodide enrichment is seen to be present in both the intrusion water types when compared with seawater ($I^-/Cl^- = 0.3 \times 10^{-5}$).

**Table 7.4** *Hydrochemical data from the Mersey Valley saline groundwaters*

| | Well 1 | Well 2 | Well 3 | Well 4 | Well 5 | Well 6 | Well 7 | Well 8 |
|---|---|---|---|---|---|---|---|---|
| $Cl^-$ (mg $l^{-1}$) | 5996 | 5145 | 4046 | 3267 | 4900 | 2136 | 3000 | 2078 |
| $Na^+/Cl^-$ | 0.52 | 0.55 | 0.51 | 0.20 | 0.36 | 0.25 | 0.26 | 0.26 |
| $NH_4^+$ (mg $l^{-1}$) | 0.25 | 2.40 | 7.55 | 0 | 0 | 0 | — | 0 |
| $I^-/Cl^-$ ($\times 10^{-5}$) | 1.6 | 3.1 | 7.4 | 0.5 | 1.5 | 1.5 | 1.3 | 1.4 |

Ratios based on mg $l^{-1}$

The source of the iodide enrichment is the estuarine muds lining the estuary channel. Chloride and iodide concentrations in the waters squeezed from these muds are given in Table 7.5 and show extreme iodide enrichment attributed to faunal remains. The reason for the lower iodide concentrations in the Type II waters is uncertain; several possibilities such as sorption can be contemplated. The range of $I^-/Cl^-$ values for Type I waters is thought to be due to the variable thickness of the estuarine muds.

**Table 7.5** *Hydrochemical data of estuarine mud pore waters from the Mersey Estuary*

| | Sample A | Sample B | Sample C | Sample D |
|---|---|---|---|---|
| $Cl^-$ (mg $l^{-1}$) | 10 000 | 1600 | 6000 | 7000 |
| $I^-$ ($\mu$g $l^{-1}$) | 2500 | 780 | 240 | 270 |
| $I^-/Cl^-$ ($\times 10^{-5}$) | 25.5 | 48.8 | 4.0 | 3.9 |

Ratios based on mg $l^{-1}$

*Modern saline groundwaters related to low flows*. In semi-arid and arid areas saline groundwaters can be encountered at the edges of fresh groundwater bodies where low transmissivity materials are in contact with the main aquifer material. Because recharge is small the low permeability materials are not adequately flushed and can thus provide quality problems outweighing their relative flow volumes. The recognition of such waters and their origins is important. In the interpretation, minor ion chemistry may be adequate or a comprehensive hydrochemistry may be necessary.

An example from Peru of the use of minor ions for distinguishing saline groundwaters is shown in Fig. 7.2. In this case three modern types of saline groundwaters entering the main Lima basin alluvial aquifer are identified. The main saline water is seawater intrusion which is partly reverse ion exchanged, while of the other two waters, one can be identified as being $Sr^{2+}$ enriched and associated with Jurassic sediments in hydraulic continuity with the alluvium and the other is $I^-$ enriched and associated with granodiorites.

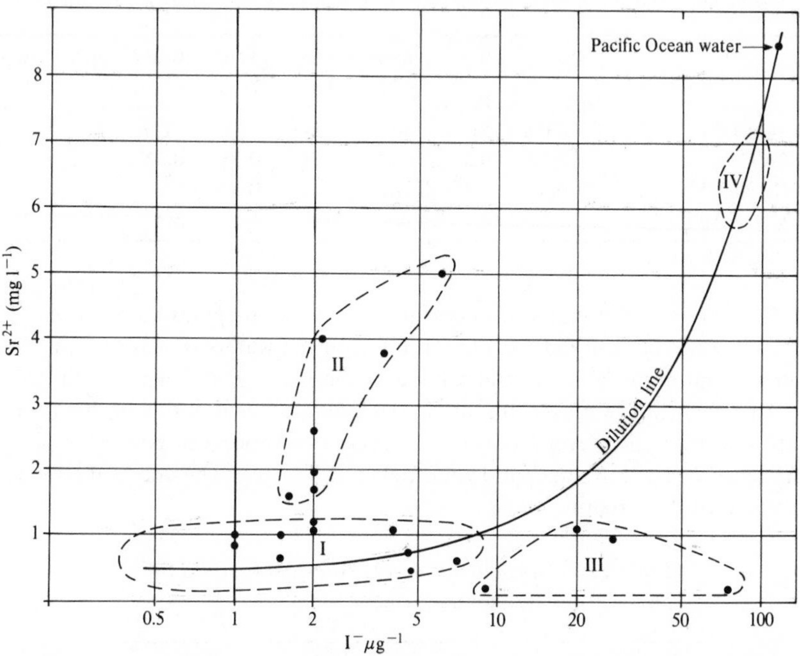


**Fig. 7.2.** Strontium–iodide relationship for saline groundwaters from the Lima Basin alluvial aquifer, Peru: I, groundwaters only associated with alluvium; II, groundwaters entering alluvium from Jurassic sediments; III, groundwaters entering alluvium from granodiorites; IV, reverse ion-exchanged groundwaters from seawater intrusion.

A more comprehensive hydrochemical interpretation of this type of 'edge' saline water condition can be illustrated for the Wadi Bisha, Saudi Arabia (Lloyd, Fritz, and Charlesworth 1980). The wadi flows north-eastwards on the Precambrian shield and contains an extensive alluvial aquifer recharged via wadi flow from mountain ranges in the southern part of the shield. As shown in

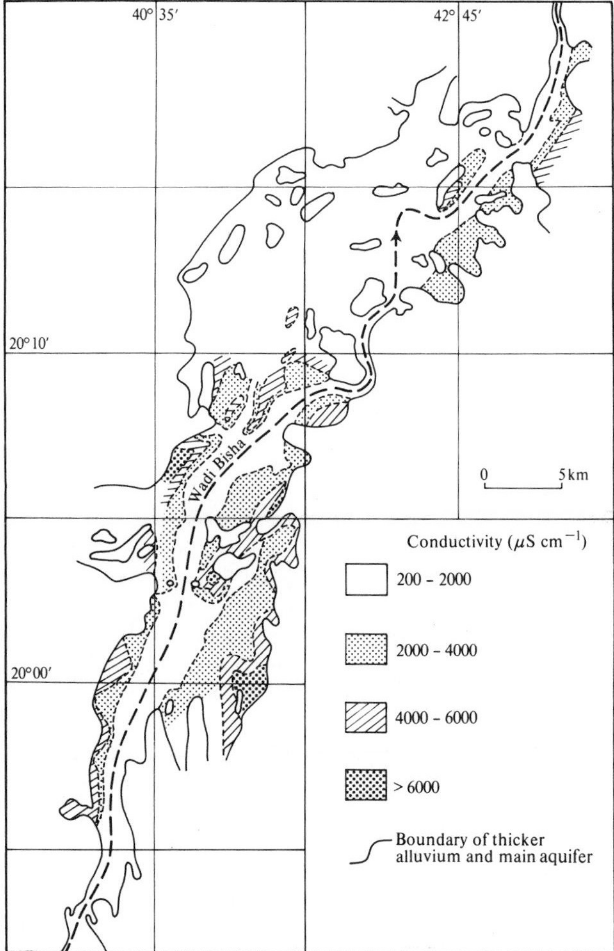


**Fig. 7.3.** Groundwater conductivity of central Wadi Bisha, Saudi Arabia.

Fig. 7.3 saline water is present at the edges of the alluvium and is a feature of the aquifer where rainfall is low ($<150$ mm). A section across the aquifer at right angles to the flow direction in the low rainfall area together with the evolution of the saline groundwaters as shown in Fig. 7.4. Initial recharge in the lateral areas is calcium bicarbonate in type and is undersaturated with respect to gypsum. Under very small shallow flows in thin alluvium and fractured basement, the groundwater progressively changes in character to calcium sulphate water and eventually sodium sulphate water because of ion exchange and sodium chloride dissolution. The lateral flow towards the main wadi alluvium is subjected to evaporation as depicted by $^{18}O$ enrichment; it is non-tritiated and 'modern'

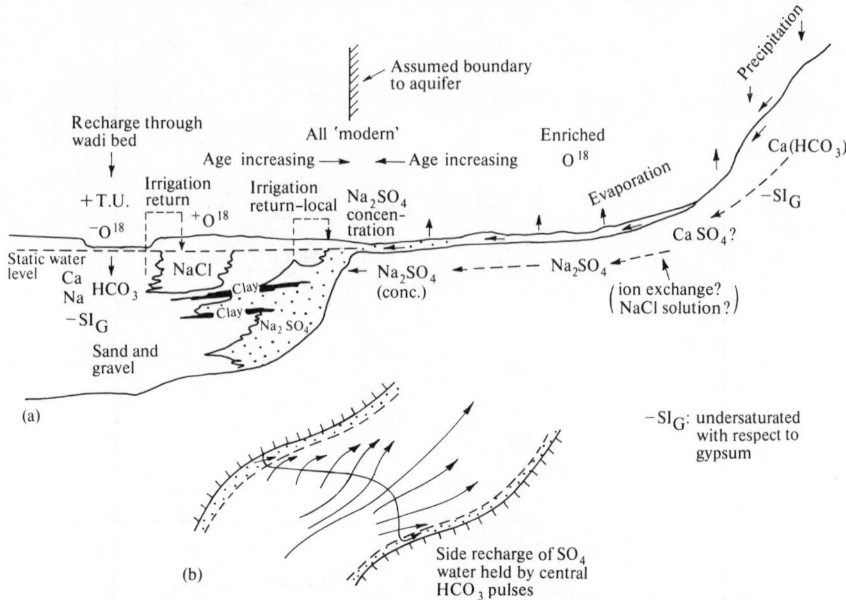


**Fig. 7.4.** Hydrochemical and flow concepts for Wadi Bisha saline groundwaters: (a) conceptual hydrochemical cross-section; (b) equipotential line and flow distribution between assumed boundaries.

in $^{14}C$ terms. The evaporation is thought to help to increase the total dissolved solids content. Upon entering the main alluvial aquifer the saline groundwater is restricted chiefly to the sides of the aquifer because of the greater flow of fresh water in the down-valley direction. The accentuation of salinity in the aquifer due to irrigation practice is shown in Fig. 7.4. Sodium chloride waters are shown evolving from calcium bicarbonate waters with sodium sulphate waters being increased in concentration.

*Salinity resulting from simple dissolution in aquifers.* The examples discussed above show that high salinity does not always relate to old groundwater origin; however, it is normal to accept that groundwater salinity increases with flow and aquifer residence chiefly because of dissolution of the type shown in Fig. 7.5. Where permeability is significant, as inherently implied in the term aquifer, high salinity due to dissolution is not a common feature in an active system as flushing will remove soluble salts. While in certain cases bicarbonates may be high, carbonate saturation restricts concentrations. An example of simple dissolution is shown in Fig. 7.5 where an active groundwater system has developed in karstified evaporites and high-concentration sulphate-chloride waters have evolved away from the recharge mound in predominantly unconfined conditions.

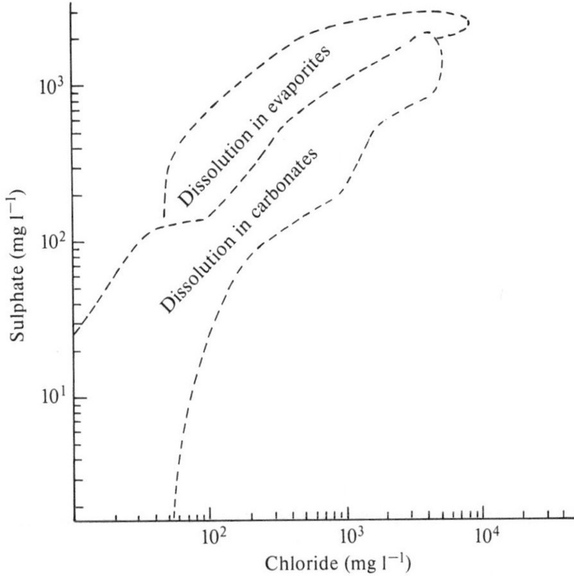


**Fig. 7.5.** Simple dissolution sequence producing high-salinity groundwater in Qatar.

In the predominant number of cases where salinity increases down the hydraulic gradient it is a function of decrease in flow and, although dissolution may play a part, the origin and certainly the age of the saline water is often very different from that in the active fresh water part of the aquifer. The increase in salinity is not necessarily a simple progression.

### *7.2.2. 'Ancient' hydrogeological influences*

*Juvenile groundwaters.* Juvenile groundwaters are those originating within the earth's crust as a part of crustal activity and usually vulcanism (White 1957). Occurrences of true juvenile water are probably rare, as much water recognized as geothermal often contains significant amounts of meteoric groundwaters. Juvenile and geothermal groundwaters tend to be saline and are not normally associated with fresh groundwater resource problems, so they are considered to be outside of the scope of this account. For information the reader is referred to Forcella (1982) and Henley and Ellis (1983).

*Connate groundwaters.* Connate groundwaters are those which existed in the aquifer when the materials forming the aquifer were deposited. The presence of truly connate groundwaters in aquifers is probably rare except in very recent deposits. Certainly unchanged connate water is unlikely; for example in the deep chalk of eastern England connate water is thought to exist in pore water form. As permeabilities are very low this may be feasible; however, diagenetic changes have influenced its composition. Chalk is almost totally of biogenic origin with the skeletal carbonate being magnesium deficient. With diagenesis the carbonate has recrystallized taking up magnesium from the pore waters which are now depleted in magnesium relative to normal seawater. With diagenesis and overburden processes, stress dissolution has occurred preferentially releasing strontium to produce strontium-enriched pore water.

Changes in connate and indeed other waters can be caused by membrane filtration effects of shales and clays (Bredehoeft, Blythe, White, and Maxey 1963). It now seems probable that connate waters are likely to be most affected because filtration would appear to be effective only below about 500–1000 m depth under very low groundwater flow conditions. Most evidence of filtration relates to brines and is not very relevant in the context of relationships with fresh groundwaters. When waters and solutes are driven under the influence of groundwater head across a shale membrane the solute movement is restricted relative to the water movement. Cation adsorption takes place on clay mineral surfaces so that further cations are repelled by the membrane. In order to maintain electrical neutrality across the membrane anions are also restricted. Membrane effects in saline waters can be recognized by ion ratios, for example the $I^- < Cl^- < Br^-$ adsorption sequence is known to be important and ratios related to the major selectivity sequences discussed in Section 3.7.2. can also be instructive. Downing and Howitt (1969), in a study of saline groundwaters in Carboniferous rocks in England, have used $Ca^{2+}/Cl^-$ and $Mg^{2+}/Cl^-$ ratios as an example of distinguishing membrane effects. They show that the ratios increase with increasing total ionic concentration and postulate selective concentrations of calcium and magnesium ions by argillaceous beds.

Because of the need for large effective stresses on clays before they can act as efficient membranes they are generally thought not to influence shallow groundwaters as noted above. However, Schwartz (1974) has raised the question of residual effective stress conditions in boulder clays and invoked membrane effects to explain hydrochemical anomalies in shallow waters beneath tills.

Graf, Friedman, and Meents (1965) have found that stable isotope fractionation occurred when waters passed through shale membranes in a number of North American sedimentary basins. The stable isotope concentrations of the saline groundwaters, however, did not indicate a direct correlation with ancient seawater and a meteoric origin has been postulated. Fractionation across argillaceous membranes does not appear to alter deuterium concentrations but certainly effects $^{18}O$ concentrations; similar conclusions have been reached by Coplen and Hanshaw (1973).

*Salinity related to hydrogeological entrapment.* Hydrogeological controls on groundwater discharge are probably the most important factors governing most occurrences of saline groundwater. Given that flow is occurring the opportunity for discharge coupled with head and permeability in the aquifer dictates the degree to which flushing may proceed. The rate of flushing or groundwater flow will determine in part the rate of removal of connate groundwater, the strength of diffusion influences, and dissolution rates (Issar 1981; Schwartz, Muehlenbachs, and Chorley 1981).

Under many hydrogeological circumstances saline groundwaters tend to occur adjacent to a discharge zone but in a position down the hydraulic gradient from the discharge zone. Such conditions are commonplace where there are low recharge mounds and large groundwater throughputs. Saline groundwaters described by Cederstrom (1946) in the eastern United States and by Lawrence, Lloyd, and Marsh (1976) and Lloyd (1981) in the United Kingdom are typical of these circumstances, as is the associated fresher groundwater evolution also described.

An excellent example of saline groundwater entrapment controlled by discharge zones occurs in eastern Saudi Arabia as depicted in Fig. 7.6. A multiple aquifer system exists with groundwater flow eastwards towards the Arabian Gulf from the outcrop areas in central Arabia. The main discharge occurs through a number of salt-dome-induced anticlinal structures and sabkhas close to the Gulf and in Bahrain and Qatar. Some discharge may also exist beneath the Gulf. High-salinity groundwaters are present extensively to the east of the main discharge zones. Salinity occurs progressively westwards and with depth. Its distribution is related to the controls and positions of upward groundwater movement across low permeability strata.

While the high-salinity waters in eastern Saudi Arabia may be partly connate as a result of poor flushing, entrapped saline groundwaters can be considerably younger than the deposition of the aquifer matrix materials. A typical example occurs in the eastern United Kingdom where saline waters have been entrapped as a result of glacial activity.

In Lincolnshire (Fig. 7.7) saline groundwaters are encountered in the Chalk aquifer adjacent to the North Sea coast. The saline water types have been distinguished initially on the basis of iodide concentrations (Howard and Lloyd 1978) as shown on Figs 7.7. and 7.8. Local modern seawater intrusion can be detected (Type I) but otherwise the bulk of saline groundwaters (Type II) are highly enriched in iodide above normal seawater concentrations. This iodide enrichment is considered to represent long aquifer residence with iodide acquired from the biogenic chalk matrix and shows that most of the saline waters are not related to any modern influences. The age of the saline waters and their mode of emplacement has been examined using isotopes. Tritium and $^{14}C$ ages for some of the groundwaters are shown in Fig. 7.9. Apart from the modern saline intrusive waters, the saline groundwaters have no significant tritium content. $^{14}C$ analyses have been carried out for brackish groundwaters at the edge of the main Type II saline groundwaters to avoid problems of possible carbon isotope fractionation and non-interpretation of the saline groundwaters (Wendt 1971). Such waters are clearly mixed and only provide a minimum age indication; however, ages of up to 21 000 years have been assessed using the Wigley (1976) dating procedure. The ages cannot be reconciled with any geological event but do indicate that the Type II saline groundwaters must

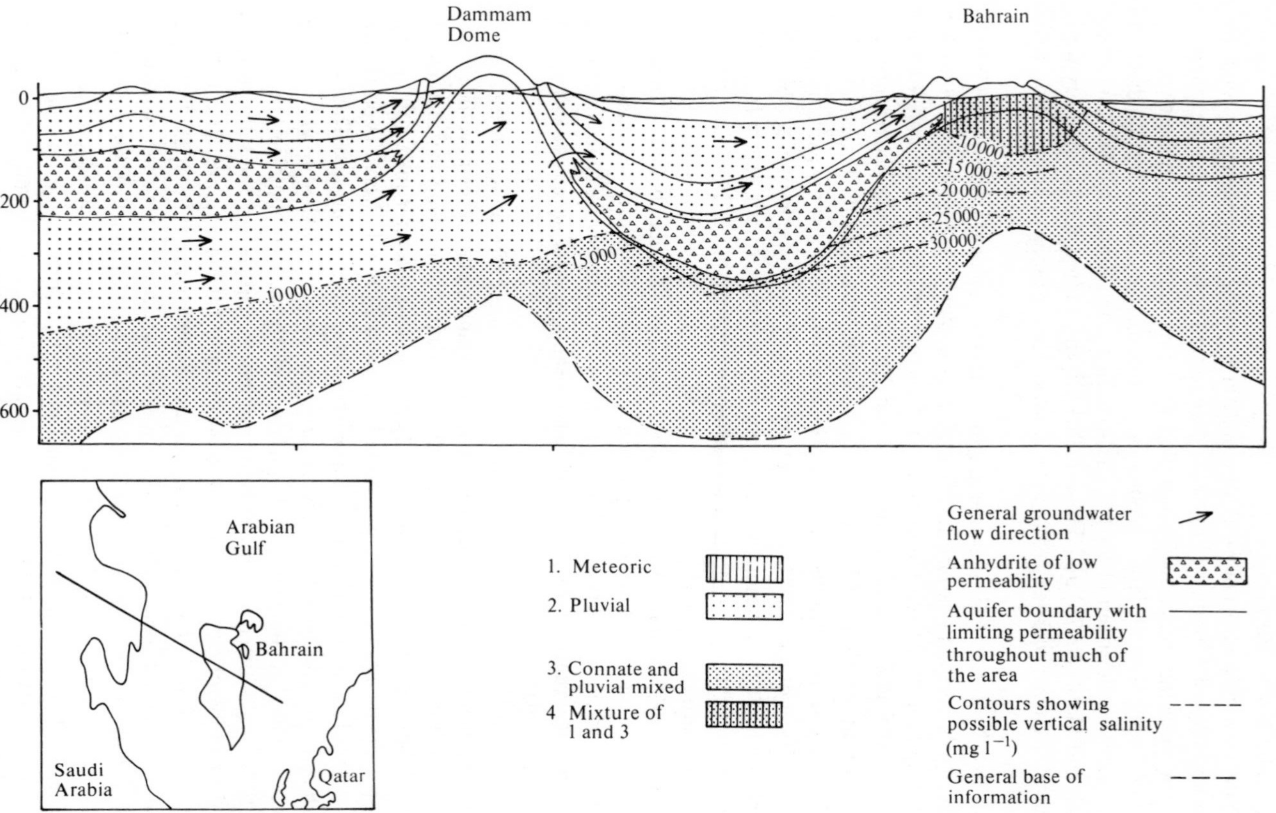


**Fig. 7.6.** Entrapped saline groundwaters in the western Arabian Gulf. (After Groundwater Development Consultants 1980.)

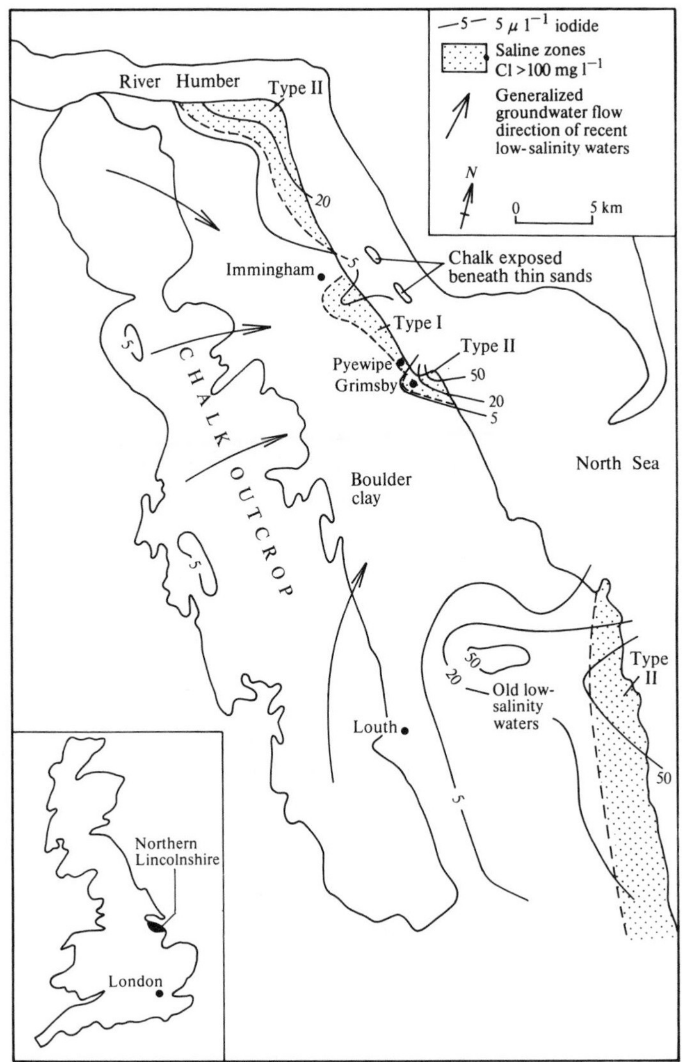


Fig. 7.7. Saline groundwater and iodide distribution in the Northern Lincolnshire Chalk aquifer, England.

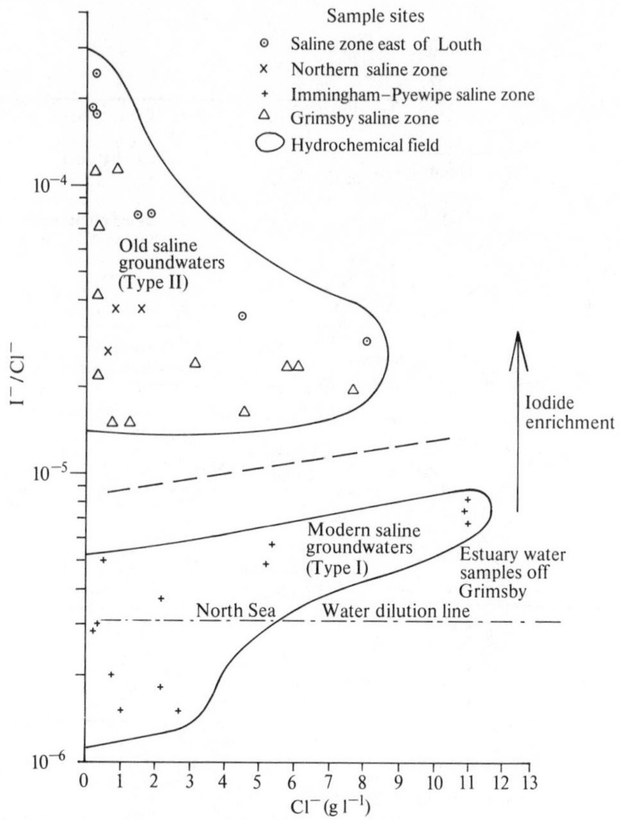


**Fig. 7.8.** Iodide and chloride relationships for saline groundwaters from Northern Lincolnshire.

be considerably older than the dates determined. The carbon isotope data bear out the iodide data conclusions. Hydrogeologically, the only feasible emplacement of the Type II saline waters relates back to glacial activity as depicted on Fig. 7.10. In the Ipswichian the Chalk was eroded by sea encroachment and saline groundwater entered the aquifer beneath a sea platform to the east of a well-developed sea cliff feature. Subsequently permafrost and glacier advancement covered the marine platform and the Pleistocene saline groundwaters were eventually entrapped by the Devensian boulder clay deposition produced by the melting of the glacier. The Type II saline groundwaters are therefore approximately 150 000 years old.

While the hydrogeological evolution described above is undoubtedly academically interesting, it also has applied significance in that the recognition of the age and disposition of the Type II saline waters allowed the incorporation of realistic boundary and very low flow conditions in the digital groundwater

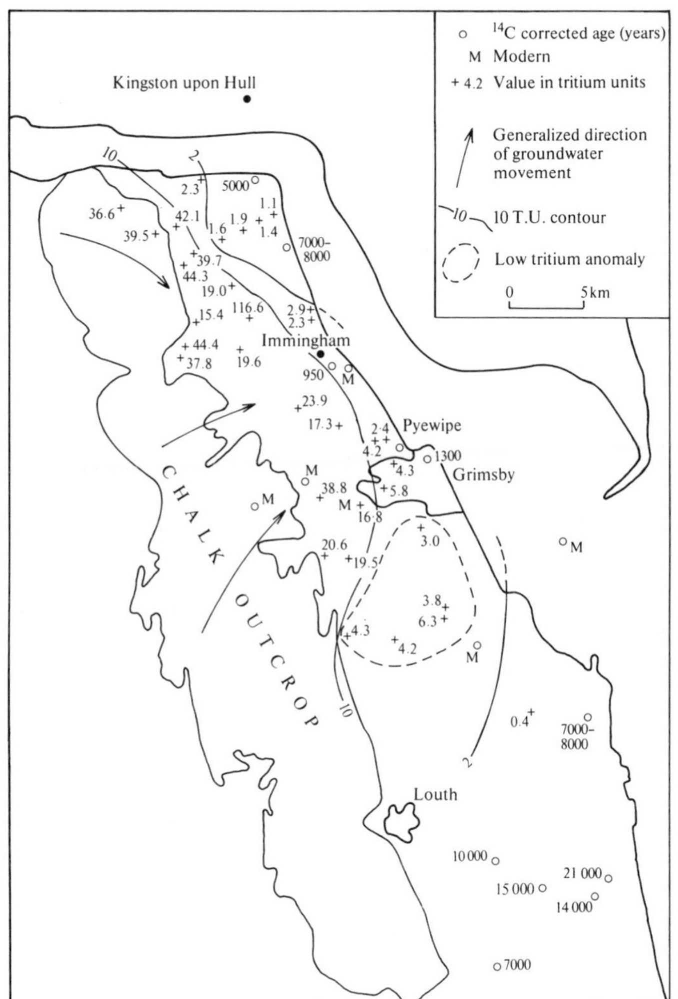


Fig. 7.9. Isotope data from the Northern Lincolnshire Chalk aquifer.

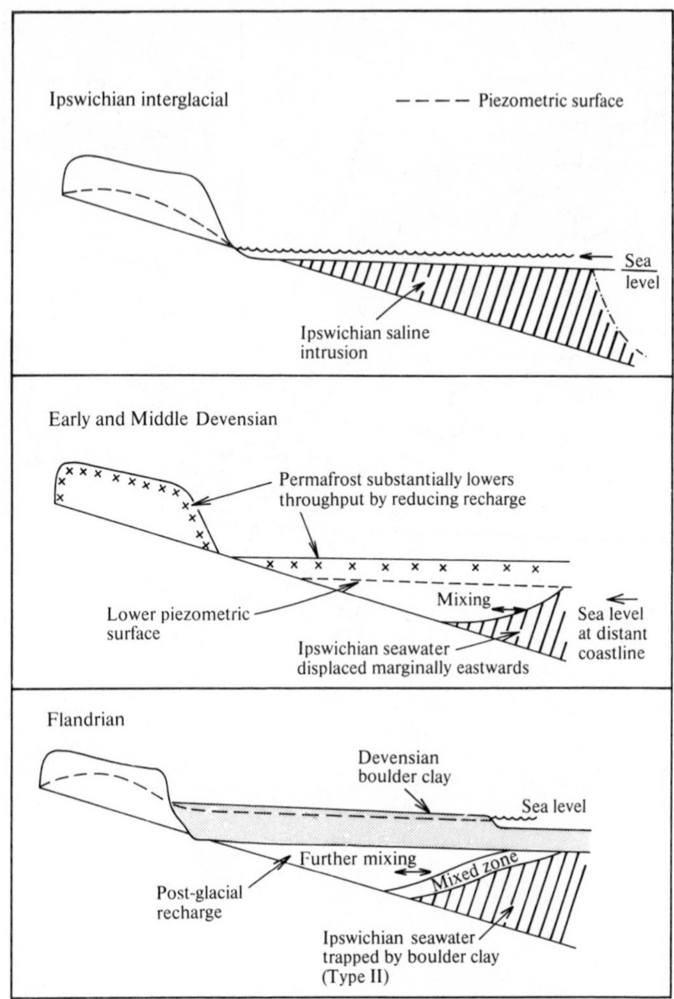


**Fig. 7.10.** Schematic representation of saline intrusion and entrapment in Northern Lincolnshire.

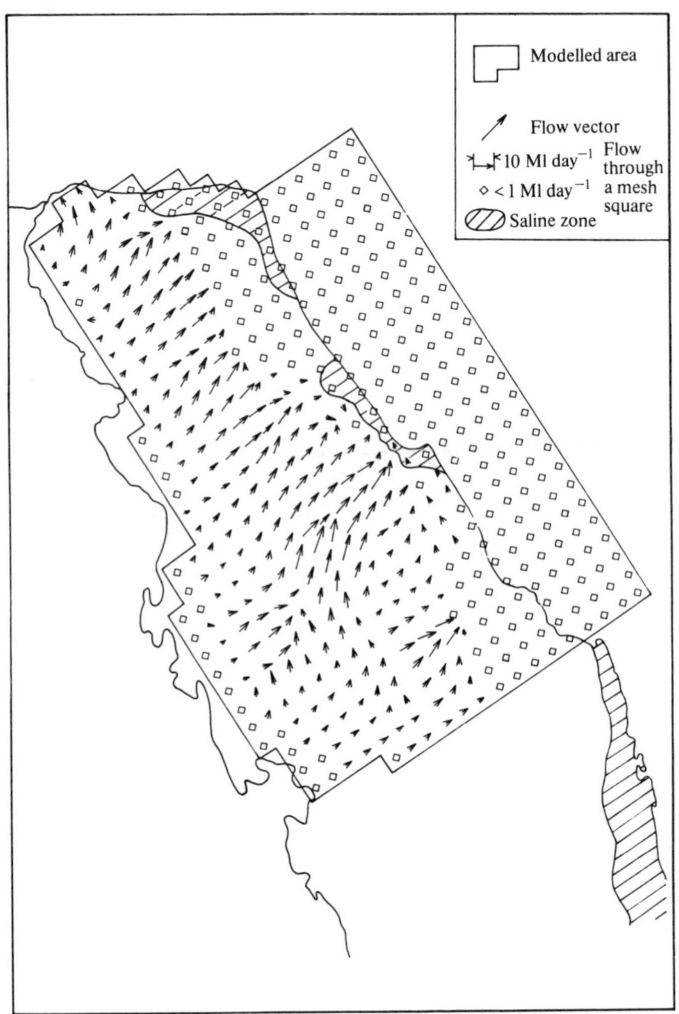


**Fig. 7.11.** Modelled groundwater flow distributions in Northern Lincolnshire. Note: no significant flow in Type II zone but seawater intrusion active at Pyewipe (see Fig. 7.7).

resources model of the area. The modelled flow distribution is shown in Fig. 7.11 with the very limited modern saline intrusion clearly identified and no significant movement of the old saline groundwaters.

Whereas the two examples quoted above attribute entrapment of saline groundwater in part to low-permeability strata distribution, entrapment can occur in areas of materials of similar permeability, particularly in flat coastal areas. Examples exist in the north Guyanan plains, the Lower Indus Basin, and the alluvial plains of Indonesia. In such areas groundwater heads and land surface gradients combine to cause major groundwater rejection in rivers some distance inland from the coast. In the coastal sections therefore groundwater flows are minimal to non-existent and saline groundwaters exist which may be partly connate in origin or old seawater intrusions because no effective modern flushing has occurred.

*Diffusion salinity*. The presence of saline groundwaters or matrix materials within or close to an aquifer, whatever their origin, may cause or further exacerbate saline conditions. Under static groundwater conditions (i.e. no-flow) diffusion could undoubtedly prove very effective, but usually groundwater flow rates are sufficiently high in areas of active groundwater resource development so that diffusion from a saline body is difficult to recognize (Lloyd, Rushton, Taylor, Barker, and Howard 1977). Undoubtedly diffusion occurs in brackish transition zones between fresh and saline groundwaters; however, flow-mixing is likely to be dominant (see Fig. 6.21) and diffusion effects are not easily recognizable.

## References

Arad, A., Kafri, U., and Fleisher, E. (1975). The Neaman springs, northern Israel. Salination mechanism of an irregular freshwater–seawater interface. *J. Hydrol.* **25**, 81–104.

Bredehoeft, J. D., Blythe, C. R., White, W. A., and Maxey, G. B. (1963). Possible mechanism for concentration of brines in subsurface formations. *Bull. Am. Assoc. Petrol. Geol.* **47**, 257–69.

Burdon, D. J. and Dounas, A. (1963). Hydrochemistry of the Parnassos–Ghiona aquifers and problems of sea-water contamination in Greece. *Int. Assoc. Sci. Hydrol.* **57**, 680–704.

Cederstrom, D. J. (1946). Genesis of groundwaters in the Coastal Plain of Virginia. *Econ. Geol.* **41**, 218–45.

Coplen, T. D. and Hanshaw, B. B. (1973). Ultrafiltration by a compacted clay membrane. Oxygen and hydrogen isotopic fractionation. *Geochim. cosmochim. Acta* **37**, 2295–310.

Desai, B. I., Gupta, S. K., Shah, M. V., and Sharma, S. C. (1979). Hydrochemical evidence of sea water intrusion along the Mangrol Chrowad Coast of Saurashtra, Gujarat. *Hydrogeol. Sci. Bull.* **24**, 71–81.

Downing, R. A. and Howitt, F. (1969). Saline groundwaters in the Carboniferous rocks of the English East Midlands in relation to geology. *Q. J. Eng. Geol.* **1**, 241–69.

Evans, W. R. (1982). Factors affecting water quality in a regional fractured rock aquifer. *Aust. Water Res. Council, Conf. on Groundwater in Fractured Rocks, Conf. Ser.* **5**, 57–69.

Forcella, L. S. (1982). Geochemistry of thermal and mineral waters in the Cascade Mountains of Western North America. *Ground Water* **20**, 39–47.

Goldberg, E. D. (1963). Chemistry in the oceans as a chemical system. In *Composition of sea water, comparative and descriptive oceanography*, pp. 3–25. Wiley, *The Sea*, Vol. 2, pp. 3–25. Wiley-Interscience, New York.

Graf, D. L. Friedman, I., and Meents, W. F. (1965). The origin of saline formation waters, II. Isotopic fractionation by shale micropore systems. *Ill. State geol. Surv., Circ.* **392**.

Groundwater Development Consultants (1980). *Umm er Rhaduma study*. Report to Ministry of Agriculture and Water, Kingdom of Saudi Arabia.

Henley, R. W. and Ellis, A. J. (1983). Geothermal systems ancient and modern; a geochemical review. *Earth Sci. Rev.* **19**, 1–50.

Hingston, F. J. and Gailitis, V. (1976). The geographic variation of salt precipitated over Western Australia. *Aust. J. Soil Res.* **14**, 319–35.

Howard, K. W. F. and Lloyd, J. W. (1978). Iodide enrichment of groundwaters of the Chalk aquifer, Lincolnshire, England. *Int. Ass. Hydrogeologists, Proc. Conf., Cieplice Spa, Warsaw*, 1978, IAH Publisher Warsaw, 87–98.

Issar, A. (1981). The rate of flushing as a major factor in determining the chemistry of water in fossil aquifers in Southern Israel. *J. Hydrol.* **54**, 285.

Jenkin, J. J. (1981). Terrain, groundwater and secondary salinity in Victoria, Australia, *Agric. Water Manage.* **4**, 143–71.

Lawrence, A. R., Lloyd, J. W., and Marsh, J. M. (1976). Hydrochemistry and groundwater mixing in part of the Lincolnshire Limestone aquifer, England. *Ground Water*, **14**, 12–20.

Lloyd, J. W. (1981). Saline groundwaters associated with fresh groundwater reserves in the United Kingdom. *A Survey of British Hydrogeology, Spec. Publ., R. Soc. London*, pp. 73–84.

—, Fritz, P., and Charlesworth, D. (1980). A conceptual hydrochemical model for alluvial aquifers on the Saudi Arabian basement shield. *Proc. advisory Group Meet., Arid-zone Hydrology, Vienna, Rep. IAEA-AG-158/15*. International Atomic Energy Agency, Vienna.

—, Howard, K. W. F., Pacey, N. R., and Tellam, J. H. (1982). The value of iodide as a parameter in the chemical characterisation of groundwaters. *J. Hydrol.* **57**, 247–65.

—, Rushton, K. R., Taylor, H. R., Barker, R. D., and Howard, K. W. F. (1977), Saline groundwater studies in the Chalk of northern Lincolnshire. *5th Int. Hydrogeological Prog. Salt Water Intrusion Meet.* Water Research Centre, Medmenham, England.

Martin, R. E. and Harris, P. G. (1982). Hydrochemical study of groundwater from an unconfined aquifer in the vicinity of Perth, Western Australia, *Aust. Water Res. Council, Tech. Paper 67*.

Schwartz, F. W. (1974). The origin of chemical variations in groundwaters from a small watershed in southwestern Ontario. *Can. J. Earth Sci.* **11** 893–904.

—, Muehlenbachs, K., and Chorley, D. W. (1981). Flow-system controls on the chemical evolution of groundwater. *J. Hydrol.* **54**, 225–43.

Van Dijk, D. C. (1969). Relict salt, a major cause of recent land damage in the Yass Valley, Southern Tablelands, N.S.W. *Aust. Geogr.* **11**, 1–12.

Wendt, I. (1971). Carbon and oxygen isotope exchange between $HCO_3$ in saline solution and solid $CaCO_3$. *Earth Planet Sci. Letter,* **12**, 439–42.

White, D. E. (1957). Magmatic, connate and metamorphic waters. *Geol. Soc. Am. Bull.* **68**, 1659–82.

Wigley, T. M. L. (1976). Effect of mineral precipitation on isotopic composition, and $^{14}C$ dating of groundwater. *Nature, Lond.*, **263**, 219–21.

# 8 Environmental isotopes

## 8.1. Introduction

In hydrogeology the term environmental isotopes is normally used to describe isotopes that occur naturally in the hydrological cycle and does not include isotopes introduced into the systems artificially for tracer purposes.

The main hydrogeological use of environmental isotopes can be summarized as follows:

(i) to provide a signature to a particular water type;
(ii) to identify the occurrence of mixing two or more water types;
(iii) to provide age information about groundwaters;
(iv) to provide information concerning travel times and groundwater velocities.

The use of environmental isotope studies in groundwater evaluation has become widespread since the early 1960s. As with most new techniques, the proponents of isotope studies initially claimed that their methods were all embracing and that isotopes could be applied to solving many of the outstanding problems of hydrogeology, e.g. recharge volume, permeability, etc. Unfortunately, but inevitably, as the work has progressed environmental isotopes have proved to be useful but limited in scope. As with all other aspects of hydrogeological study isotopes are only contributary to the solution of our problems and can only be considered in the framework of the rest of our hydrogeological knowledge. For example, quite simply, it is essential that the isotope data correlate with the hydraulic data. Equally, however, it is essential that the hydraulic data correlate with the isotope data and that the latter data are not overlooked but are used as a worthwhile contribution to understanding a system. The environmental isotopes commonly in use in groundwater studies are:

| *Radioactive* | *Stable* |
|---|---|
| Tritium (T or $^{3}H$) | Oxygen-18 ($^{18}O$) |
| Carbon-14 ($^{14}C$) | Deuterium ($^{2}H$ or D) |
| | Carbon-13 ($^{13}C$) |

Less commonly, research programmes have been carried out and are in progress using the following isotopes:

| *Radioactive* | *Stable* |
|---|---|
| Silicon-32 ($^{32}Si$) | Sulphur-34 ($^{34}S$) |
| Uranium ($^{234}U/^{238}U$) | Nitrogen ($^{14}N/^{15}N$) |
| Thorium ($^{230}Th/^{234}U$) | |

The radioactive isotopes are those that are unstable and undergo nuclear transformation emitting radioactivity; decay is spontaneous and is not changed by external influences. The stable isotopes undergo no radioactive decay. However, those which are studied exclusively on their own merits (i.e. $^{18}O$ and $^{2}H$) form part of the water molecule and are influenced by changes in the $H_2O$ state.

Isotope studies can be expensive and are frequently inconclusive. Individual isotopes do not normally provide adequate information so that multiple isotope studies tend to be more advantageous. As they represent only another aspect of hydrochemistry they cannot be sensibly studied without a thorough understanding of major ion hydrochemistry and controlling hydrochemical processes.

## 8.2. The δ notation

For instrumental reasons, it is difficult to determine isotope abundances absolutely by techniques that are suitable for routine analysis. However, relative isotope abundances can be determined easily using double-inlet double-collector mass spectrometers. By using an agreed standard, the relative measurements are as useful as absolute abundances. The relative difference is designated by the symbol δ and is normally expressed in parts per thousand (permil °/₀₀).

$$\delta_x = \frac{R_X - R_{std}}{R_{std}} \times 1000. \tag{8.1}$$

Isotope ratios ($R$) for the elements of hydrochemical interest are quoted with the light isotope as the denominator, e.g. $^{18}O/^{16}O$, as in all cases the heavy isotope is the rare isotope. Thus an increasing value of δ implies an increasing proportion of the heavy isotope. If one sample is *heavy* compared with another, it has a more *positive* δ value. $R_X$ relates to the sample and $R_{std}$ to the standard.

The isotope ratios of coexisting species are of great geochemical interest, and these are expressed by fractionation factors:

$$\alpha_{A\text{-}B} = \frac{R_A}{R_B}. \tag{8.2}$$

Substituting the definition of δ gives

$$\alpha_{A\text{-}B} = \frac{1000 + \delta_A}{1000 + \delta_B}. \tag{8.3}$$

Fractionation may occur either as a result of kinetic effects or at equilibrium. The equilibrium fractionation factor is equal to the equilibrium constant for the exchange reaction written so that one atom is exchanged. For example, for the exchange of oxygen between carbon dioxide and water

$$\tfrac{1}{2}\,C^{16}O_2 + H_2\,^{18}O \rightleftharpoons \tfrac{1}{2}\,C^{18}O_2 + H_2\,^{16}O \tag{8.4}$$

$$K = \frac{(C\,^{18}O_2)^{\frac{1}{2}}\,(H_2\,^{16}O)}{(C\,^{16}O_2)^{\frac{1}{2}}\,(H_2\,^{18}O)} = \frac{(^{18}O/^{16}O)_{CO_2}}{(^{18}O/^{16}O)_{H_2O}} = \alpha_{CO_2\text{-}H_2O}. \tag{8.5}$$

Most equilibrium isotope fractionation factors are close to unity. For convenience, the enrichment factor $\epsilon$ is used, where

$$\epsilon_{A\text{-}B} = (\alpha_{A\text{-}B} - 1) \times 1000. \tag{8.6}$$

The following approximation is valid for small $\epsilon$

$$\epsilon \approx 1000 \ln \alpha. \tag{8.7}$$

This is useful in so far as $1000 \ln \alpha$ is proportional to $1/T$ for small ranges at low temperatures.

Various useful interrelationships involving the $\delta$ notation are presented in Table 8.1.

**Table 8.1** *Relationships involving $\alpha$, $\delta$ and $\epsilon$*

| Given | Required | Relationship | Approximation |
|---|---|---|---|
| $\delta_A$, $\delta_B$ | $\alpha_{A\text{-}B}$ | $\dfrac{1000 + \delta_A}{1000 + \delta_B}$ | |
| $\alpha_{A\text{-}B}$ | $\epsilon_{A\text{-}B}$ | $1000\,(\alpha_{A\text{-}B} - 1)$ | $1000 \ln \alpha_{A\text{-}B}$ |
| $\delta_{A\text{-}B}$, $\delta_{B\text{-}C}$ | $\delta_{A\text{-}C}$ | $\delta_{A\text{-}B} + \delta_{B\text{-}C}\left(1 + \dfrac{\delta_{A\text{-}B}}{1000}\right)$ | $\delta_{A\text{-}B} + \delta_{B\text{-}C}$ |
| $\delta_{A\text{-}B}$ | $\delta_{B\text{-}A}$ | $-1000\left(\dfrac{\delta_{A\text{-}B}}{1000 + \delta_{A\text{-}B}}\right)$ | $-\delta_{A\text{-}B}$ |
| $\delta_A$, $\epsilon_{A\text{-}B}$ | $\delta_B$ | $1000\left(\dfrac{\delta_A - \epsilon_{A\text{-}B}}{1000 + \epsilon_{A\text{-}B}}\right)$ | $\delta_A - \epsilon_{A\text{-}B}$ |
| $\delta_B$, $\epsilon_{A\text{-}B}$ | $\delta_A$ | $\delta_B + \epsilon_{A\text{-}B}\left(1 + \dfrac{\delta_B}{1000}\right)$ | $\delta_B + \epsilon_{A\text{-}B}$ |
| $\alpha_{A\text{-}B}$, $\alpha_{B\text{-}C}$ | $\alpha_{A\text{-}C}$ | $\alpha_{A\text{-}B}\,\alpha_{B\text{-}C}$ | |
| $\alpha_{A\text{-}B}$ | $\alpha_{B\text{-}A}$ | $1/\alpha_{A\text{-}B}$ | |

## 8.3. Carbon isotopes

### *8.3.1. Standards*

Three isotopes of carbon occur naturally. Their abundances are given in Table 8.2.

**Table 8.2** *Abundance of carbon isotopes*

| Isotope | Average terrestrial abundance (per cent) |
|---|---|
| $^{12}C$ | 98.89 |
| $^{13}C$ | 1.11 |
| $^{14}C$ | $10^{-10}$ ($t\frac{1}{2}$ = 5730 years)† |

† $t\frac{1}{2}$ is half-life.

The internationally agreed standard for $^{13}C$ measurements is carbon dioxide prepared from a calcite sample separated from a specimen of *Belemnitella americana* from the Pee Dee Formation (Cretaceous) of South Carolina, U.S.A., which is referred to as PDB. This has long been unavailable so the International Atomic Energy Agency (I.A.E.A.) now distributes NBS-19 where $\delta^{13}C_{NBS\text{-}19/PDB} = -2.26 \pm 0.04$ parts per thousand. The absolute abundance of $^{13}C$ in PDB has been measured by Craig (1957): $^{13}C/^{12}C = 11237.2 \times 10^{-6}$. Measurements of $^{14}C$ are normally reported in terms of per cent modern carbon (pmc) where

$$A = \frac{(^{14}C/^{12}C)\text{ sample}}{(^{14}C/^{12}C)\text{ modern}} \times 100 \text{ pmc}. \tag{8.8}$$

The activity of modern carbon is very variable as a consequence of the combustion of fossil fuels and the testing of thermonuclear devices, and therefore an artificial oxalic acid standard is distributed by the National Bureau of Standards (U.S.A.) whereby the $^{14}C$ activity of modern carbon was equal to $0.95 \times {}^{14}C$ activity of NBS oxalic acid in 1950.

An activity of 100 pmc is close to the natural rate of $^{14}C$ generation in the troposphere and corresponds to 13.56 dpm $g^{-1}$ of carbon. $^{14}C$ measurements can also be quoted in terms of an apparent age. This is calculated by normalizing the $^{14}C$ result to $\delta^{13}C = -25$ parts per thousand, assuming $\epsilon^{14}C$ (per cent) = 0.2 $\epsilon^{13}C$ (parts per thousand) and by using the 'Libby' half-life of 5568 years. This practice is to be discouraged when the data are required for hydrogeological purposes. Because the $^{14}C$ standard is defined for vegetable carbon and because there are fractionations between various carbon reservoirs, the $^{14}C$ activities of the various reservoirs differ, as illustrated in Fig. 8.1. Usually these differences can be neglected in groundwater studies and it is assumed that the original modern input was 100 pmc.

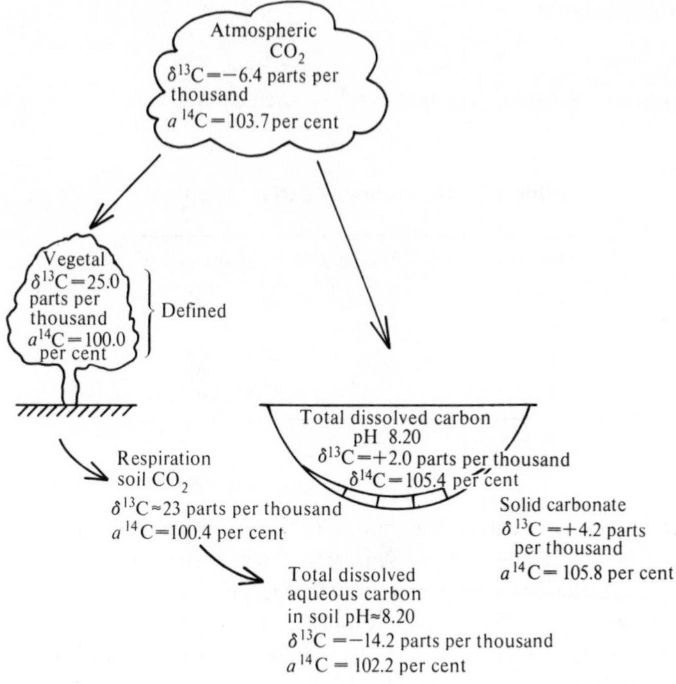


**Fig. 8.1.** Range of carbon isotopes in carbonates and related materials. (After Degens 1967.)

### *8.3.2. The occurrence of carbon isotopes in groundwater*

The occurrence of carbon isotopes in groundwater is intimately related to the carbonate chemistry of groundwaters discussed in Chapter 4. The sole natural source of $^{14}C$ is cosmic ray bombardment of the upper atmosphere, where it is produced as a result of neutron capture by atmospheric nitrogen:

$$^{14}N + n \rightarrow {}^{14}C + p.$$

The $^{14}C$ nuclei so produced oxidize to $CO_2$ and mix with the inactive carbon dioxide of the lower atmosphere. Some carbon dioxide enters the groundwater system by direct absorption in rainwater, but the most important source in most regions is via plant carbon. During their life, plants maintain an equilibrium with atmospheric carbon dioxide and therefore contain predictable amounts of $^{13}C$ and $^{14}C$. After death this is no longer maintained and the $^{14}C$ content decreases by radioactive decay. Vegetable carbon becomes assimilated in the soil organic matter, where its decay produces carbon dioxide whose isotopic composition is determined by the isotopic composition of the vegetable matter.

The $^{14}C$ content of vegetable matter was close to 100 pmc until the nineteenth century, when large-scale combustion of fossil fuels reduced the 'modern' activity to near 90 pmc. Since 1963 this effect has been eliminated by the contribution of $^{14}C$ from nuclear testing, which has caused 'modern' values of up to 200 pmc in the northern hemisphere (Tamers and Scharpenseel 1970). In hydrogeological studies, an initial input value of 100 pmc is usually assumed. Failure of the models used to produce positive ages, thus indicating that an input value greater than 100 pmc is needed, shows that the water is very recent (post 1963). The $^{13}C$ content of vegetable matter is rather variable as revealed by Deines (1980). Land plants can be divided into three groups on the basis of their photosynthetic mechanism and these groups have characteristic $^{13}C$ contents. The majority of temperate plants use the Calvin ($C_3$) cycle which produces a mean plant carbon composition close to $\delta^{13}C = -26$ parts per thousand. The Hatch-Slack ($C_4$) cycle used by some tropical plants, notably maize and sugar cane, produces a mean plant carbon composition around $\delta^{13}C = -13$ parts per thousand. A few plants with crassulacean acid metablism use both cycles, therefore producing carbon spanning the range $\delta^{13}C = -26$ to $-13$ parts per thousand.

It is usually assumed that no further fractionation of carbon takes place in the decay to carbon dioxide of soil organic matter, few direct measurements having been made. Galimov (1966) made a study of various soil types near Moscow, finding an average soil carbon dioxide isotopic composition of $\delta^{13}C = -24.8$ parts per thousand with a range of $-22.6$ to $-28$ parts per thousand. Hendy (1971) found an average of $-23.8$ parts per thousand for subtropical rain forest soils in New Zealand. The influence of Hatch–Slack plants can be seen in the data of Pearson and Hanshaw (1970) for arid soils in Texas, with a carbon dioxide composition in the range $\delta^{13}C = -15$ to $-19$ per cent. Interdiffusion with atmospheric carbon dioxide ($-7$ to $-10$ parts per thousand) may alter the carbon dioxide composition of sandy soils with low organic productivity (Galimov 1966).

Soil zone carbon dioxide reacts with percolating water and with soil zone carbonate as described in Section 3.4. The reaction can be summarized by the (non-balanced) equation

$$CO_2 + H_2O + CaCO_3 \rightleftharpoons Ca^{2+} + H_2CO_3^* + HCO_3^- + CO_3^{2-}. \tag{8.9}$$

Thus this step may involve the addition of further dissolved carbon, whose isotopic composition should be known to evaluate the isotopic composition of the input to the aquifer. If direct measurement is impossible, the $^{13}C$ content of the carbonate can be estimated by assuming it to be detrital and using the value of the parent rock. The $^{13}C$ content of sedimentary carbonates may range between $-55$ and $+7$ parts per thousand. (Degens 1967) (see Fig. 8.2) but marine carbonates are confined to a fairly narrow range around $+1$ parts per thousand. The $^{14}C$ content of such detrital carbonate is often assumed to be zero, although Geyh (1970) has shown that the $^{14}C$ content of the soil zone carbonate may be as high as 78 pmc but is more usually in the range 2–15 pmc. Such $^{14}C$ contents may arise from solution and precipitation of carbonate in the soil zone in response to seasonal temperature changes (Wendt, Stahl, Geyh, and Fauth 1967).

After the carbonate bearing water enters the saturated aquifer, further changes may modify the isotopic composition of the dissolved carbon. Some of these changes also modify the amount of dissolved carbon and are thus fairly easily detected. Such changes include further dissolution of carbonate to achieve saturation, precipitation of carbonate as part of an incongruent dissolution process (see Section 3.4), dissolution of carbonate as a result of ion exchange (see Section 3.7.2), and addition of carbon dioxide produced by the oxidation of organic matter, often associated with nitrate or sulphate reduction. It is also possible for the water to exchange carbon with the aquifer carbonate, a process which does not change the dissolved carbon content of the water. Direct exchange is negligible at normal groundwater temperatures, although it may become important in geothermal areas. Apparent exchange in calcite-saturated waters is produced by an incremental solution–precipitation process in response to small changes in physical conditions (Wigley 1976). In the relatively constant conditions of a deep aquifer this process is much slower than in the soil zone. Reaction with aquifer carbonate will clearly have very similar effects on the isotopic composition to reaction with soil zone carbonate, except that it can usually be assumed that the $^{14}C$ content of aquifer carbonate is zero. The isotopic effect of the addition of carbon dioxide by organic matter is highly variable, and depends on the isotope composition of the organic matter.

### *8.3.3. Models of isotope composition*

The qualitative discussion has indicated the main sources of dissolved carbon in groundwater and the ranges of their isotopic composition. A full interpretation of the isotopic composition of dissolved carbon can only be achieved by a quantitative treatment of the various steps. Ideally, from a study of carbon isotopes it is possible to decipher the mechanism controlling carbonate chemistry and also to estimate the age of the dissolved carbon.

The dissolution of carbonate (in the soil zone or the aquifer) can be described by the stoichiometric equation

$$\begin{aligned}&(x + 0.5y)CO_2 + 0.5yCaCO_3 + (x + 0.5y)H_2O \\ &\rightleftharpoons 0.5yCa^{2+} + yHCO_3^- + xH_2CO_3^*.\end{aligned} \tag{8.10}$$

assuming that the pH is such that $(CO_3^{2-})$ can be neglected. From mass balance it is possible to evaluate the isotopic composition of the resulting dissolved carbon using the following nomenclature:

| | |
|---|---|
| $\delta_g, a_g$ | $^{13}C$ (parts per thousand) and $^{14}C$ (pmc) content of gas phase; |
| $\delta_s, a_s$ | $^{13}C$ (parts per thousand) and $^{14}C$ (pmc) content of soil or aquifer carbonate; |
| $\Sigma$ | total dissolved carbon = $[HCO_3^-] + [H_2CO_3^*]$; |

$\delta_\Sigma, a_\Sigma$ $^{13}C$ (parts per thousand) and $^{14}C$ (pmc) content of total dissolved carbon;

The $^{13}C$ and $^{14}C$ contents are given by

$$\delta_\Sigma = \frac{[H_2CO_3^* + 0.5HCO_3^-]}{\Sigma}\delta_g + \frac{0.5[HCO_3^-]}{\Sigma}\delta_s \tag{8.11}$$

$$a_\Sigma = \frac{[H_2CO_3^* + 0.5HCO_3^-]}{\Sigma}a_g + \frac{0.5[HCO_3^-]}{\Sigma}a_s. \tag{8.12}$$

Two models of carbonate solution were discussed in Section 3.4, the closed-system and the open-system models. The carbon isotope chemistry associated with these models has been explored by Deines, Langmuir, and Harmon (1974). In the open system solution takes place at constant $P_{CO_2}$ and a three-phase equilibrium betweeen gas, solid, and solution is reached. When considering the isotopes, the additional constraint of isotopic equilibrium is imposed. Deines *et al.* (1974) assumed that isotopic equilibrium would be obtained between gas and solution, but not between solution and solid. On this basis, the isotopic composition of the dissolved carbon is governed only by $\delta_g$ and the fractionation factor $\epsilon_g{}^{13}$ (Table 8.3), and is

$$\delta_\Sigma = \delta_g + \epsilon_g{}^{13}\left(1 + \frac{\delta_g}{1000}\right) \tag{8.13}$$

$$a_\Sigma = a_g\frac{(1000 + 2\epsilon_g{}^{13})}{1000}. \tag{8.14}$$

**Table 8.3** *Enrichment factors for* $^{13}C$

| | | |
|---|---|---|
| $\epsilon H_2CO_3^*\text{-}CO_2$ | $= -0.373 \times 10^3/T + 0.19$ ‰ | |
| $\epsilon HCO_3^-\text{-}CO_2$ | $= +9.654 \times 10^3/T - 24.40$ ‰ | |
| $\epsilon HCO_3^-\text{-}H_2CO_3^*$ | $= +10.063 \times 10^3/T - 24.70$ ‰ | |
| $\epsilon CaCO_3\text{-}HCO_3^-$ | $= -4.232 \times 10^3/T + 15.10$ ‰ | |
| $\epsilon CaCO_3\text{-}H_2CO_3^*$ | $= +5.383 \times 10^3/T - 9.16$ ‰ | |
| $\epsilon_g = \epsilon\Sigma - CO_2$ | $= \dfrac{\epsilon H_2CO_3^*\text{-}CO_2\ [H_2CO_3^*] + \epsilon HCO_3^-\text{-}CO_2[HCO_3^-]}{\Sigma}$ | at pH < 9 |
| $\epsilon_s = \epsilon CaCO_3\text{-}\Sigma$ | $= \dfrac{\epsilon CaCO_3\text{-}H_2CO_3^*\ [H_2CO_3^*] + \epsilon CaCO_3\text{-}HCO_3^-\ [HCO_3^-]}{\Sigma}$ | |

After Mook 1980.
*T* is in kelvins.

In the closed system, after an initial step described by eqns. (8.13) and (8.14), further solution in the absence of a gas phase is described by

$$\Sigma'\delta'_{\Sigma} = \Sigma\delta_{\Sigma} + \Delta_s\delta_s \tag{8.15}$$

where $\Delta_s$ is the increment of carbonate dissolved and $\Sigma$ and $\Sigma'$ are the dissolved carbon concentrations before and after dissolution respectively. The results produced by these two models are illustrated in Figs. 8.2 and 8.3. A computer implementation of this model has been presented by Reardon and Fritz (1978).

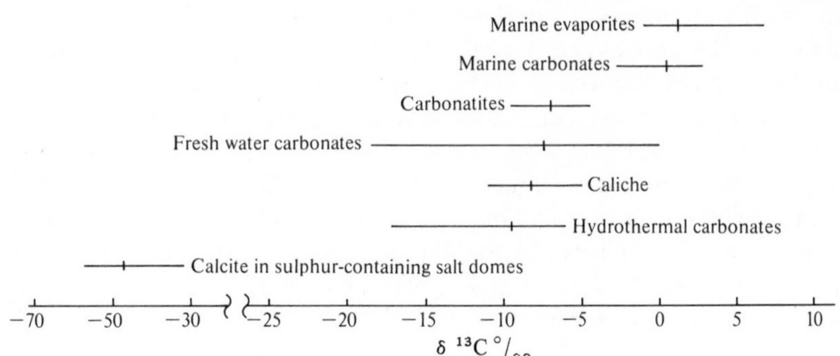


**Fig. 8.2.** Carbon isotope content of natural carbon reservoirs. (After Dengens 1967.)

In practice it is often found that this approach does not correctly predict the $\delta^{13}C$ values observed experimentally, suggesting that isotopic equilibrium with the gas phase is not fully achieved. An equation which compares measured and calculated $\delta^{13}C$ values to estimate exchange and thence to evaluate $a_{\Sigma}$ has been presented by Mook (1976):

$$a_{\Sigma} = \frac{([H_2CO_3^*] + 0.5[HCO_3^-])a_g + 0.5[HCO_3^-]a_s}{\Sigma} + \frac{0.5(a_g - a_s + 0.2\epsilon_g{}^{13})}{\Sigma}$$

$$\times \frac{\Sigma\delta_{\Sigma} - ([H_2CO_3^*] + 0.5[HCO_3^-])\delta_g - 0.5[HCO_3^-]\delta_s}{0.5(\delta_g - \delta_s) + \epsilon_g{}^{13}}. \tag{8.16}$$

When typical values for temperate zone groundwaters, where $\delta_g = -25$ parts per thousand, and $\delta_s = +1$ parts per thousand, are used, the stoichiometric reaction (eqn 8.11) predicts $\delta_{\Sigma}$ values around −14 parts per thousand. Open-system conditions involving isotopic equilibrium with the gas phase produce $\delta_{\Sigma}$ values around − 18 parts per thousand. Waters for which eqn (8.16) is valid should fall within this range. However, it is not uncommon for groundwaters to have more positive $\delta_{\Sigma}$ values than those predicted by eqn (8.11), implying exchange with a source of heavy carbon, usually aquifer carbonate. As mentioned above, such exchange proceeds via solution-precipitation steps and

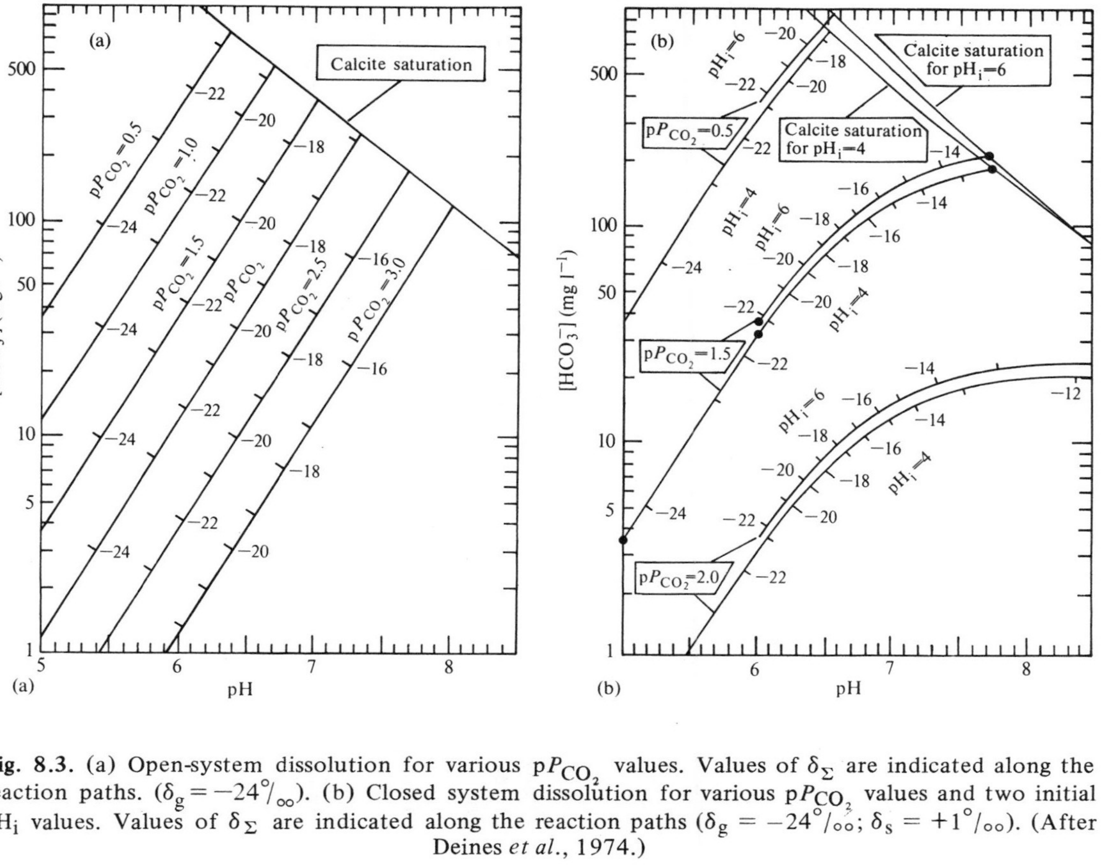


**Fig. 8.3.** (a) Open-system dissolution for various $pP_{CO_2}$ values. Values of $\delta_\Sigma$ are indicated along the reaction paths. ($\delta_g = -24°/_{oo}$). (b) Closed system dissolution for various $pP_{CO_2}$ values and two initial $pH_i$ values. Values of $\delta_\Sigma$ are indicated along the reaction paths ($\delta_g = -24°/_{oo}$; $\delta_s = +1°/_{oo}$). (After Deines *et al.*, 1974.)

therefore can be considered only for calcite-saturated waters. This produces a limiting $\delta_\Sigma$ value at equilibrium given by

$$\delta_\Sigma = 1000 \frac{\delta_s - \epsilon_s^{13}}{1000 + \epsilon_s^{13}}. \tag{8.17}$$

This value is near +0.8 parts per thousand for the input values assumed above. Partial exchange with aquifer carbonate, in the absence of a gas phase, produces the result

$$a_2 = \frac{a_1 \{\delta_s - \delta_2 - \epsilon_s^{13}(1 + \delta_2/1000)\} + a_s(\delta_2 - \delta_1)}{\delta_s - \delta_1 - \epsilon_s^{13}(1 + \delta_2/1000) + (2\epsilon_s^{13}/1000)(\delta_2 - \delta_1)}. \tag{8.18}$$

where $a_1$, $\delta_1$, $a_2$, and $\delta_2$ are the compositions of the dissolved carbon before and after the exchange step respectively. If the substitution $a_s = 0$ is made and second-order terms are neglected, it can be seen that eqn (8.18) is equivalent to the result of Wigley (1976):

$$a_2 = a_1 \frac{(\delta_s - \delta_2 - \epsilon_s^{13})}{(\delta_s - \delta_1 - \epsilon_s^{13})}. \tag{8.19}$$

Fontes and Garnier (1979) produced an equation which attempted to allow for exchange with either the gas or solid phases; unfortunately their derivation does not specify the isotopic composition of the water, which is the only parameter accessible to measurement in groundwater systems. It is possible for exchange with the gas phase and exchange with the solid phase to proceed simultaneously, in which case it is not possible to determine uniquely the amount of each exchange that has taken place. An estimate of the amount of carbon contributed by gas and solid reservoirs can be obtained from the isotope mixing model of Pearson (1965):

$$a_\Sigma = \frac{\delta_\Sigma - \delta_s}{\delta_g - \delta_s}(a_g - a_s) + a_s. \tag{8.20}$$

This is only an approximation to $a_\Sigma$ since no account can be taken of fractionation factors.

A general treatment of the isotopic changes in a water body which has several sources and sinks of carbon has been given by Wigley, Plummer, and Pearson (1978). The carbon content and isotopic composition of the solution are defined by differential equations, where $\mathscr{I}$ and $\mathscr{O}$ represent sources and sinks of carbon respectively. The carbon content is given by

$$\mathrm{d}\Sigma = \sum_{i=1}^{m} \mathrm{d}\mathscr{I}_i - \sum_{i=1}^{m} \mathrm{d}\mathscr{O}_i. \tag{8.21}$$

The isotopic composition of the solution is given by

$$\mathrm{d}(R\Sigma) = \sum_{i=1}^{m} R_i \mathrm{d}\mathscr{I}_i - \sum_{i=1}^{m} \alpha_i R \mathrm{d}\mathscr{O}_i \tag{8.22}$$

where $R$ is the isotope ratio of the solution (i.e. $^{13}C/C_{total}$) and $R_i$ is the isotope ratio of the $i$th input.

Solutions of these equations are possible for certain conditions. In the general case numerical solutions, which must be combined with numerical models of the carbonate chemistry, are necessary, the whole requiring the use of large computers. Often the result is seen to be insensitive to the reaction path taken, and also the chemical data available may provide insufficient control; therefore the simpler methods of correction described above may be preferred.

The $^{14}C$ content of the dissolved carbon is reduced by radiometric decay independently of chemical changes. The activity $a$ is related to the time $t$ by a first-order rate equation:

$$\frac{\mathrm{d}a}{\mathrm{d}t} = -\lambda a \tag{8.23}$$

whence

$$a_t = a_0 \exp(-\lambda t) \tag{8.24}$$

where $a_0$ is the activity at $t = 0$, and $\lambda$ is the decay constant.

Rearranging this equation and substituting numerical values for time in years gives

$$t = -19\,035 \log_{10} \left(\frac{a_t}{a_0}\right). \tag{8.25}$$

Thus by comparing the measured $^{14}C$ activity with the activity estimated from the chemical models discussed above, it is possible to determine the age of the water.

### *8.3.4. $^{14}C$ age determination of water*

In using $^{14}C$ to determine the age of a water body, the best that can be done is to determine when the dissolved carbon in the water became isolated from the source of atmospheric $^{14}C$, which is usually decaying plant matter. The discussion above has shown that there are several ways in which the $^{14}C$ content of the dissolved carbon can be modified, only one of which, radioactive decay, is of any use in determining the age of the carbon.

All approaches to $^{14}C$ age determination in water aim to estimate the $^{14}C$ content the water samples would have had at zero age, and then use eqn (8.25) with the measured $^{14}C$ content to determine time elapsed. The simplest method is to determine the carbon isotopic composition after recharge at the beginning of the aquifer flow system. Water samples from further along the flow system can then be assigned relative ages using eqn (8.25). Use of this technique assumes that no solution of aquifer carbonate and no exchange with aquifer carbonate have occurred, which must be checked by measurements of $\Sigma$ and $\delta^{13}C$.

Dissolutions of carbonate without exchange between points 1 and 2 situated on a flowline produces a change in $^{14}C$ activity given by (Pearson and Hanshaw 1970):

$$a_2 = \frac{(\Sigma_2 - \Sigma_1)\, a_s + \Sigma_1\, a_1}{\Sigma_2}. \tag{8.26}$$

This produces a corresponding change in the $\delta^{13}C$ value. If this does not agree with the measured value exchange is indicated and the equations of Wigley *et al.* (1978) must be used. Exchange with no change in $\Sigma$ is corrected using eqn (8.19).

It may be impossible to measure the isotopic composition of the recharge water, or there may be other reasons for needing an absolute rather than a relative age. In this case, the models of carbon dissolution discussed earlier are used.

### *8.3.5. Case histories of $^{14}C$ age determination*

*Carbonate aquifer covered by calcareous soil.* The major aquifer of south-east England is the Chalk, a fine-grained limestone of upper Cretaceous age. In many areas recharge to the Chalk aquifer is derived from overlying Pleistocene formations consisting of highly calcareous till and its derivatives (Lloyd, Harker, and Baxendale 1981). The Chalk groundwaters are uniformly calcite saturated and most waters in the upper 30 m of the aquifer have a $pP_{CO_2}$ value close to 1.5. This has been interpreted as showing that the recharge water reaches chemical equilibrium with soil gas and soil calcite under open-system conditions in the soil zone, where the carbon dioxide partial pressure is raised by organic activity.

Isotope data from Heathcote (1981) for part of the Chalk aquifer near Ipswich are presented in Table 8.4. Samples 1 and 2 originate near outcrop and the remainder originate from down-gradient locations. Before attempting to interpret these data in terms of $^{14}C$ age it is necessary to understand the processes operating.

**Table 8.4** *Chemical and isotopic data for Chalk groundwaters*

| Sample number | $M_{H_2CO_3^*}$ | $M_{HCO_3^-}$ | $\delta^{13}C$ (PDB) ‰ | $a$ (pmc) | Model age (years) |
|---|---|---|---|---|---|
| 1 | 1.22 | 5.82 | −13.3 | 67.1 | — |
| 2 | 1.32 | 5.63 | −12.6 | 62.3 | — |
| 3 | 1.42 | 5.99 | −12.2 | 49.2 | 858 |
| 4 | 1.68 | 6.87 | −11.6 | 39.8 | 2140 |
| 5 | 0.83 | 4.83 | − 6.0 | 8.6 | 9045 |

The open-system hypothesis must be checked using the isotopic data for the recharge samples 1 and 2. The open system $\delta^{13}$C values of these samples, calculated from eqn (8.13) are −17.1 parts per thousand and −17.2 parts per thousand respectively, assuming a soil gas $\delta^{13}$C value of −25 parts per thousand. Clearly the system does not behave as though isotopically open. Detailed study suggests that kinetic considerations prevent the system from being isotopically open, and that the initial reaction is stoichiometric (eqn 8.11). Since measured $\delta^{13}$C values are heavier than those calculated for stoichiometric reaction, exchange with a source of heavy carbon, is this case soil zone calcite, is indicated. Such exchange is described by eqn (8.18). No data for the $^{14}$C content of the soil zone calcite are available and therefore a value of 10 pmc is assumed on the basis of data collected by Geyh (1970). The initial and final $^{13}$C values for this exchange are taken to be the stoichiometric and measured values respectively for the recharge waters, producing measured $^{14}$C contents in excess of those calculated, and therefore substitution into eqn (8.25) will give future ages. This impossible result implies an error either in the input parameters or in the choice of model. The most probable error is in assuming that $a_g$ = 100 pmc. The use of $a_g$ values of up to 113 pmc produces zero ages for both recharge waters but this assumes the presence of 'post-bomb' carbon. Thus it can safely be said that samples 1 and 2 are very recent, as might be expected on aquifer outcrop.

The down-gradient waters have heavier $\delta^{13}$C values than the outcrop waters, showing that assimilation of aquifer carbonate ($\delta^{13}C = +1.2$ parts per thousand; $a = 0$ pmc) occurs. This must be by solution–precipitation exchange since major-ion data contradict the hypothesis of further net calcite dissolution once the water moves away from the recharge zone. Since two stages of exchange with solid carbonate have been proposed, firstly with $^{14}$C-containing soil carbonate and secondly with $^{14}$C-free aquifer carbonate, it is necessary to separate them because their effects on the calculated $^{14}$C actively are different. Using the data from the outcrop samples, it is assumed that the $\delta^{13}$C value after the soil zone step is no heavier than −13 parts per thousand ($\delta_W$ in Fig. 8.4), with subsequent exchange taking place in the aquifer. Exchange in the aquifer is adequately described by eqn (8.19). The overall scheme of the calculation is summarized in Fig 8.4 and the results are included in Table 8.4. This calculation scheme is appropriate to many limestone aquifer situations.

In the area studied these data have significant resource implications. Samples 3 and 4 originate from a flow system active at present whose turnover time is therefore characterized as at least several hundred years. It is therefore probable that it will not be affected by modern nitrate pollution in the near future. Sample 5 originated from a body of high quality water in the confined zone of the aquifer. Its age suggests that it originated during special hydrogeologic conditions at the end of the Devensian (Weichsel) glaciation of Britain. These special conditions no longer obtain, implying that this body of high quality water is fossil. Its development will almost certainly constitute mining.

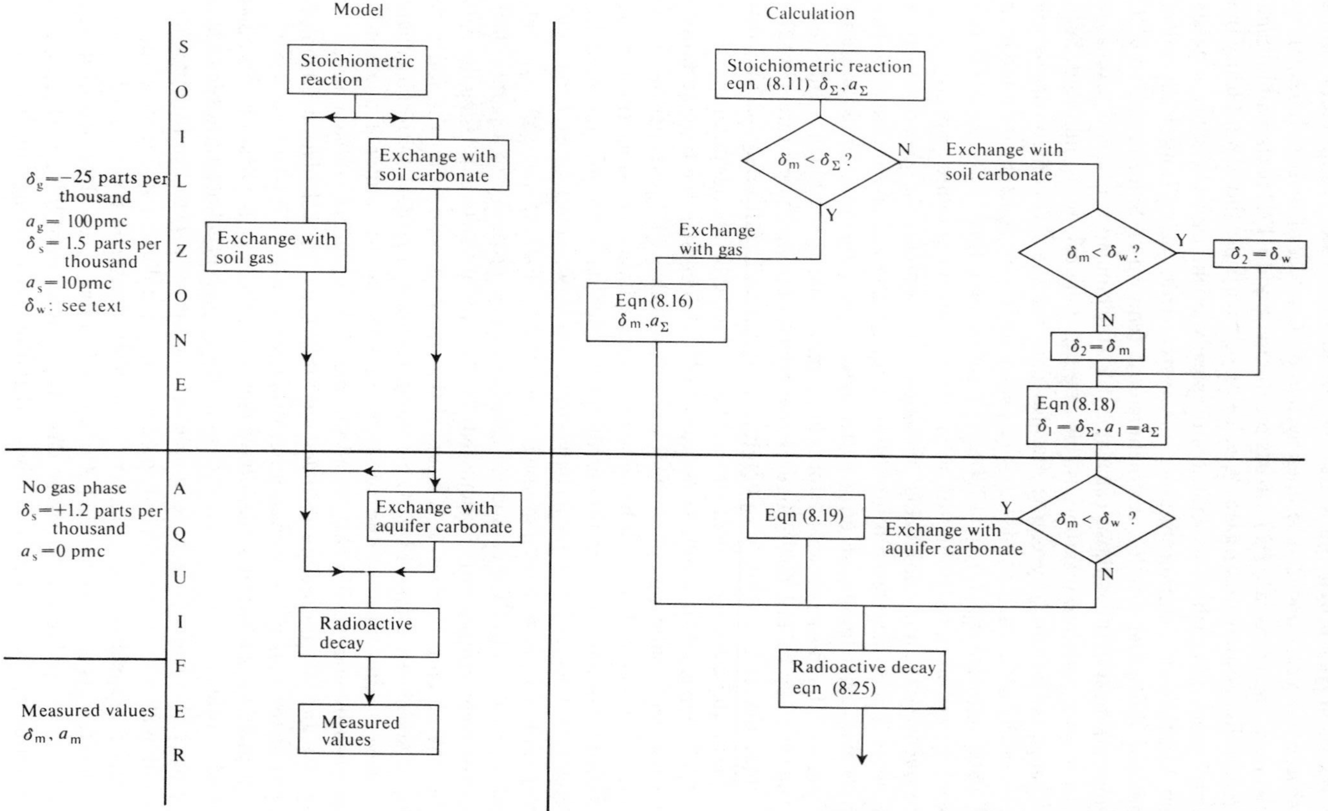


**Fig. 8.4.** Model of carbon isotope evolution.

*Arid zone aquifers*. The dating of groundwaters in arid zones poses special problems concerning the values of the input parameters to the equations. In particular, the sparse vegetation may consist of plants with both Hatch–Slack and Calvin photosynthetic pathways. In addition the low organic productivity of the soil means that contamination of soil air by atmospheric carbon dioxide may be significant; thus the value of $\delta_g$ may be between −26 and −7 parts per thousand. Soil and aquifer carbonate may consist predominantly of caliche ($\delta^{13}C \approx -5$ parts per thousand; may contain modern carbon) instead of marine limestone. The nature of the carbonate dissolution mechanism may be obscure and may not be operating under the present climatic conditions. Despite these problems, $^{14}C$ dating is of considerable importance in arid areas in determining whether a groundwater body is fossil, originating from some post-pluvial period, or whether it is a renewable resource receiving modern recharge.

An investigation of carbonate dissolution processes currently operating in southern Arizona has been made by Wallick (1976). The Tucson Basin aquifer consists of alluvial materials and has a cover of dry scrub and sparse grass. Widespread caliche has resulted from the chemical weathering of detrital silicates. As a consequence of the solutional remobilization of the caliche calcite during successive erosion cycles, it was assumed that the calcite had attained isotopic equilibrium with soil carbon dioxide via the medium of surface water. The closed-system process described in Section 3.4 was modelled chemically and isotopically to find the value of $\delta_g$ that produces the $\delta^{13}C$ value for total dissolved carbon observed in the groundwaters.

The results of the model were then compared with measured $\delta^{13}C$ values for soil gas (about −12 parts per thousand) and caliche (about −3.7 parts per thousand). The fit is good. Because the equilibrium between soil carbon dioxide and soil calcite is not a true dynamic equilibrium, but takes place at discrete times possibly separated by thousands of years, the model cannot predict the $^{14}C$ content of the soil calcite; the measured values (about 25 pmc) differ considerably from the expected values assuming equilibrium (about 102 pmc).

The use of dissolved calcium to calculate the soil carbonate contribution to the total dissolved carbonate was proposed by Wallick (1976) according to eqn (8.10). After corrections for gypsum dissolution etc., this is an alternative to the use of eqn (8.11) and should therefore produce the same results. Significant differences are present, thus showing that eqn (8.10) does not account for all the dissolved carbon. Dissolution of natron ($Na_2CO_3$) is one possibility not explored by Wallick (1976), and therefore there must be doubt concerning the validity of this method.

An interesting approach to the problems of arid zone aquifers is illustrated by Pearson and Swarzenki (1974). The study concerned an aquifer of semi-consolidated fine sand with silt and clay in the Northeastern Province of Kenya. The long-term average rainfall of the area is around 250 mm per year compared with a potential evaporation of 2500 mm per year. It was therefore important to determine whether recharge to the aquifer was still occurring during locally intense storms.

No data on the composition of the soil gas or the soil and aquifer carbonate in the area were available, and therefore it was necessary to estimate these parameters from the composition of the water. The dilution of the initial dissolved carbon by the dissolution of the soil zone and aquifer carbonate was modelled by a simple linear binary mixing model. If the end memebers of the mixture (m) are rock (r) and water (w) components with total carbon contents $C_m$, $C_r$, $C_w$ and isotopic compositions $I_m$, $I_r$, $I_w$, the composition of any mixture is given by

$$I_m = \frac{C_w\,(I_w - I_r)}{C_m} + I_r. \tag{8.27}$$

This is a straight line if $I_m$ is plotted against $1/C_m$, with slope $C_w(I_w - I_r)$ and intercept $I_r$. This equation can be sensibly applied to groundwater only if the radioactive decay of $^{14}C$ can be neglected; thus it is only applicable to individual recharge episodes where the duration of the episode is short on the $^{14}C$ time scale. This restriction is clearly unacceptable in a temperate climate with annual recharge.

Plotting of $^{13}C$ and $^{14}C$ data from Pearson and Swarzenki (1974) (Table 8.5) produces Fig. 8.5. The $^{14}C$ data were considered by Pearson and Swarzenki

**Table 8.5** *Carbon isotope data*

| Sample number | Total dissolved carbon (mmol $l^{-1}$) | $\delta^{13}C_{HCO_3^-}$ (‰) | $^{14}C$ (pmc) | Age group (Pearson and Swarzenki) | Age, years (this account) |
|---|---|---|---|---|---|
| 1 | 7.11 | −12.0 | 38.4 | III | 7251 |
| 2 | 7.19 | −13.1 | 39.4 | III | 8429 |
| 3 | 8.20 | −11.3 | 37.9 | III | 6333 |
| 4 | 6.05 | −11.7 | 48.4 | III | 4913 |
| 5 | 9.85 | −10.2 | 58.2 | II | 864 |
| 6 | 9.49 | −11.4 | 12.8 | IV | 15 462 |
| 7 | 10.89 | −10.7 | 57.6 | II | 1880 |
| 8 | 7.38 | −12.0 | 42.1 | III | 6490 |
| 9 | 10.23 | −9.8 | 64.1 | II | −761 |
| 10 | 10.05 | — | 87.7 | I | — |
| 11 | 12.76 | −9.5 | 52.4 | II | 225 |
| 12 | 9.88 | −11.0 | 18.6 | IV | 11 736 |
| 13 | 12.95 | −9.5 | 21.0 | III | 7784 |
| 14 | 18.21 | −11.8 | 35.0 | II | 7737 |
| 15 | 8.46 | −9.4 | 2.8 | IV | 24 201 |
| 16 | 11.09 | −10.1 | 13.3 | IV | 12 868 |
| 17 | 30.97 | −10.4 | 7.3 | III | 18 411 |
| 18 | 55.15 | −8.0 | 6.7 | III | 12 062 |
| 19 | 38.74 | −8.8 | 28.9 | I | 3300 |
| 20 | 22.40 | −9.2 | 4.4 | IV | 19 964 |

From Pearson and Swarzenki 1974.

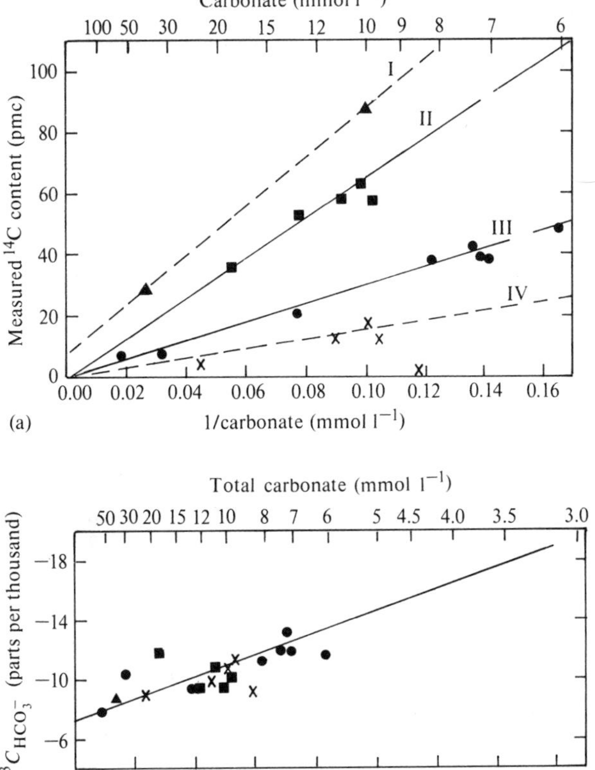


**Fig. 8.5.** Graph of (a) measured $^{14}C$ content and (b) $\delta^{13}C$ (bicarbonate) versus reciprocal dissolved carbonate content (the symbols correspond to linear groups). (From Pearson and Swarzenki 1974.)

(1974) to fall on four straight lines which can be interpreted as corresponding to distinct recharge periods, each with its characteristic $C_W(I_W - I_r)$ value as a consequence of subsequent radioactive decay. The lines pass close to the origin showing that the present $^{14}C$ content of the soil carbonate that reacted with past recharge water is small, but it is necessary to be able to correct for the radioactive decay to estimate the past $^{14}C$ content.

The $^{13}C$ data produce a single straight line (Fig. 8.5(b)), implying that the controlling parameters have remained constant over the time concerned. The intercept of the line gives a mean $\delta^{13}C$ value for the soil and aquifer carbonate of $-7.3$ parts per thousand, which is within the range reported for caliche carbonates. Derivation of the soil gas $\delta^{13}C$ value from the slope of the line ($-36.7$) requires a value for $C_W$. On the assumption that the minimum observed $C_m$ value ($\sim 7$ mmol $l^{-1}$) represents the input value before dissolution of soil zone carbonate, the value of $\delta^{13}C_W$ is $-12.5$ parts per thousand. The line of steepest slope (line I) in Fig. 8.5(a), which represents the most modern water, has a slope of 870. If it is assumed that the input $^{14}C$ content of the water did not exceed 100 pmc, the value of $C_w$ is greater than 8.7 mmol $l^{-1}$, giving $\delta^{13}C_w = -11.5$ parts per thousand. The two methods of estimating $\delta^{13}C_W$ therefore produce comparable results. At the pH and temperature of the waters in this study, the value of $\epsilon_g$ is approximately 7.8 parts per thousand; therefore the equilibrium gas phase has $\delta^{13}C \approx -20$ parts per thousand, i.e. between the values for Calvin and Hatch–Slack plants.

The isotope chemistry and the major ion chemistry provide complementary information useful in understanding the carbonate dissolution process. The dissolved carbon concentrations of even the most dilute waters exceed those that might be expected for a three-phase equilibrium between soil gas, water, and calcite, i.e. open-system conditions. The high sodium content of the waters suggests that ion exchange is contemporaneous with calcite dissolution in the soil zone, thus removing the constraint of calcite saturation. The soil carbon dioxide isotopic compositions deduced from the dissolved carbon isotope chemistry are plausible, thus suggesting isotopic as well as chemical equilibrium. Since the fractionation factor used is for $7 < pH < 9$, this confirms the occurrence of carbonate dissolution and ion exchange in contact with the soil air. (If carbonate dissolution did not occur, the equilibrium pH would be approximately 5 whence $\epsilon_g$ is approximately zero.) Further carbonate dissolution must take place out of contact with the soil air, otherwise the observed correlations between carbonate content and isotopic composition would not be seen. This continued solution is probably drawn by continued replacement of calcium ions by sodium ions, removing the calcite saturation constraint.

The relative ages of the waters can be deduced from Fig. 8.5(a) by assuming a constant initial dissolved carbon content, whence the differences in $^{14}C$ content can be ascribed to radioactive decay. Pearson and Swarzenki (1974) used an approximate initial dissolved carbon content of 10 mmol $l^{-1}$ which gives the following ages relative to group I: group II, 2600 years; group III, 8900 years; group IV, 14100 years. These ages are considered hydrogeologically plausible. Of the alternative methods of interpreting $^{14}C$ data, only the isotope mixing model (eqn (8.20)), is appropriate for an area about which so little is known. Substitution of the parameters derived above ($\delta_g = -12.5$ parts per thousand, $a_g = 100$ pmc, $\delta_s = -7.3$ parts per thousand, and $a_s = 25$ pmc (Wallick 1976)) gives the ages listed in Table 8.5. This treatment confirms the general results deduced above but differs for a few samples. Groups can be distinguished with ages modern, 2000 years, 7000 years, and greater than 12 000 years. Further study of the processes involved may clear up these anomalies. From the point of view of groundwater resources, the work shows that restricted recharge in the recent past has occurred but that most of the water is very old, having been recharged sporadically at intervals of several thousand years. The prospects for a large-scale renewable resource are poor.

## 8.4. Hydrogen isotopes

### *8.4.1. Standards*

Three isotopes of hydrogen occur naturally. Their abundances are given in Table 8.6.

**Table 8.6** *Abundances of hydrogen isotopes*

| Isotope | Average terrestrial abundance (%) |
|---|---|
| $^1H$ | 99.984 |
| $^2H(D)$ | 0.015 |
| $^3H(T)$ | $\approx 10^{-4}$ ($t\frac{1}{2} = 12.35$ years) |

The internationally agreed standard for $^2H$ (deuterium) measurements is V-SMOW (Vienna Standard Mean Ocean Water). The absolute isotope ratio of V-SMOW has been measured by Hageman, Neif, and Roth (1970) as $^2H/^1H = (155.76 \pm 0.05) \times 10^{-6}$. Large systematic errors are possible in the mass spectrometric analysis for $^2H$, and therefore a secondary standard SLAP (Standard Light Antarctic Precipitation) is also used, where $\delta^2H_{SLAP/V\text{-}SMOW} = -428$ parts per thousand. Measured values are corrected to the normalized V-SMOW/SLAP scale, defined by

$$\delta = \frac{R_{sample} - R_{V\text{-}SMOW}}{R_{V\text{-}SMOW}} \frac{\delta_{SLAP}}{(R_{SLAP} - R_{V\text{-}SMOW})/R_{V\text{-}SMOW}} \tag{8.28}$$

Although deuterium is a hydrogen isotope it is not discussed in this section but is considered with $^{18}O$ in Section 8.5 as the two isotopes are normally studied together.

There is no reference standard for measurements of $^3H$ (tritium). Absolute concentrations are determined radiometrically and reported as tritium units (TU), where 1 TU equals $10^{-18}$ atoms of $^3H$ per atom of $^1H$. The activity corresponding to 1 TU is 7.2 dpm per litre of water or 3.2 pCi $l^{-1}$.

### *8.4.2. Tritium in precipitation*

Tritium is produced in the atmosphere at a rather low rate through cosmic high energy radiation and enters the hydrological cycle in precipitation. Estimates of the concentration of the isotope in precipitation resulting from this source vary, but are of the order of 10 TU. Unfortunately few data are available for natural levels although some have been published by Kaufmann and Libby (1954), Von Buttlar and Libby (1955), and Begemann and Libby (1957).

The natural levels of tritium have been drastically disturbed since 1952 by the detonation of thermonuclear devices. The hydrogen bomb was found to introduce tritium to levels that swamped any natural backgrounds and the intermittent occurrence of detonations also introduced periodic pulses of tritium into the precipitation. More recently the international agreements to curb atmospheric detonations has resulted in a decrease in tritium levels so that, providing no bomb activity occurs in the future, the use of tritium as a signature will have been transient and its future use will greatly diminish. The present precipitation data, however, indicate that the stratosphere must still contain a reasonably high tritium concentration. The tritium concentrations in precipitation in the United Kingdom are shown in Fig. 8.6.

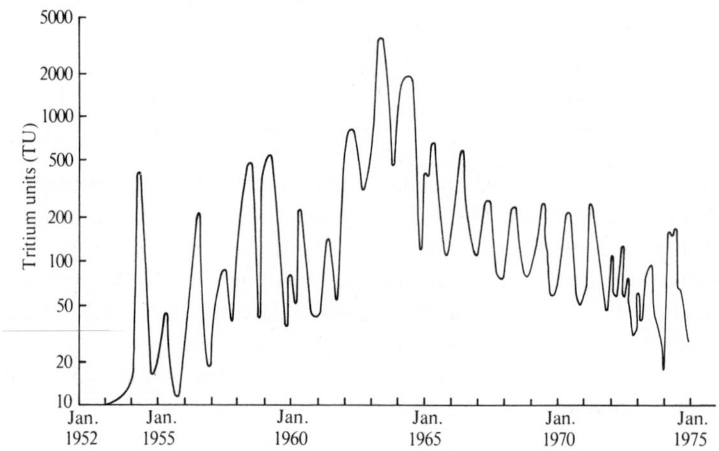


**Fig. 8.6.** Tritium concentration in United Kingdom rainfall.

The circulation pattern of tritium depends on the motion of the atmosphere and the eddy diffusion processes. The isotope enters the troposphere from the stratosphere and is incorporated in the water molecule. The infusion occurs mainly in the temperate latitudes and is dependent upon seasonal factors. Maximum concentrations in the precipitation are found in May and June in the northern hemisphere.

Tritium input to the earth's surface is also variable depending upon the nature of the surface. In oceanic areas vapour exchange between the atmosphere and the surface of the sea will be far more effective than precipitation in removing tritium from the atmosphere. Over land areas the removal of tritium from the atmosphere is essentially via precipitation. However, this is very variable and is related to the various climatic, land distance, and orographic influences affecting precipitation.

From the tritium concentration data available for precipitation it is readily apparent that significant variations in the level of the isotope input to the earth's surface can occur in both time and space. Consequently the recharge levels of tritium to groundwater in both direct and indirect recharge can be very variable.

### *8.4.3. Tritium in groundwater*

Tritium concentrations in groundwaters have frequently been studied to provide indications of age of the water (Libby 1953; Nir 1964) and the quantitative replenishment of the resource. Before the value of tritium in these contexts can be discussed, however, some account has to be taken of tritium input from rainfall and any modifications to concentrations of tritium in the unsaturated zone.

Tritium mass balance calculations have been carried out in many aquifers and have often concentrated upon the peak input periods of the early 1960s (Smith, Wearn, Richards, and Rowe 1970; Bredenkamp *et al.* 1974). Unfortunately rainfall tritium concentrations for the period are limited so that transposed data tend to have been used in the balances. Foster and Smith-Carrington (1980) demonstrate that significant errors can be introduced in this manner and emphasize the need to use local tritium rainfall data in any mass balance study.

While there is no isotopic enrichment of water during transpiration through plants (Zimmerman, Ehhalt, and Munnich 1966), a degree of fractionation involving preferential evaporation of non-tritiated water occurs during evaporation from bare soil. McFarlane, Rodgers, and Bradley (1978) also point out that rainfall is not the only source of tritium entering the ground in that soil micro-organisms can take up tritium directly from the atmosphere and introduce it into the soil water. These effects, however, are likely to be small.

Once the tritium enters the ground it is generally considered to move within the water molecule (as does $^{1}H$) and not to enter preferentially into any hydrochemical process. This would not appear to be completely true. The degree to which modifications of tritium concentration occur in quantifiable terms, however, is not easy to assess. Matthess, Pekdeger, Schultz, and Bauet (1979) record reductions in soil moisture tritium due to exchange with low tritiated interstitial water in simple intergranular permeability materials. Under certain conditions, such as the unsaturated zones of the British Chalk aquifer where the matrix has a high porosity and low permeability and where very high pore water retention occurs, exchange due to diffusion gradients between infiltrating and stored water can cause marked changes in tritium concentrations in the infiltrating waters. Kaufmann and Todd (1962) and Stewart (1967) have demonstrated that tritium can be sorbed on clay minerals with a resultant reduction in the concentration in the infiltrating water; again such reductions are likely to be small.

In infiltration experiments tritium has been shown to have similar concentration profiles to nitrate, so it has been examined as a means of assessing rates of nitrate movement in the unsaturated zone and the ensuing nitrate pollution potential. Unfortunately, conclusive results have not as yet been forthcoming because, although many of the controls on tritium movement are understood, the inhomogeneity of the unsaturated zone lithologies do not easily allow the application of simplistic models to obtain detailed tritium profile simulations.

Recharge estimates using tritium have been applied in a number of countries where more rigorous methods cannot be applied through lack of hydrological data. Whereas hydrological analyses of recharge input are moving towards daily calculations, tritium recharge assessments relate to annual input. A number of approaches have been adopted including those of Brown (1960), Allison and Holmes (1973), and Sukhifa and Shah (1976). The models assume simple unconfined shallow homogeneous aquifers with complete mixing under steady state conditions. Allison and Holmes provide a reasonable model as depicted in Fig. 8.7. Flow past any section of unit thickness is given as

$$-K_h z \frac{dz}{dx} = qx \tag{8.29}$$

where $q$ is the annual recharge and $K_h$ is the hydraulic conductivity. Integration of (8.29) gives

$$z^2 + \frac{qx^2}{K_h} = z_m^2. \tag{8.30}$$

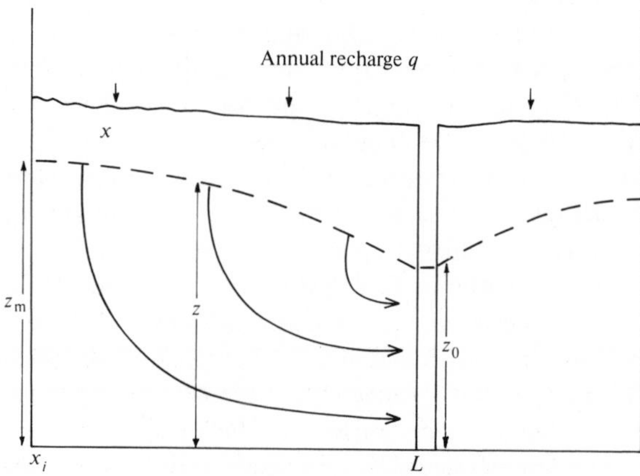


**Fig. 8.7.** Geometry of the system used to express the tritium recharge balance.

The laminar flow velocity $u$ can be expressed as

$$\phi u = K_h \frac{dz}{dx} \tag{8.31}$$

where $\phi$ is the porosity, and, if it is assumed that $u$ is independent of $z$, the substitution of $dx/dt = u$ in eqn (8.31) and combination with eqn (8.29) gives

$$dt = \frac{\phi z \, dx}{qx}. \tag{8.32}$$

If the time taken for a particle of water to move from $x = x_i$ to $x = L$ (Fig. 8.7) is $t_i$, then

$$t_i = \frac{\phi}{q} \int_{x_i}^{L} \frac{z\,dx}{x} \tag{8.33}$$

which gives on integration

$$t_i = \frac{z_m \phi}{q} \left\{ \frac{z_0 - z_i}{z_m} + \ln\left(\frac{z_m + z_i}{z_0 + z_m}\right) + \ln \frac{L}{x_i} \right\} \tag{8.34}$$

Combining this equation with the radioactive law of decay

$$T_{t_i} = T_0 \exp(-t_i\lambda) \tag{8.35}$$

where $T_0$ is the initial tritium concentration input, $T_{t_i}$ is the tritium concentration after time $t_i$, and $\lambda = 0.0565$ year$^{-1}$ is the decay constant for tritium, gives the tritium concentration $T$ for flow in $n$ throughput paths as

$$T = \frac{1}{n} \sum_{i=1}^{n} T_0 \exp\left[ \frac{-z_m \phi\lambda}{q} \left\{ \frac{z_0 - z_i}{z_m} + \ln\left(\frac{z_m + z_i}{z_0 + z_m}\right) + \ln \frac{L}{x_i} \right\} \right]. \tag{8.36}$$

As Allison and Holmes (1973) have demonstrated, if $T_0$ is known for an area the tritium load can be distributed with respect to porosity for the calculation of mean annual recharge. In practical groundwater resources terms the method provides a means of assessing overall recharge under very specific groundwater conditions and when hydrogeological data are sparse; unfortunately where such a dearth of data occurs it is also unlikely that tritium data, particularly long-term tritium rainfall data, will be available.

Although under specific conditions tritium has been shown to provide meaningful information, it has never really fulfilled its expected potential as a dating tool (Nir 1964). As thermonuclear explosions decrease so will the value of tritium as a signature. In real groundwater terms the value of tritium lies in its verification or otherwise of modern recharge. While this is useful in all climatic zones, it is perhaps of most value in the semi-arid and arid areas where recharge can be a matter of pure conjecture. Normally it is accepted that if a concentration of 4 TU or greater is present in a groundwater system modern recharge has occurred. This would be justified under the simplistic hydrogeological conditions discussed above for tritium balance calculations, but unfortunately lithological and hydraulic layering effects are so dominant in groundwater bodies that without very careful tritium sampling from discrete aquifer zones the data can be totally misleading. Where tritium samples are taken from pumping wells mixing may frequently reduce tritium concentrations and mask the presence of modern recharge water.

The tritium concentrations in groundwaters from a sandstone aquifer in southern Jordan (Fig. 8.8(a)) indicate that although recharge is occurring it is localized and relates to recharge from run-off. A more regional tritium distribution in an arid area of Saudi Arabia (Fig. 8.8(b)) shows modern recharge influences. A good example of a semi-arid area where recharge assessment poses a problem is provided by the Kalahari Desert. Although there is much scepticism about recharge because of the presence of thick sand layers and evaporation removing infiltration water in the Kalahari, Mazor (1982) and Foster, Bath, Farr, and Lewis (1982) have recorded local soil moisture at depths greater than 6m, and groundwater tritium concentrations of 10–24 TU, indicating that without doubt modern recharge is active despite indications to the contrary by classical soil moisture balance calculations.

In such cases as those quoted above the value of tritium is purely qualitative but it does help in pointing to areas where recharge may occur which subsequently allows the mechanisms to be determined.

## 8.5. Oxygen isotopes

### *8.5.1. Standards*

Three isotopes of oxygen are found naturally. Their abundances are given in Table 8.7. The isotope $^{17}O$ is not normally measured in hydrochemical studies

**Table 8.7** *Abundance of oxygen isotopes*

| Isotope | Average terrestrial abundance (%) |
|---|---|
| $^{16}O$ | 99.76 |
| $^{17}O$ | 0.037 |
| $^{18}O$ | 0.2 |

because in almost all cases $\delta^{17}O = 0.5\ \delta^{18}O$ and in addition the measurement is technically difficult.

Three internationally agreed standards are available at present and a fourth is in the process of calibration. The primary oxygen standard, distributed by the International Atomic Energy Agency, is the water standard V-SMOW (Vienna Standard Mean Ocean Water) whose isotopic composition is close to that of average ocean water. The absolute isotope ratio of V-SMOW has been measured by Baertschi (1976): $^{18}O/^{16}O = (200\ 5.2 \pm 0.45) \times 10^{-1}$. Gas for analysis is prepared by equilibrating carbon dioxide with the water at 25 °C, the fractionation factor being $\alpha_{CO_2-H_2O} = 1.0412$ (Friedman and O'Neil 1977). The range of $^{18}O/^{16}O$ ratios encountered in natural waters is rather wide and it is therefore useful to have secondary standards spanning the range since the precision of a measurement decreases as the difference between the sample and the standard

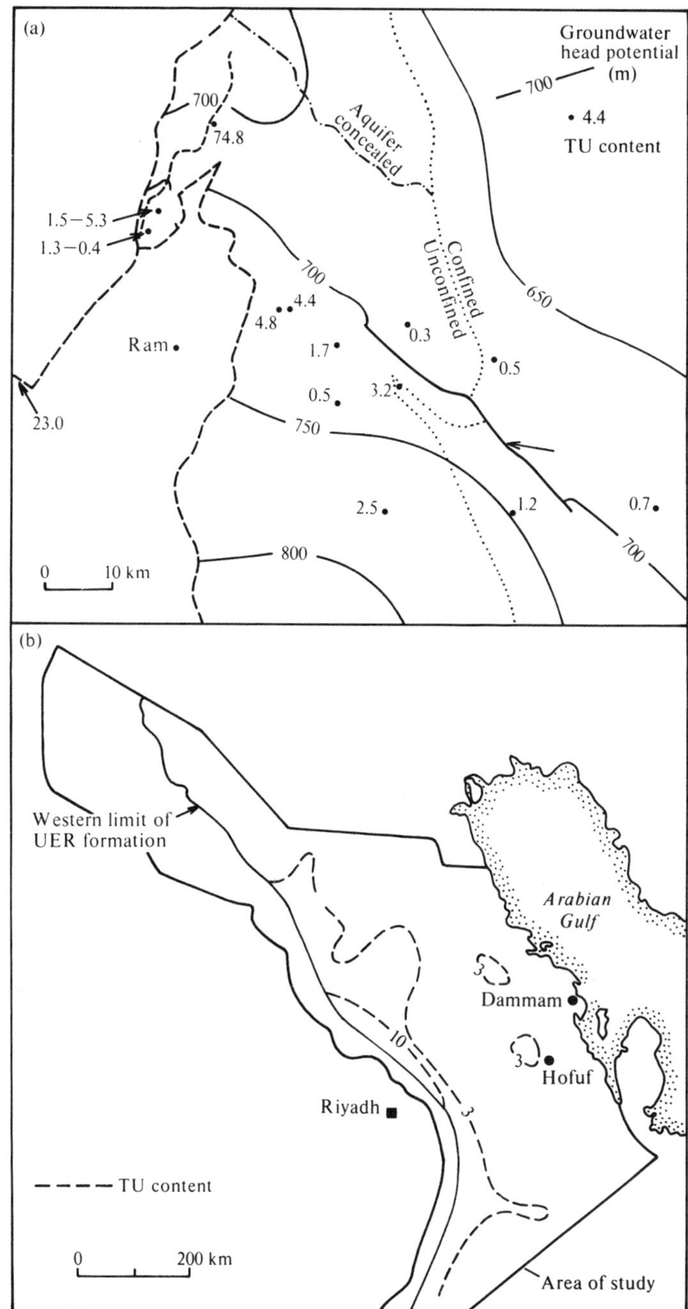


**Fig. 8.8.** Tritium distributions in groundwater in arid areas: (a) typical distribution indicating recharge from localized run-off in sandstones in southern Jordan (after Lloyd 1980); (b) contoured tritium concentrations from the Upper Umm Er Rhaduma (UER) calcareous aquifer of central Saudi Arabia (after Bakiewicz, Milne, and Noori 1982).

increases. For this reason, the International Atomic Energy Agency also distributes SLAP (Standard Light Antarctic Precipitation), where $\delta^{18}O_{SLAP/V\text{-}SMOW}$ = −55.5 parts per thousand (Gonfiantini 1978). A further secondary standard **GISP** (Greenland Ice Sheet Precipitation), is in preparation ($\delta^{18}O_{GISP/V\text{-}SMOW} \approx$ −25 parts per thousand). Results are quoted normalized to the V-SMOW/SLAP scale.

The other standard is PDB calcite (see Section 8.3.1), from which carbon dioxide is prepared by reaction with 100 per cent $H_3PO_4$. Because only two of the three oxygen atoms in calcite appear in the carbon dioxide, there is a fractionation of oxygen between calcite and carbon dioxide. The accepted value of the fractionation factor at 25 °C (Sharma and Clayton 1965) is $\alpha_{CO_2\text{-calcite}}$ = 1.01025. The PDB oxygen standard is primarily used in palaeotemperature studies of carbonates and is little used in hydrochemical studies.

The V-SMOW and PDB oxygen scales are interrelated by

$$\delta_{V\text{-}SMOW} = 1.03086\,\delta_{PDB} + 30.86 \tag{8.37}$$

$$\delta_{PDB} = 0.97006\,\delta_{V\text{-}SMOW} - 29.94. \tag{8.38}$$

### 8.5.2. $^{18}O$ *and deuterium in precipitation*

The oxygen and hydrogen isotope concentrations in precipitation are to a considerable extent controlled by vapour pressures. The vapour pressures of the various isotopic molecules of water are inversely proportional to their masses. Therefore $H_2{}^{16}O$ has a significantly higher vapour pressure than $D_2{}^{18}O$, $D_2{}^{17}O$, $D_2{}^{16}O$, $H_2{}^{18}O$, etc.

The difference in the vapour pressures is described by the fractionation factor $\alpha^1$ (Dansgaard 1964; Faure 1977). In oxygen isotopes for example

$$\alpha^1 = P_L/P_S \tag{8.39}$$

(for equilibrium between vapour and liquid) where $P_L$ is the vapour pressure of the light component ($H_2{}^{16}O$), $P_S$ is the vapour pressure of the heavy component ($H_2{}^{18}O$), and $P_L > P_S$. If the relative concentration of the heavy component is $a_d$ in a vapour and $a_f$ in a liquid, then the concentration of the lighter component is $1 - a_d$ in the vapour and $1 - a_f$ in the liquid.

If both the vapour and liquid are assumed to be ideal mixtures, the partial pressure of the heavy component is

$$a_d P = a_f P_S \tag{8.40}$$

where $P$ is the total vapour pressure. The partial pressure of the light component is

$$(1 - a_d)P = (1 - a_f)P_L. \tag{8.41}$$

Since both $a_f$ and $a_d$ are much less than unity $P = P_L$. Therefore from eqn (8.40)

$$a_f/a_d = P_L/P_S = \alpha^1. \tag{8.42}$$

The fractionation factor is temperature dependent and tends towards unity with increasing temperature. At 20 °C under equilibrium conditions the following values apply for $^{18}O$–$^{16}O$ and D–$^{1}H$:

| | $^{18}O$ | D |
|---|---|---|
| $\alpha^1$ | 1.0092 | 1.079. |

Therefore the relationship between $\delta^{18}O$ and $\delta D$ has a gradient of 79/9.2 which is approximately 8.

In practice equilibrium conditions are realized at a phase boundary if the process of exchange proceeds slowly. If non-equilibrium conditions exist, a kinetic fractionation occurs which is a consequence of different rates of molecular transport for the different components. The corresponding fractionation factor depends upon the square root of the molecular weight ratio (Craig and Gordon 1965). It is also sensitive to the relative humidity of the atmosphere (Ehhalt and Knott 1965). Kinetic fractionation influences $^{18}O$ and D so that at 20 °C under non-equilibrium conditions the following values apply at 30 per cent relative humidity:

| | $^{18}O$ | D |
|---|---|---|
| $\alpha^1$ | 1.018 | 1.087. |

This produces a relationship between $\delta^{18}O$ and $\delta D$ with a gradient of 87/18 which is approximately 5.

Water vapour which produces precipitation is chiefly derived from the ocean. When condensation occurs to form precipitation the isotopic concentration changes according to what is known as the Rayleigh process. The condensate (precipitation) is immediately removed from the vapour which becomes isotopically lighter producing raining out of isotope in a cloud mass as it crosses a continent for example (Fig. 8.9(a)). Kinetic effects are demonstrated in Fig. 8.9(b) showing that in fast evaporation the vapour becomes isotopically lighter than in the equilibrium Rayleigh process and the residual fluid becomes more enriched in isotopes. Clearly this has most significance in hydrogeology where recharging waters are subject to evaporation.

The equilibrium process controlling $^{18}O$ and D in precipitation originating from a large ocean mass has been confirmed by Craig (1961) who has defined a world meteoric water relationship of

$$\delta D = 8\delta^{18}O + 10. \qquad (8.43)$$

Where sea water tends to be restricted by land masses, a different precipitation relationship may exist, as is the case in the Mediterranean Sea:

$$\delta D = 8\delta^{18}O + 22. \qquad (8.44)$$

Such a relationship shows equilibrium transfer processes but indicates a different initial sea isotope composition from the large oceans.

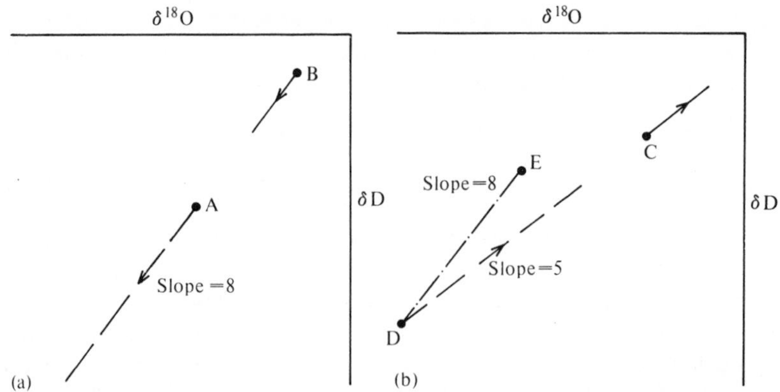


**Fig. 8.9.** $^{18}$O/D relationships due to fractionation by (a) the Rayleigh process and (b) kinetic effects (evaporation): A, original vapour composition; B, first condensate; C, original composition of the water; D, first small amount of vapour; E, first condensate from D; – – –, vapour line: ——, condensate line; — · —, effect of equilibrium fractionation.

As temperature is a dominant effect on isotopic fractionation, variations in isotopic concentration in precipitation can relate to climate and altitude. The initial vapour source can be important, as can the distance of travel of vapour over major land masses. Because of the atmospheric complexities related to precipitation, isotope concentrations at any one locality can vary significantly and do not necessarily correlate with precipitation amount (Fig. 8.10).

In climatic areas where rapid and high levels of evaporation occur rainfall can evaporate as it descends through the atmosphere under what is termed virgis effects. This is most significant in low intensity rainfall so that in desert rainfall one can occasionally see evaporative influences as shown in Fig. 8.11 for the Jordan desert. In the southern hemisphere, Hughes and Allison (1982) record that Australian rainfall accords with the Craig meteoric lines but that deuterium depletion is evident in summer rainfall.

Because evaporation effects are severe in arid climates, difficulties exist in accurate sampling of the stable isotopes in precipitation and indeed in run-off. Rainfall can evaporate on a rain gauge surface and within containers, while flood run-off samples need protection. The types of installations that have been found to be efficient are shown in Fig. 8.12 (Levin, Gat, and Issar 1980).

### *8.5.3. $^{18}$O and D in groundwater*

Although quite marked variations occur in stable isotope concentrations in precipitation the limited data available indicate that where direct recharge is significant $^{18}$O/D ratios in recharging waters approximate to the average weighted precipitation ratios. As a result, in many areas groundwater isotopic ratios tend to correlate closely with precipitation and thus provide little information. This is particularly so in western continental environments. In groundwater

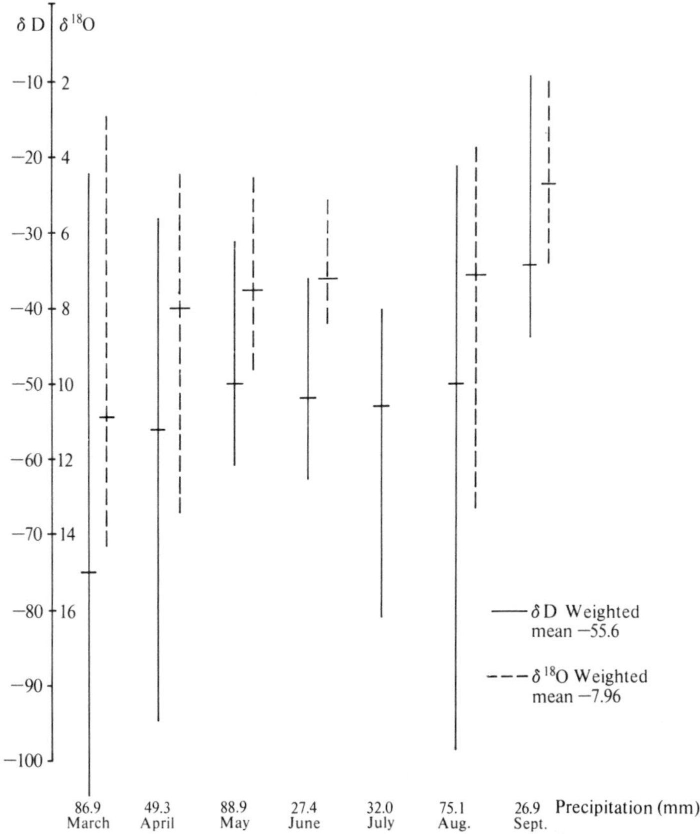


**Fig. 8.10.** Example of monthly D and $^{18}O$ concentrations (in parts per thousand) in precipitation from Lincolnshire, eastern England.

resources terms in such areas, the stable isotopes are therefore of limited value. They do, however, provide some information with respect to climatic controls as for example is shown in Fig. 8.13 for the Triassic Sandstone aquifer of Merseyside in western England (see Section 9.3.3). Deep saline groundwaters in the aquifer have relatively negative stable isotope concentrations indicating recharge under colder conditions than the modern fresh groundwaters. Their position within the aquifer is consistent with their age being older than the fresh groundwaters; however, the isotopic composition does not necessarily imply anything about the origin of the salinity. The stable isotope information in this case is interesting but not very positive.

Many attempts have been made to relate $^{18}O$ and D to groundwater temperature and thence age and to broaden the understanding of past recharge mechanisms. Dansgaard (1964) derived the empirical relationship

$$\delta^{18}O = 0.69\ T_a - 13.6 \text{ parts per thousand} \tag{8.45}$$

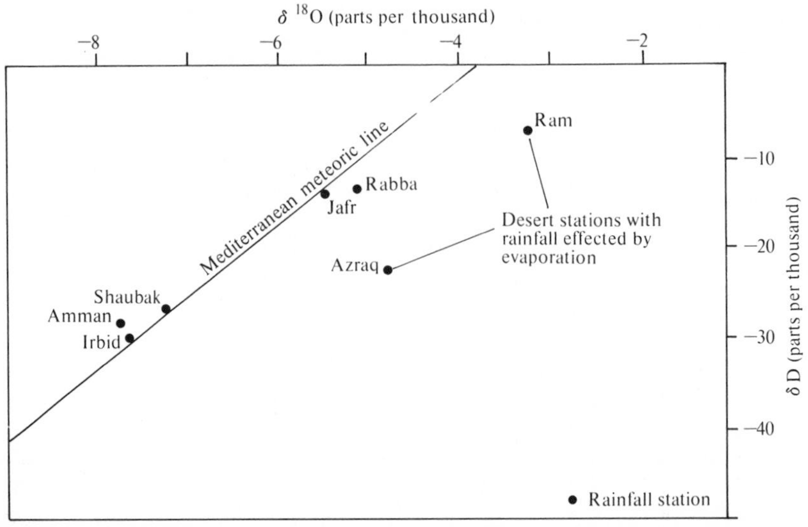


**Fig. 8.11.** $^{18}$O and D relationships for precipitation in the eastern Mediterranean and the Jordan Desert.

where $T_a$ is mean air temperature in degrees Celsius, to provide a relationship between $\delta^{18}$O and temperature for North Atlantic areas. This equation has been applied, for example, to data from the Chalk aquifer of the London Basin (Smith, Downing, Monkhouse, Otlett, and Pearson 1976) and gives a mean recharge temperature range of 1 °C spanning a 20 000 year period. As the $\delta^{18}$O range is −7.1 to −7.9 parts per thousand and bearing in mind a measurement error limit of ± 0.2 parts per thousand such calculations appear to be of little value.

Evans, Otlet, Downing, Monkhouse, and Rae (1978) have revised the Dansgaard relationship to provide the following equation for north-west European stations:

$$\text{mean annual} \quad (T_a\ -20\ ^\circ\text{C to } 18\ ^\circ\text{C})\ \delta^{18}\text{O} = 0.23T_a - 8.62 \tag{8.46}$$

$$\text{mean winter} \quad (T_a\ -20\ ^\circ\text{C to } 14\ ^\circ\text{C})\ \delta^{18}\text{O} = 0.22T_a - 8.26 \tag{8.47}$$

While eqns (8.46) and (8.47) may be more accurate than eqn (8.45) their application in terms of past recharge temperatures does not appear very meaningful.

Although it can be concluded that $^{18}$O and D do not provide much useful information for resources where direct recharge occurs and has been a long-term feature, in semi-arid and arid areas the stable isotope application may be more worthwhile.

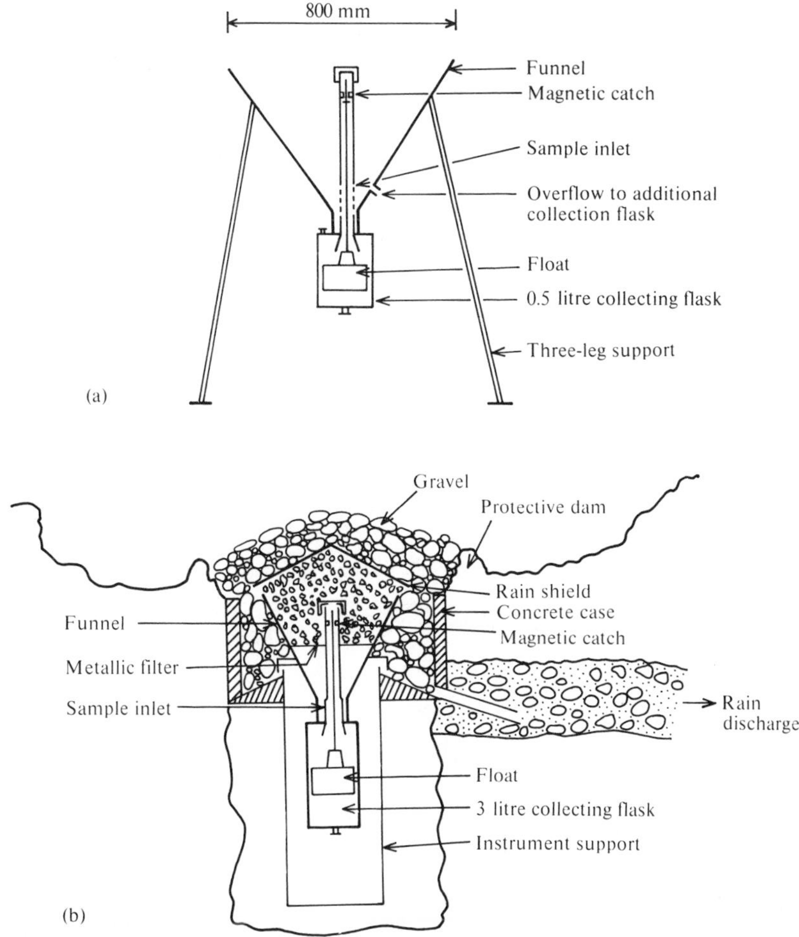


**Fig. 8.12.** Gauges for collecting isotopes in (a) precipitation and (b) flood run-off. (After Levin *et al.* 1980.)

Two aspects of $^{18}O$ and D can be studied to limited advantage in semi-arid and arid hydrogeology:

(i) the implication of the isotopes with respect to modern recharge;
(ii) the inferences of the isotopes with respect to overall resources and long-term recharge.

Particularly under arid circumstances recharge is very intermittent and almost exclusively indirect in character, resulting from run-off. Quantitative assessments pose considerable problems and, where feasible, consist of measuring river bed losses between flood gauging stations (Lloyd 1981). In a desert environments

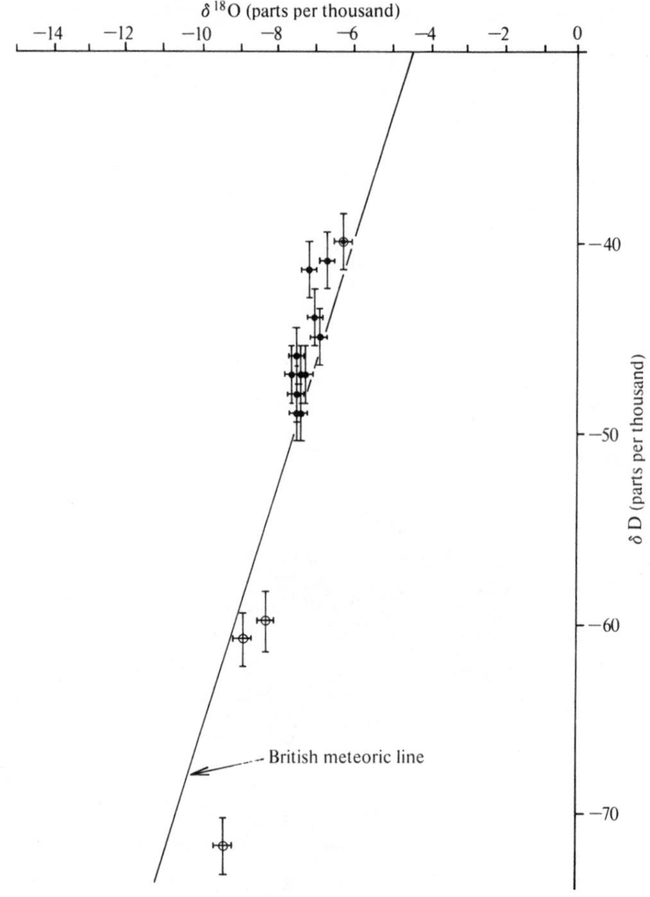


**Fig. 8.13.** $^{18}O/D$ relationship for groundwaters from the Triassic Sandstone aquifer in the Lower Mersey Valley, England: ○, saline groundwaters; ●, fresh groundwaters.

where water enters the ground in such a manner it can be subject to evaporation from the wadi bed. In Fig. 8.14 flood water in the Negev (Levin *et al.* 1980) is compared with modern recharge groundwater in the adjacent part of Jordan (Lloyd 1980). The groundwaters, which are highly tritiated (up to 75 TU), however, show distinct displacement from the flood waters and demonstrate distinct evaporative affects which must be taken into account in recharge calculations.

The problem of modern recharge sustaining groundwater flow and gradients in the major arid area sedimentary basins is a difficult hydrogeological problem and has been discussed at length by Burdon (1977), Lloyd and Farag (1978), and Bakiewicz *et al.* (1982). $^{14}C$ ages suggest that modern recharge is very limited and this concept is supported to some extent by $^{18}O$ and D data.

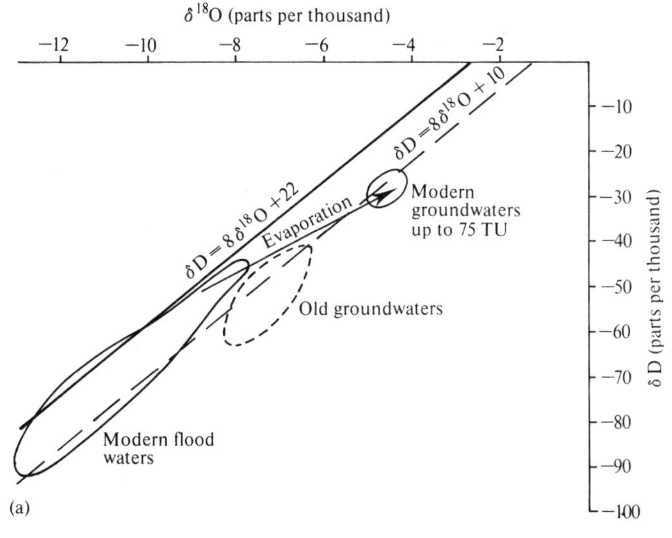


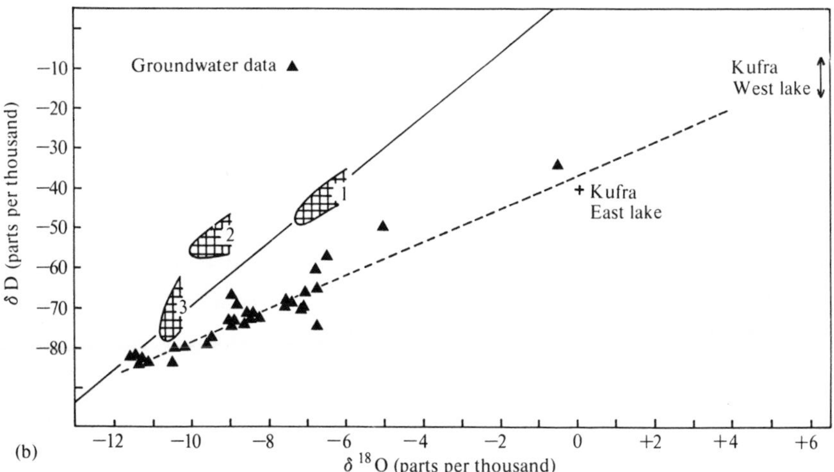


**Fig. 8.14.** (a) Stable isotopic composition of groundwaters in southern Jordan: (b) stable isotopic composition of groundwaters from the Kufra and Sirte basins of Libya with surface data from (1) Tunis, (2) Fort Lamy and Kano, and (3) Bamako. (After Edmunds and Wright 1979.)

Figure 8.14(a) shows that old groundwaters with $^{14}C$ ages of 5000–30 000 years in southern Jordan plot distinctly away from the modern flood waters and the Mediterranean meteoric line. The groundwaters adhere more to the world meteoric line, suggesting that their recharge was associated with storms tracking through the area from either the Atlantic or the Indian Ocean during the late Pleistocene and early Holocene. A similar situation is shown in Fig. 8.14(b) from

Libya where groundwaters up to 30 000 years old from the Kufra and Sirte basins relate to an evaporation line with the slope $\delta D = 4.5\ \delta^{18}O - 35$. Comparison with present-day rainfall data from North African stations suggests that the groundwater recharge source was from isotopically more negative rainfall. The interception of the evaporation line on the meteoric line is at $\delta^{18}O = -12$ parts per thousand and $\delta D = -88$ parts per thousand, indicating a probably colder precipitation than today. A close relationship to present rainfall exists with Bamako, which may indicate that the groundwater recharge originated from monsoonal storms tracking northwards from the Gulf of Guinea.

In both the cases given above, the stable isotopes provide substantive data for the $^{14}C$ isotope interpretation. The stable isotope data tend to confirm that modern recharge is limited but in no way do they provide conclusive evidence.

In understanding the arid zone flow system, the discharge conditions are equally as important as the recharge condition. In some systems discharge zones occur in the form of sabkhas from which the discharge usually occurs as evaporation. Quantification of the evaporation is difficult and has been attempted using energy budgets (Idrotecneco 1979). Recent work has shown that stable isotope profiles can be modelled and interpreted to provide evaporation estimates (Allison 1982).

## 8.6. Other isotopes

As can be deduced from Table 4.7 a number of other environmental isotopes are under study. A certain amount of intriguing data are available from such studies; however, as yet the results are even less conclusive than the more conventional isotope results discussed above and are unlikely to have much real impact on groundwater resource studies for some considerable time. The reader is referred to Andrews and Kay (1978), Osmond and Cowart (1982), and Rightmire, Pearson, Bach, Rye, and Hanshaw (1974).

## References


Allison, G. B. (1982). The relationship between $^{18}O$ and deuterium in water in sand columns undergoing evaporation. *J. Hydrol.* **55**, 163–9.

— and Holmes, J. W. (1973). The environmental tritium concentration of underground water and its hydrological interpretation. *J. Hydrol.* **19**, 131–43.

Andrews, J. N. and Kay, R. L. F. (1978). The evolution of enhanced $^{234}U/^{238}U$ activity ratios for dissolved uranium in groundwater dating. In R. E. Zartman (ed.), *Short papers of the 14th Int. Conf. on Geochronology, Cosmochronology, and Isotope Geology, Colarado. U.S. Geol. Sur. Open File Rep 78-701, 11-13.*

Baertschi, P. (1976). Absolute $^{18}O$ content of standard mean ocean water. *Earth Planet. Sci. Lett.* **31**, 341–4.

Bakiewicz, W., Milne, D. M., and Noori, M. (1982). Hydrogeology of the Umm Er Rhaduma aquifer, Saudi Arabia, with reference to fossil gradients. *Q. J. eng. Geol.* **15**, 105–26.

Begemann, F. and Libby, W. F. (1957). Continental water balance, groundwater inventory and storage times, surface ocean mixing rates and worldwide circulation patterns from cosmic ray and bomb tritium. *Geochim. cosmochim, Acta* **12**, 277–87.

Bredenkamp, D. B., Schutte, J. M., and Du Toit, G. J. (1974). Recharge of a dolomite aquifer as determined by tritium profile. *Proc. Symp. on Isotope Hydrology, International Atomic Energy Agency, Vienna, pp. 65–75.*

Brown, R. M. (1960). Hydrology of tritium in the Ottawa valley. *Geochim. cosmochim. Acta* **21**, 199–216.

Burden, D. J. (1977). Flow of fossil groundwater. *Q. J. eng. Geol.* **10**, 97–124.

Craig, H. (1957). Isotopic standards for carbon and oxygen and correction factors for mass spectrometric analysis of carbon dioxide. *Geochim cosmochim. Acta* **12**, 133–49.

— (1961). Isotopic variations in meteoric water, *Science* **133**, 1702–3.

— and Gordon, L. (1965). Deuterium and oxygen-18 variations in the ocean and marine atmosphere. *Proc. Congr. on Stable Isotopes in Oceanography Studies and Palaeotemperatures.* pp. 9–13. Lab. Geol. Nucl., Pisa, Italy.

Dansgaard, W. F. (1964). Stable isotopes in precipitation. *Tellus* **16**, 436–49.

Degens, E. T. (1967). Stable isotope distribution in carbonates. In G. V. Chilinger, H. J. Bessall, and R. W. Fairbridge (eds.), *Developments in sedimentology; carbonate rocks*, pp. 193–208. Elsevier, Amsterdam.

Deines, P. (1980). The isotopic composition of reduced organic carbon. In *Handbook of environmental isotope chemistry* (ed. P. Fritz and J. Ch. Fontes), Vol. 1, *The terrestrial environment*, pp. 329–406. Elsevier, Amsterdam.

—, Langmuir, D., and Harmon, R. S. (1974). Stable carbon isotope ratios and the existence of a gas phase in the evolution of carbonate groundwaters. *Geochim. Cosmochim. Acta* **38**, 1147–64.

Edmunds, W. M. and Wright, E. P. (1979). Groundwater recharge and palaeoclimate in the Sirte and Kufra Basins, Libya. *J. Hydrol.* **40**, 215–41.

Ehhalt, D. and Knott, K. (1965). Kinetische isotopentrennung bei der Verdampfung von Wasser. *Tellus* **17**, 389–97.

Evans, G. V., Otlet, R. L., Downing, R. A., Monkhouse, R. A., and Rae., G. (1978). Some problems in the interpretation of isotope measurements in British aquifers, *Proc. Symp. on Isotope Hydrology, Neuherberg*, pp. 679–708. *Rep. SM-228/34* (International Atomic Energy Agency, Vienna).

Faure, G. (1977) *Principles of isotope geology*. Wiley, New York.

Fontes, J. C. and Garnier, J. M. (1979). Determinations of the initial activity of the total dissolved carbon: a review of existing methods and a new approach. *Water Resources Res.* **15**, 399–413.

Foster, S. S. D., Bath, A. H., Farr, J. L., and Lewis, W. J. (1982). The likelihood of active groundwater recharge in the Botswana Kalahari. *J. Hydrol.* **35**, 113–36.

— and Smith-Carrington, A. (1980). The interpretation of tritium in the Chalk unsaturated zone. *J. Hydrol.* **46**, 343–64.

Friedman, I. and O'Neil, J. R. (1977). Compilation of stable isotope fractionation factors of geochemical interest. In M. Fleischer (ed.), *Data of Geochemistry, U.S. Geol. Surv. Prof. Pap. 440-KK*, pp. 1–12.

Fritz, P. and Fontes, J. Ch. (1980). *Handbook of environmental chemistry*, Vol. 1, *The terrestrial environment*. Elsevier, Amsterdam.

Galimov, E. M. (1966). Carbon isotopes of soil $CO_2$. *Geochem. Int.* **3**, 889–97.

Geyh, M. A. (1970). Carbon-14 concentration of lime in soils and aspects of the carbon-14 dating of groundwater. *Proc. Symp. on Isotopes in Hydrology*, pp. 215–23. International Atomic Energy Agency, Vienna.

Gonfiantini, R. (1978). Standards for stable isotope measurements in natural compounds. *Nature (Lond.)* **271**, 534–5.

Hageman, R., Neif, G., and Roth, E. (1970). Absolute isotopic scale for deuterium analysis of natural water. absolute D/H ratio for SMOW. *Tellus* **22**, 712–15.

Heathcote, J. A. (1981). Hydrochemical aspects of the Gipping Chalk. Salinity investigation. Ph.D. Thesis. Department of Geological Sciences, University of Birmingham.

Hendy, C. H. (1971). The isotopic chemistry of speleothems. *Geochim. Cosmochim. Acta* **35**, 801–24.

Hughes, M. W. and Allison, G. B. (1982). Stable isotope concentrations in Australian rainfall. *Australian Acad. Sci. Workshop on Stable Isotopes in the Environment, Canberra*. Australian Academy of Sciences, Melbourne.

Idrotecneco (1979). Hydrogeology study of Wadi Ash Shati Al Jufrak and Jabal Fezzan area, Libya. *Report*, Libyan Secretariat of Dams and Water Resources, Tripoli.

Kaufmann, S. and Libby, W. F. (1954). The natural distribution of tritium. *Phys. Rev.* **93**, 1327–43.

Kaufmann, W. J. and Todd, D. K. (1962). Application of tritium tracer to canal seepage measurements. *Proc. Symp. on Tritium in the Biological and Physical Sciences*, pp. 84–9. International Atomic Energy Agency, Vienna.

Levin, M., Gat, J. R., and Issar, A. (1980). Precipitation, flood and groundwaters of the Negev highlands: An isotopic study of desert hydrology. *Proc. Advisory Group Meeting, Arid Zone Hydrology, Investigations with Isotope Techniques. Rep. Ag-158/1*, pp. 3–22. International Atomic Energy Agency, Vienna.

Libby, W. F. (1953). The potential usefulness of natural tritium. *Proc. natl. Acad. Sci. U.S.* **35**, 245–7.

Lloyd, J. W. (1980). Aspects of environmental isotope chemistry in groundwaters in eastern Jordan. *Proc. Advisory Group Meeting, Arid Zone Hydrology, Investigations with Isotope Techniques, Rep. AG-158/14*, pp. 193–204. International Atomic Energy Agency, Vienna.

— (1981). A review of groundwater recharge assessment. *Proc. Groundwater Recharge Conf., Australian Water Resources Council Conf. Ser.* **3**, 1–25.

— and Farag, M. H. (1978). Fossil groundwater gradients in arid regional sedimentary basins. *Groundwater* **6**, 59–68.

—, Harker, D., and Baxendale, R. A. (1981). Recharge mechanisms and groundwater flow in the Chalk and drift deposits of East Anglia. *Q. J. eng. Geol.* **14**, 87–96.

McFarlane, J. C., Rodgers, R. D., and Bradley, D. V. (1978). Environmental tritium oxidation in surface soil. *Environ. Sci. Technol.* **12**, 590–3.

Matthess, G. A. Pekdeger, H. D., Schultz, H. R., and Bauet, W. (1979). Tritium tracing in hydrogeological studies using model lysimeters. Proc. Symp. on *Isotope Hydrology*, Vol. 2, pp. 769–85. International Atomic Energy Agency, Vienna.

Mazor, E. (1982). Rain recharge in the Kalahari: a note on some approaches to the problem. *J. Hydrol.* **55**, 137–44.

Mook, W. G. (1976). The dissolution exchange model for dating groundwater with $^{14}$C. *Interpretation of environmental isotope hydrochemical data in groundwater*, pp. 213–25. International Atomic Energy Agency, Vienna.

— (1980). Carbon-14 in hydrogeological studies. In *Handbook of environmental isotope geochemistry* (ed. P. Fritz and J. Ch. Fontes), Vol. 1, *The terrestrial environment*, pp. 49–74. Elsevier, Amsterdam.

Nir, A. (1964). On the interpretation of tritium age measurements of groundwater. *J. geophys. Res.* **69**, 2589–95.

Osmond, J. K. and Cowart, J. B. (1982) Groundwater. In M. Ivanovich and R. S. Harmon (eds.), *Uranium series disequilibrium applications and environmental problems*, pp. 202–45. Clarendon, Oxford.

Pearson, F. J. (1965). Use of $^{13}C/^{12}C$ ratios to correct radiocarbon ages of material initially diluted by limestone. *Proc. 6th int. Conf. on Radiocarbon and Tritium Dating.* pp. 357–64. *Pulmann.*

—and Hanshaw, B. B. (1970). Sources of dissolved carbonate species in groundwater and their effects on carbon-14 dating. *Proc. Symp. on Isotope Hydrology*, pp. 271–86. International Atomic Energy Agency, Vienna.

— and Swarzenki, W. V. (1974). $^{14}C$ evidence for the origin of arid region groundwater, Northwestern Province, Kenya. *Isotope Techniques in Groundwater Hydrology*, Vol. 1, pp. 95–108. International Atomic Energy Agency, Vienna.

Reardon, E. J. and Fritz, P. (1978). Computer modelling of groundwater $^{13}C$ and $^{14}C$ isotope composition. *J. Hydrol.* **36**, 201–224.

Rightmire, C. T., Pearson, F. J., Bach, W., Rye, R. O., and Hanshaw, B. (1974). Distribution of sulphur isotopes of sulphates in groundwater from the principal artesian aquifer of Florida and the Edwards aquifer of Texas, U.S.A. *Proc. Symp.*, pp. 32–41. International Atomic Energy Agency, Vienna.

Sharma, T. and Clayton, R. N. (1965). Measurement of $^{18}O/^{16}O$ ratios of total oxygen of carbonates. *Geochim. Cosmochim. Acta* **25**, 1347–53.

Smith, D. B., Downing, R. A., Monkhouse, R. A., Otlett, R. L., and Pearson, F. J. (1976). The age of groundwater in the Chalk of the London Basin. *Water Resources Res.* **12**, 392–404.

—, Wearn, P. L., Richards, H. J., and Rowe, P. C. (1970). Water movement in the unsaturated zone of high and low permeability strata by measuring natural tritium. *Proc. Symp. on Isotope Hydrology*, pp. 73–87. International Atomic Energy Agency, Vienna.

Stewart, G. L. (1967). Fractionation of tritium and deuterium in soil water. *Proc. Am. Geophys. Union Symp. on Isotope Techniques in the hydrologic Cycle, Geophys. Monogr., Am. Geophys. Union* **11**, 159–68.

Sukhifa, B. S. and Shah, C. R. (1976). Conformity of groundwater recharge rate by tritium method and mathematical modelling. *J. Hydrol.* **30**, 167–78.

Tamers, M. A. and Scharpenseel, H. W. (1970). Sequential sampling of groundwater. *Proc. Symp. on Isotope Hydrology*, pp. 241–56. International Atomic Energy Agency, Vienna.

Von Buttlar, H. and Libby, W. F. (1955). Natural distribution of cosmic-ray produced tritium. *J. inorg. nucl. Chem.* **1**, 75–87.

Wallick, E. I. (1976). Isotopic and chemical considerations in radiocarbon dating of groundwater within the semi-arid Tucson Basin, Arizona. *Interpretation of environmental isotope and hydrochemical data in groundwater hydrology*, pp. 15–24. International Atomic Energy Agency, Vienna.

Wendt, I., Stahl, W., Geyh, M., and Fauth, F. (1967). Model experiments for $^{14}C$ water-age determinations. *Proc. Symp. on Isotopes in Hydrology*, pp. 321–37. International Atomic Energy Agency, Vienna.

Wigley, T. M. L. (1976). Effect of mineral precipitation on isotopic composition and $^{14}C$ dating of groundwater. *Nature (Lond.)* **263**, 219–21.

—, Plummer, L. N., and Pearson, F. J. (1978). Mass transfer and carbon isotope evolution of natural water systems. *Geochim. Cosmochim. Acta* **42**, 1117–39.

Zimmerman, U., Ehhalt, D. H., and Munnich, K. O. (1966). Soil–water movement and evapotranspiration: changes in the isotopic composition of the water. *Proc. Symp. on Isotopes in Hydrology*, pp. 567–85. International Atomic Energy Agency, Vienna.

# 9 Chemical processes in the hydrogeological context

## 9.1. Introduction

The theory of the hydrochemical processes introduced in Chapter 2 was examined in Chapter 3. The purpose of this chapter is to examine the main hydrochemical processes in real systems, thus showing the extent to which the simple systems of Chapter 3 can explain the chemistry of the complex systems that occur in most aquifers. Firstly, the main processes are illustrated by examples where a single process dominates and thus its action can be seen clearly.

Graphical techniques useful in the interpretation of these systems are illustrated. Graphical techniques are important in deciding the operating processes, but the detailed understanding of the system is often best obtained using a hydrochemical model to simulate the proposed processes. Particularly in the complex carbonate system, interactions between the various processes are so strong that it is difficult to perceive the overall effect of the main process without such a model. Caution must be exercised, however, because the model is usually far more accurate than the field data and conversely the real system is far more complex than the modelled system; therefore the fit of the model will seldom be perfect. Only experience can judge whether it is adequate.

On the scale of a regional groundwater investigation it is usual to find several hydrochemical processes acting and interacting. The approach which has been found most successful in handling these complex systems is based on classification, described in the second part of this chapter. The classification approach based on the methods of Chapter 6 can be used to divide the large amount of data acquired during a regional study into manageable parts in which the techniques of the first part of this chapter are applicable. The synthesis of the parts into a whole can then provide insight into the hydrogeological mechanisms of the region.

## 9.2. Process examples

### *9.2.1. Input from precipitation*

Before considering the implications of a reactive aquifer, it is useful to consider firstly the chemistry of the rainwater input. As can be seen in Table 9.1, rainwater is an extremely dilute solution. Much of the sodium, chloride, and magnesium in rain derives from seawater via the entrainment of spray in the atmosphere. As a consequence, these constituents are highest in coastal areas where they may reach tens of milligrams per litre. Carbonates are partly derived from atmospheric carbon dioxide and also from fine rock dust and usually

**Table 9.1** *Typical analyses of rainwater*

| Constituent | Analysis (mg $l^{-1}$) | | | | | |
|---|---|---|---|---|---|---|
| $Ca^{2+}$ | 0.9 | 1.20 | 0.77 | 0.53 | 1.42 | 0.42 |
| $Mg^{2+}$ | 0.0 | 0.5 | 0.43 | 0.15 | 0.39 | 0.09 |
| $Na^{+}$ | 0.4 | 2.46 | 2.24 | 0.35 | 2.05 | 0.26 |
| $K^{+}$ | 0.2 | 0.37 | 0.35 | 0.14 | 0.35 | 0.13 |
| $Cl^{-}$ | 0.2 | 4.43 | 3.75 | 0.22 | 3.47 | 0.38 |
| $SO_4^{2-}$ | 2.0 | — | 1.76 | 0.45 | 2.19 | 3.74 |
| $HCO_3^-$ | 2.0 | — | 1.95 | — | — | — |
| $NO_3^-$ | — | — | 0.15 | 0.41 | 0.27 | 1.96 |
| $NH_4^+$ | — | — | — | 0.6 | 0.41 | 0.48 |
| $SiO_2$ | 0.1 | — | 0.29 | 0.6 | — | 0.9 |
| pH | — | — | 5.9 | 5.3 | 5.5 | 4.1 |

After Freeze and Cherry 1979.

buffer rainwater at pH 5-6. Some sulphate is derived from seawater but much results from the oxidation of $H_2S$ and $SO_2$ in the atmosphere, the latter often being of man-made origin:

$$H_2S + 2O_2 \rightarrow H_2SO_4 \tag{9.1}$$

$$2H_2O + 2SO_2 + O_2 \rightarrow 2H_2SO_4. \tag{9.2}$$

It should be noted that both these reactions produce sulphuric acid, which is a strong acid, and therefore the pH of rain containing significant quantities may be low (<3). Nitrates in rain are produced as a consequence of the oxidation of nitrogen during lightning and in most vehicle engines:

$$N_2 + 2O_2 \rightleftharpoons 2NO_2 \tag{9.3}$$

$$O_2 + 2H_2O + 4NO_2 \rightarrow 4HNO_3. \tag{9.4}$$

Even though rainwater has a very low dissolved solids content, this may be highly significant in arid zones where only a very small proportion of precipitation becomes run-off or groundwater recharge. In this case the rainwater salts are concentrated in the soil and can produce concentrations dangerous to plants. This evaporation effect has been used to estimate groundwater recharge; if the concentration in an amount $P$ of precipitation is $C_P$ and the concentration in an amount $R$ of recharge is $C_R$, the by mass balance for zero run-off

$$R = \frac{C_P}{C_R} P. \tag{9.5}$$

In very old semi-arid environments rainwater-borne salts may accumulate to the extent that hypersaline brines are produced, as is now occurring in southern Australia (Lawrence 1975), see Section 7.2.1.

### *9.2.2. Solution processes*

Solution processes are of prime importance in determining the chemistry of groundwater, since the composition of rainwater is far from chemical equilibrium with plausible aquifer matrix minerals. Aquifer matrix solution is most pronounced with more soluble lithologies such as carbonates and evaporites, but is significant even with silicates and quartz. In many sandstone aquifers solution of traces of carbonate present either as cement or detrital grains may swamp any chemical contribution from the silicate minerals.

The normal solution processes in sedimentary aquifers can be illustrated by the chemistry of groundwaters from Qatar. The main aquifer developed here is the Rus Formation together with parts of the underlying Umm er Rhaduma Formation (both of early Tertiary age). Two different facies of the Rus Formation are present, the carbonate facies comprising chalky limestones and dolomites, and the evaporite facies compirising anhydrite ($CaSO_4$) and gypsum ($CaSO_4.2H_2O$) with shales. In such an aquifer dissolution processes are usually very important, producing groundwater in which calcium and magnesium are the dominant cations and bicarbonate and sulphate are the dominant anions. Typical analyses are given in Table 9.2. Within these two different facies the chemistry of the groundwater is distinctly different, as illustrated in Fig. 9.1.

Before considering these analyses further, it should be noted that the highly supersaturated $SI_C$ values are unlikely, confirmed by unusually high pH values. In the hot climate it is almost certain that these waters will have lost $CO_2$ before measurement. As explained in Section 3.4, this does not affect the alkalinity because dissolved $CO_2$ is the reference measure for alkalinity. Thus the missing $CO_2$ can be restored by subtracting $SI_C$ from pH, if it is assumed that the water was originally calcite saturated. The new $pP_{CO_2}$ value is given by $pP_{CO_2} - SI_C$.

Because of the constraint of the pH data, it is not possible to recognize definitely calcite-undersaturated waters that might form the start of the evolutionary sequence; however, this can be remedied by modelling the solution process in reverse. By removing dolomite from Sample 1, a possible recharge water 1a is produced. The changes in chemistry as this water equilibrates with calcite can be seen in Fig. 9.2. In the real aquifer dolomite solution is clearly also important, as it produces the high magnesium concentrations, but the simultaneous dissolution of two carbonate minerals is difficult to model.

To explore the chemistry of the water in the evaporite facies, the same recharge water is allowed to dissolve gypsum. Saturation with gypsum is achieved at around 1200 mg $l^{-1}$ $SO_4^{2-}$ and 500 mg $l^{-1}$ $Ca^{2+}$, and this water is strongly undersaturated with respect to calcite, despite its high calcium content. Comparison with the observed water shows the modelled water to be deficient in calcium, magnesium, and particularly alkalinity. Clearly calcite and dolomite are available for solution in the evaporite facies. Considering the simpler case of calcite and gypsum only, modelling confirms in general the hypothesis of simultaneous calcite and gypsum solution. The presence of calcium contributed by gypsum (or anhydrite) dissolution reduces the solubility of calcite (Section 2.6), though

**Table 9.2** *Typical analyses of Qatar groundwaters*

| | No. | $Ca^{2+}$ ($mg\ l^{-1}$) | $Mg^{2+}$ ($mg\ l^{-1}$) | $Na^{+}$ ($mg\ l^{-1}$) | $K^{+}$ ($mg\ l^{-1}$) | $Cl^{-}$ ($mg\ l^{-1}$) | $SO_4^{2-}$ ($mg\ l^{-1}$) | Alkalinity ($mg\ l^{-1}\ CaCO_3$) | pH | $SI_C$ | $pP_{CO_2}$ | Temperature (°C) |
|---|---|---|---|---|---|---|---|---|---|---|---|---|
| Carbonate facies | 1 | 72 | 32.5 | 66.5 | 40.1 | 80 | 15.6 | 364 | 7.9 | 0.88 | 2.62 | 30 |
| | 1a | 24 | 2.8 | 66.5 | 40.1 | 80 | 15.6 | 80 | 6.0 | −2.1 | 0.97 | 30 |
| | 2 | 65 | 54.0 | 56.0 | 6.5 | 92 | 65.0 | 342 | 6.6 | −0.50 | 0.98 | 30 |
| | 3 | 173 | 85.0 | 278.5 | 27.0 | 420 | 487 | 252 | 7.8 | 0.80 | 2.36 | 29 |
| Evaporite facies | 4 | 80 | 44.0 | 52.0 | 12.5 | 68 | 190 | 200 | 8.1 | 0.81 | 2.73 | 30 |
| | 5 | 150 | 62.0 | 85.0 | 11.0 | 150 | 280 | 299 | 7.5 | 0.60 | 1.97 | 29 |
| | 6 | 344 | 122 | 295.0 | 43.0 | 420 | 1078 | 231 | 7.6 | 0.76 | 2.21 | 29 |

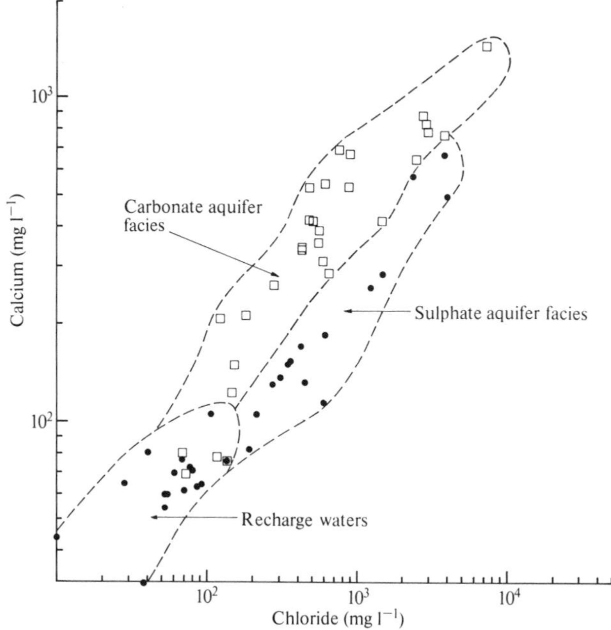


**Fig. 9.1.** Groundwater chemistries from Qatar indicating dissolution.

by less than the expected amount because of the formation of the $CaSO_4^0$ ion pair, and therefore calcite-saturated groundwaters in the evaporite facies have lower alkalinities, pH values, and $pP_{CO_2}$ values than analogous waters in the carbonate facies, as illustrated in Fig. 9.2. It should be noted that the final composition of the water is independent of the path taken, i.e. whether calcite or gypsum dissolution occurs first.

Fissured crystalline rocks are important as sources of groundwater in many areas. Glacial deposits derived from crystalline rocks are important aquifers at high latitudes. In these carbonate-free rocks the chemistry of the contained groundwater is dominated by silicates, predominantly feldspars, pyroxenes, amphiboles, and quartz. The mineral chemistry of these rocks is greatly complicated by atomic substitution within the minerals. The minerals often dissolve incongruently in several stages, for example the dissolution of albite, a common feldspar (Table 9.3). Using thermodynamic data, these equations can be used to plot stability diagrams. For example the albite-kaolinite reaction can be derived from

$$K = \frac{[Na^+]\,[Si(OH)_4]^2}{[H^+]}. \tag{9.6}$$

Similar equations can be derived for the other reactions, which when plotted together give Fig. 9.3. The same approach can be used for the other important

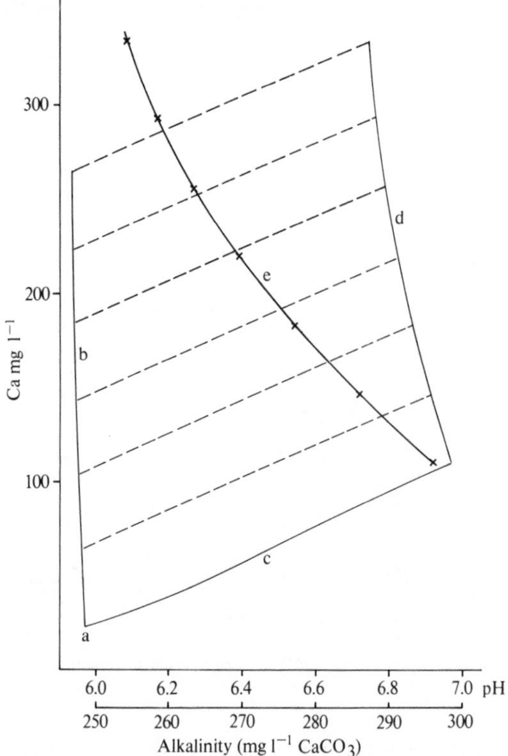


**Fig. 9.2.** Composition of groundwater controlled by calcite and gypsum dissolution: curve a, initial water; curve b, gypsum solution; curve c, calcite solution; curve d, gypsum solution in equilibrium with calcite; curve e, alkalinity corresponding to d.

feldspars, orthoclase ($KAlSi_3O_8$) and anorthite ($CaAl_2Si_2O_8$). Ferromagnesian minerals involve redox equilibria and are thus more difficult to handle.

Using such diagrams, Tardy (1971) found that most waters draining granitic rocks plotted in the kaolinite field and that waters draining basic rocks straddled the kaolinite and montmorillonite fields, from which it is deduced that reactions (2) and (3) described the dissolution process. Typical analyses are presented in Table 9.4. It will be noted that by comparison with waters from most sedimentary rocks, waters from crystalline rocks are extremely dilute. This is in part a consequence of extremely sluggish reaction rates caused by the formation of insoluble surface coatings which prevent the waters reaching equilibrium with the minerals. Saline waters sometimes encountered in crystalline rocks, for example the Paraiba Basin of Brazil (Schoff 1972), cannot be explained by matrix dissolution processes. Salts of marine origin cycled and concentrated in the arid environment are responsible in this case, as is shown by the close correspondence of ionic ratios to the seawater values. Dissolution of all silicates except

**Table 9.3** *Reactions for incongruent dissolution of sodium aluminosilicates (solid phases underlined)*

| | | |
|---|---|---|
| (1) | Kaolinite | $\underline{Al_2Si_2O_5(OH)_4} + 5H_2O \rightarrow \underline{Al_2O_3.3H_2O} + 2Si(OH)_4$ Gibbsite |
| (2) | Albite | $\underline{NaAlSi_3O_8} + H^+ + (9/2)H_2O \rightarrow \underline{\tfrac{1}{2}Al_2Si_2O_5(OH)_4} + Na^+ + 2Si(OH)_4$ Kaolinite |
| (3) | Albite | $\underline{NaAlSi_3O_8} + (6/7)H^+ + (20/7)H_2O \rightarrow \underline{(3/7)Na_{0.33}Al_{2.33}Si_{3.67}O_{10}(OH)_2} + (6/7)Na^+ + (10/7)Si(OH)_4$ Na montmorillonite |
| (4) | Na montmorillonite | $\underline{Na_{0.33}Al_{2.33}Si_{3.67}O_{10}(OH_2)} + (1/3)H^+ + (23/6)H_2O \rightarrow \underline{(7/6)Al_2Si_2O_5(OH)_4} + (1/3)Na^+ + (4/3)Si(OH)_4$ Kaolinite |
| (5) | Quartz | $\underline{SiO_2} + 2H_2O \rightarrow Si(OH)_4$ |

**Table 9.4** *Composition of waters from crystalline rocks*

| | Location | pH | $Ca^{2+}$ (mg $l^{-1}$) | $Mg^{2+}$ (mg $l^{-1}$) | $Na^+$ (mg $l^{-1}$) | $K^+$ (mg $l^{-1}$) | $Cl^-$ (mg $l^{-1}$) | $SO_4^{2-}$ (mg $l^{-1}$) | $HCO_3^-$ (mg $l^{-1}$) | $SiO_2$ (mg $l^{-1}$) |
|---|---|---|---|---|---|---|---|---|---|---|
| Granitic rocks | Norway | 5.4 | 1.7 | 0.6 | 2.6 | 0.4 | 5.0 | 4.6 | 4.9 | 3.0 |
| | Vosges | 6.1 | 5.8 | 2.4 | 3.3 | 1.2 | 3.4 | 10.9 | 15.9 | 11.5 |
| | Sahara | 6.9 | 40 | — | 30 | 1.8 | 4.0 | 20 | 30.4 | 9 |
| | Chad | 7.9 | 8.0 | 2.5 | 15.7 | 3.4 | <3 | 1.4 | 54.4 | 85 |
| Basic rocks | Norway | 5.5 | 1.7 | 0.9 | 3.5 | 0.5 | 3.5 | 4.8 | 5.0 | 3.4 |
| | Vosges | 6.1 | 4.5 | 1.5 | 2.3 | 0.6 | 2.2 | 5.8 | 19.3 | 9.2 |
| | Sahara | 6.8 | 31.6 | 3.9 | 6.9 | 4.4 | 10.6 | 16.1 | 30.0 | 14.5 |
| | Senegal | 7.0 | 24 | 23 | 24 | 2.5 | 4.1 | 1.6 | 198 | 55 |

From Tardy 1971.

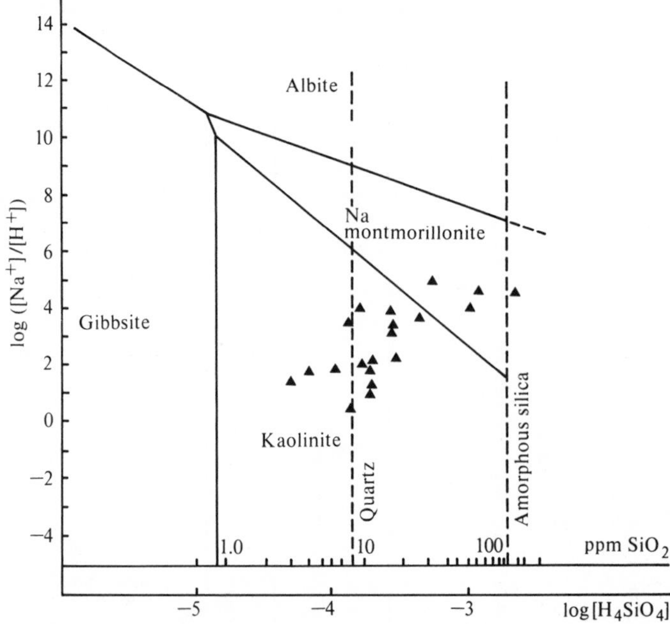


**Fig. 9.3.** Composition of surface waters from silicate terrains with reference to the stability fields of sodium aluminium silicates. (From Tardy 1971.)

quartz consumes hydrogen ions, usually supplied from $H_2CO_3$; thus in the absence of acid rain pollution the dominant cation is $HCO_3^-$. In arid areas the products of weathering are concentrated at the top of the weathering profile, i.e. the soil, where sodium and calcium carbonate form caliche. These carbonates can be incorporated in later cycles of erosion, producing dominantly calcium-bicarbonate and sodium bicarbonate waters in unreactive aquifers.

Detailed modelling of the composition of water in crystalline rocks using a stoichiometric reaction-mass balance approach has been used by Garrels and MacKenzie (1967) and Cleaves, Godfrey, and Bricker (1970) among others.

### *9.2.3. Ion exchange*

The adsorptive properties of solids responsible for ion exchange properties have been described in Sections 2.7 and 3.6. Ion exchange in natural water systems predominantly affects calcium and sodium ion concentrations, the usual ion exchange media being clay minerals. Certain criteria for the manifestation of ion exchange can be developed: there must be a significant ion exchange capacity accessible to the groundwater and there must be a disequilibrium between the adsorbed ions and the groundwater. The most important geological ion exchange media are clays, which tend to have extremely low permeabilities. Thus massive clay formations seldom cause ion exchange of resource significance. Optimum exchange conditions occur in 'dirty' sands with a significant clay content where the permeability is still sufficiently high for the water to be developed as a resource. Glauconitic sands are good ion exchange materials because the pelletized nature of the clay mineral glauconite maintains the permeability. Zeolites within basalts also have a high ion exchange capacity.

Since the dominant cation in most recharge waters is calcium and the dominant cation in seawater is sodium, disequilibrium between groundwater and adsorbed ions is a common occurrence where saline water has formed part of an aquifer's chemical history. This may occur as calcium-rich recharge water moves through an aquifer of marine origin containing sodium-charged clays, or where seawater begins to intrude into a fresh water aquifer with calcium clays.

The former case is well illustrated by the Lincolnshire Limestone aquifer of eastern England, described by Edmunds (1973) and Lawrence, Lloyd, and Marsh (1976). The aquifer comprises 10–30 m of oolitic limestone with a variable content of dispersed clay and organic matter and it is confined down-dip by thick marine clays. The chemical variation along the hydraulic gradient is summarized in Fig. 6.4. The variation of calcium and sodium in the central zone is of particular interest, showing a decrease in calcium with a concomitant rise in sodium. A confusing factor is a rise in chloride caused by the presence of a mixing front with type III (sodium–chloride) water. The presence of ion exchange against the background of this mixing can be revealed by studying the equivalent $Na^+/Cl^-$ ratio, which is expected to be unity for a sodium–chloride water. Values of $Na^+/Cl^-$ for type I waters vary somewhat because of the small concentrations concerned but lie in the range 1–1.5, while type II waters have $Na^+/Cl^-$ ratios in the range 2.3–3.8, clearly showing the influence of ion exchange. The $Na^+/Cl^-$ ratio falls to about 1.2 in the more saline type III waters because of the influence of NaCl.

An alternative approach is to plot the exchange ions against each other. In this case, examination of the data shows that magnesium as well as calcium is being exchanged for sodium. To allow for the increase in sodium contributed by mixing with sodium chloride water, $[Cl^-]$ is subtracted from $[Na^+]$. The plot of $[Ca^{2+} + Mg^{2+}]$ against $[Na^+ - Cl^-]$ is shown in Fig. 9.4. The ion exchange relationship is revealed clearly in the higher calcium samples as a line with slope $-1$. To explain the departure from this line in the high sodium samples, it is necessary to study the carbonate chemistry in more detail.

Type I waters moving away from the recharge area are calcite saturated, i.e. their calcium concentrations are determined by the equilibrium

$$CaCO_3 \rightleftharpoons Ca^{2+} + CO_3^{2-}. \tag{9.7}$$

The removal of calcium by ion exchange disturbs this equilibrium, dissolving more calcite and tending to restore the calcium concentration. This additional calcium may also be removed by ion exchange. The process is eventually limited by the rising $CO_3^{2-}$ concentration which raises the overall pH and reduces calcite solubility. The process can be followed in detail using a numerical model, the results from which are shown in Fig. 9.4. The fit with the observed data is very

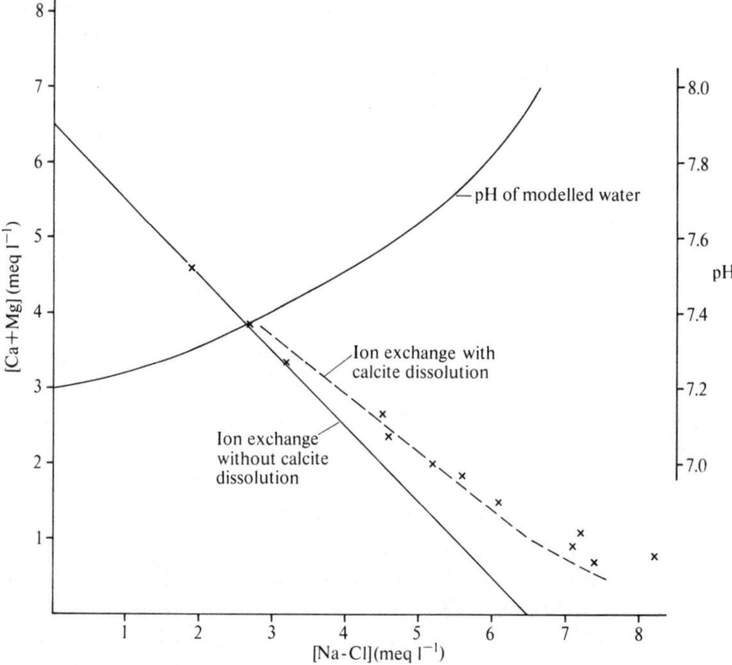


**Fig. 9.4.** Comparison of Lincolnshire Limestone groundwater chemistry with ion exchange models.

good until the Na parameter exceeds 7 meq $l^{-1}$. It is clear that a continuation of this ion exchange process cannot produce the very high alkalinities at moderate pH values (~ pH 8) observed in the saline type III waters, an additional source of carbon dioxide within the aquifer being necessary to produce this chemistry.

The ion exchange front has a width of only 1.5 km, indicative of a fast process although the exact rate cannot be determined in the absence of groundwater flow data. The lack of ion exchange up-dip of the ion exchange zone, is explained by the exhaustion of the limited exchange capacity of the aquifer.

Reverse ion exchange is currently occurring beneath the bed of the River Mersey in north-west England. Here modern seawater is infiltrating through estuarine muds into a Triassic sandstone aquifer containing fresh water. In order to resolve the ion exchange process against the varying ionic background produced by the mixing of fresh water and seawater, the expected cation concentrations based on linear mixing measured by chloride are considered. The differences between observed and expected concentrations are plotted in Fig. 9.5. The inverse relationship between $\Delta[Ca^{2+} + Mg^{2+}]$ and $\Delta[Na^+ + K^+]$ can be seen clearly, although the points are somewhat scattered by other hydrochemical processes occurring in the estuarine sediments. The occurrence of reversed ion exchange indicates the effect of the very high Na/Ca ratio in seawater, which

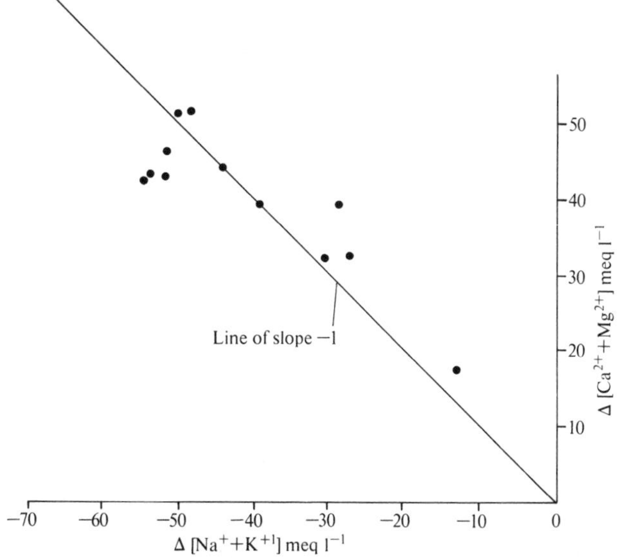


**Fig. 9.5.** Reverse ion exchange relationship occurring during saline intrusion from the Mersey Estuary.

overrides the selectivity of most clays towards calcium adsorption. It should also be remembered that the selectivity of clay minerals is considerably reduced at seawater ionic strengths (see Section 3.6).

### *9.2.4. Redox phenomena*

Rainwater at the moment of infiltration is a strongly oxidizing solution; it is almost saturated with atmospheric oxygen and in addition contains oxyanions, particularly sulphate and nitrate. Despite the consumption of oxygen by biological respiration in the soil zone, in most soils these oxidizing characteristics persist when the rainwater becomes groundwater recharge. In contrast, many aquifers contain reducing substances, notably organic carbon, iron sulphides, and iron silicates. The complex interactions between the oxidizing recharge water and the reducing aquifer matrix are responsible for the redox phenomena observed in many aquifers. The $E_H$ and pH controls imposed by these major reactants influence the chemistry of trace elements, particularly transition metals.

The major $E_H$ controls in natural water systems arising from the reactants mentioned above are given in Table 9.5.

Using these data, with the addition of some acid-base equilibria, it is possible to construct a composite $E_H$-pH diagram (Fig. 9.6) in which the interrelations for the various reactions can be seen. These are best illustrated by describing the evolution of oxygenated recharge water containing nitrate and sulphate as it moves through an aquifer containing traces of organic matter, e.g. the coastal

**Table 9.5** *Reactions controlling the $E_H$ of natural waters*

| Reaction | $E_0$ (V) |
|---|---|
| $O_2 + 4H^+ + 4e^- \rightleftharpoons 2H_2O$ | 1.23 |
| $2H^+ + 2e^- \rightleftharpoons H_2$ | 0.00 |
| $Fe(OH)_3 + 3H^+ + e^- \rightleftharpoons Fe^{2+} + 3H_2O$ | 0.94 |
| $SO_4^{2-} + 8H^+ + 8e^- \rightleftharpoons S^{2-} + 4H_2O$ | 0.16 |
| $NO_3^- + 6H^+ + 5e^- \rightleftharpoons 3H_2O + \frac{1}{2}N_2$ | 1.21 |
| $CH_2O + 4H^+ + 4e^- \rightleftharpoons CH_4 + H_2O$ | 0.41 |
| $CO_2 + 4H^+ + 4e^- \rightleftharpoons CH_2O + H_2O$ | −0.07 |
| $CO_2 + 8H^+ + 8e^- \rightleftharpoons CH_4 + 2H_2O$ | 0.17 |

$CH_2O$ (formaldehyde) is taken as a representative carbohydrate

plain sequence of eastern North America (Back 1960). The water entering the aquifer will have a pH near 7 because of buffering by the carbonate system, the corresponding $E_H$ for water in equilibrium with atmospheric oxygen being about 800 mV. The final $E_H$, which will be stable in contact with organic matter such as cellulose ($C_6H_{10}O_5$), is about −400 mV. (N.B. Both these potentials are hypothetical; they would not be measured by a platinum electrode.) Reduction proceeds sequentially; therefore dissolved oxygen is reduced first, followed by nitrate. The $E_H$ can then fall to a value such that organic matter can reduce itself, a type of reaction known as *disproportionation*. Equal amounts of carbohydrate are oxidized and reduced during this reaction, which therefore proceeds until the organic matter is entirely converted to methane and carbon dioxide. Reduction of insoluble iron(III) oxides to soluble $Fe^{2+}$ takes place at a similar $E_H$. The equilibrium between methane and carbon dioxide requires a slightly lower $E_H$ value than that at which sulphate is stable so this is reduced to sulphide before the $E_H$ reaches a stable value of about −250 mV at pH 7, which is determined by the $CO_2$-$CH_4$ couple. It should be noted that water is not reduced except in the presence of unusual agents such as metallic iron.

This sequence describes the general behaviour of many groundwater systems rather well in a qualitative way, but attempts to interpret measured $E_H$ values quantitatively in terms of chemical equilibria are usually unsuccessful (Champ, Gulkens, and Jackson 1979). In particular, redox potentials below −100 mV are seldom observed, yet sulphide species are present in many groundwaters. Two reasons for this can be proposed: firstly slow electron transfer for the reactions concerned produces chemical disequilibrium and the resulting mixed potentials are not amenable to quantitative interpretation (Morris and Stumm 1967), and secondly the reactions actually occur as the result of bacterial processes, probably in local micro-environments atypical of the whole aquifer.

The classical sequence of redox reactions is followed in the Lincolnshire Limestone (Edmunds 1973). Waters at outcrop contain dissolved oxygen, nitrate, and sulphate, and have an $E_H$ value of approximately +400 mV. Where the aquifer becomes confined the concentrations of dissolved oxygen and nitrate

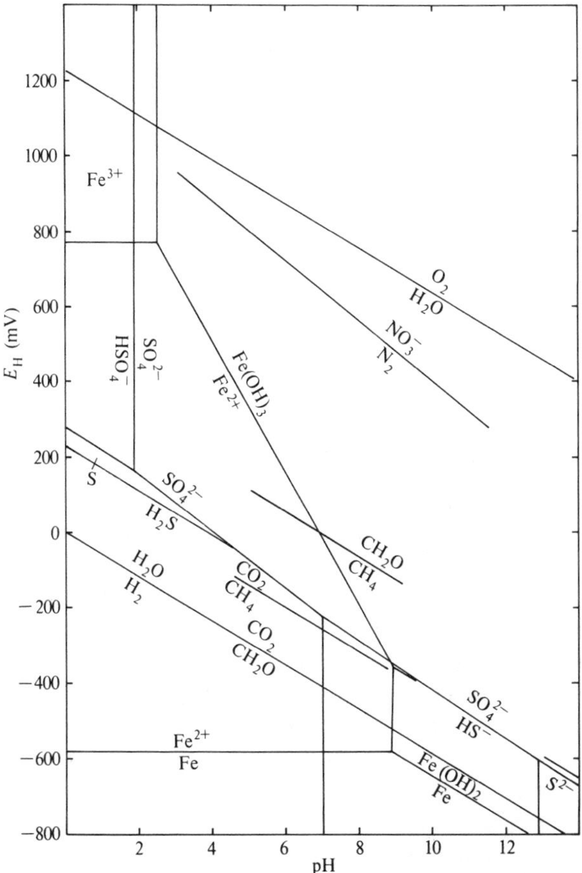


**Fig. 9.6.** $E_H$-pH relationships for important hydrochemical species.

fall to zero and the smell of hydrogen sulphide is apparent at well heads. The $E_H$ in this region is approximately +125 mV and steadily falls to −100 mV further into the confined zone (Fig. 9.7). The cause of the reduction in this aquifer is almost certainly the dispersed organic carbon (0.2 per cent) within the aquifer matrix. The more reducing waters in this aquifer are saline, and are thought to be caused by mixing with ancient seawater. If this were true, it would be possible to estimate the sulphate content of these saline waters using chloride to follow the dilution. However, the measured values (about 30 mg $l^{-1}$ $SO_4^{2-}$) are very different from the expected values (about 150 mg $l^{-1}$). One possible explanation is that the sulphate has been reduced by bacteria using the organic matter as a substrate:

$$SO_4^{2-} + \underset{\text{organic matter}}{2H_2CO} \rightarrow 2HCO_3^- + H_2S. \qquad (9.8)$$

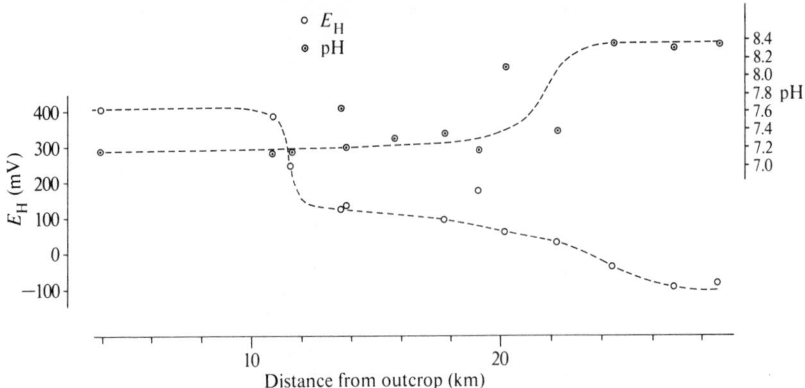


**Fig. 9.7.** Variation of $E_H$ and pH away from outcrop in the Lincolnshire Limestone.

This reaction contributes dissolved inorganic carbon which accounts for the rise in bicarbonate shown in Fig. 6.4 for type III waters beyond the ion exchange zone. Small numbers of sulphate-reducing bacteria have been detected in this water, and this supports the hypothesis. The only problem is the fate of the sulphide; the reduction of 100 mg $l^{-1}$ $SO_4^{2-}$ will produce 35 mg $l^{-1}$ $H_2S$, an amount far in excess of that detected in the water. Some sulphide is undoubtedly lost through precipitation of insoluble sulphides, particularly those of iron, but there is insufficient dissolved iron in the water to account for more than a small proportion of the sulphide. The problem remains unsolved.

Marsh (1978) has studied the effect that the $E_H$ and pH constraints imposed by the major ion groundwater chemistry has on the chemistry of trace elements in the water, the major change being from low pH (about 7), high $E_H$ (about 400 mV) water at outcrop to high pH (8-8.5), low $E_H$ (−100 mV) water in the confined zone. The variation of concentration of selected trace elements is shown in Fig. 9.8. Marsh used measured pH and $E_H$ values and the chemistry of the water to calculate the equilibrium concentration of each element in contact with the most stable solid phase, making due allowance for ion pairing. Apart from iron, it was found that the concentration of the metals was below that expected from equilibrium considerations and thus it was concluded that availability of the elements in the aquifer matrix was an important control on their concentrations. It was found difficult to explain the shape of the curves in Fig. 9.8 in detail. The calculations effectively ignored the presence of sulphide species since their equilibrium concentration is around $10^{-32}$ mol $l^{-1}$, even for the most reducing waters, but the detection of hydrogen sulphide by smell at well heads suggests that disequilibrium exists. Edmunds (1973) measured total sulphide concentrations of up to $10^{-4}$ mol $l^{-1}$ in the more reduced waters, implying that $S^{2-}$ concentrations of up to $10^{-9}$ mol $l^{-1}$ are present. Using this $S^{2-}$ concentration in the equilibrium calculations for the confined zone decreases the equilibrium $Pb^{2+}$ and $Ni^{2+}$ concentrations considerably

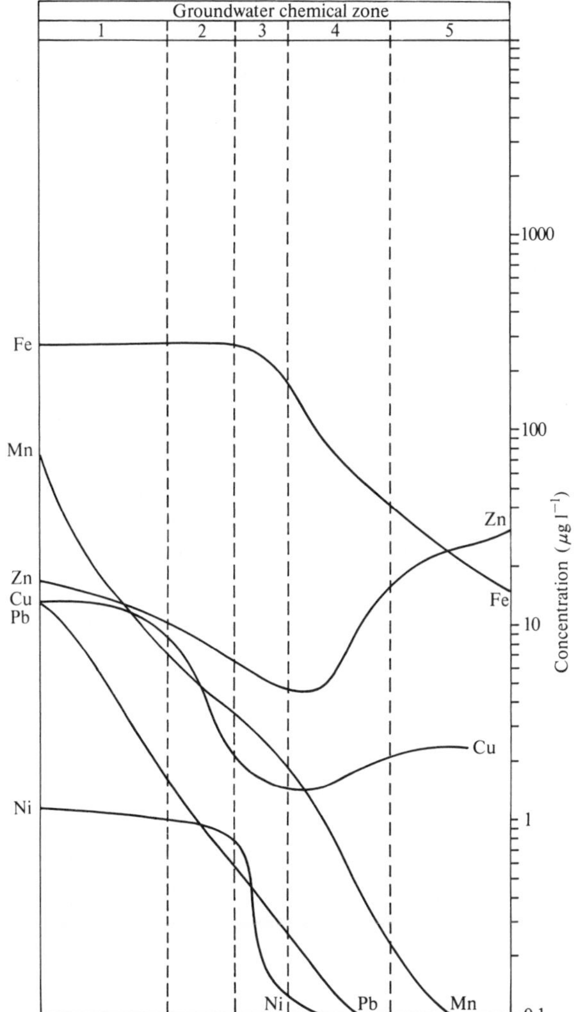


**Fig. 9.8.** Generalized hydrochemical section of the Lincolnshire Limestone showing trace heavy metal concentrations (see also Fig. 6.4).

because of the formation of highly insoluble sulphides. $Zn^{2+}$ and $Cu^{2+}$ also form insoluble sulphides but the total Zn and Cu concentrations are increased because of the formation of hydroxide and carbonate ion pairs in the high pH water. Little difference is made to the concentration of iron in the confined zone, since under the conditions assumed FeS and $FeCO_3$ (the least soluble phase in the absence of sulphide) have similar solubilities. The high and variable concentration of iron in the oxidizing part of the aquifer is explained by the presence of suspended $Fe(OH)_3$ in these waters. The concentration of Mn falls steadily, despite being undersaturated. A possible explanation is that $Mn^{2+}$ is coprecipitated with the more insoluble sulphides. Thus in the Lincolnshire Limestone it can be seen that the development of reducing conditions at high pH is effective in removing a number of heavy metals from the groundwater.

The behaviour of iron in the Atlantic coastal plain aquifers studied by Barnes and Back (1964) is very different. The aquifer comprises a sequence of continental sands of Cretaceous age, with abundant dispersed iron oxide ('limonite'; not identified more specifically). In this aquifer the $E_H$ at outcrop is about 700 mV, falling to −20 mV at down-gradient locations. The pH varies in the range 3.5-7.5. Most of the waters studied were extremely dilute with total dissolved solids in the range 10-200 mg $l^{-1}$. The lower TDS waters are almost carbonate free and thus are very poorly buffered, which accounts for the variability of the pH. Sulphate reduction cannot be quantitatively important in this aquifer because the sulphate input to the system is extremely low, although sulphide was detected by smell at many locations.

When the experimental data are plotted on an $E_H$-pH diagram (Fig 9.9), a clear correspondence is seen between the iron content of the water samples and the position of the plotted point with respect to the $Fe^{2+}$-$Fe(OH)_3$ line. This suggests that the reaction

$$Fe(OH)_3 + 3H^+ + 3e^- \rightleftharpoons Fe^{2+} + 3H_2O$$

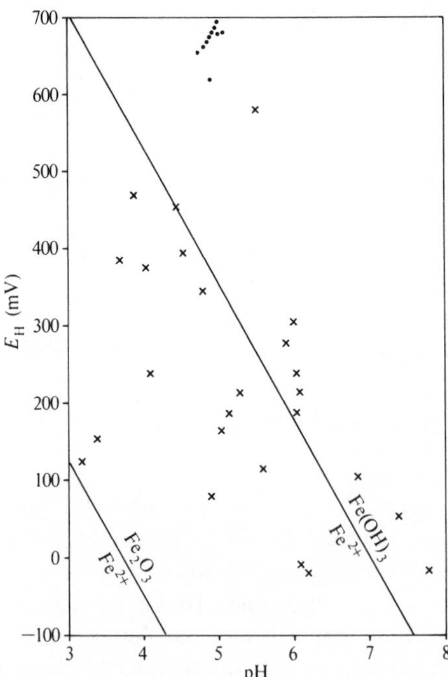


**Fig. 9.9.** Dissolved iron in the Atlantic plain of North America compared with $E_H$–pH controls: •, $Fe_{tot} < 0.05$ mg $l^{-1}$; ×, $Fe_{tot} > 0.1$ mg $l^{-1}$.

is the source of the dissolved iron. The exact position of the line drawn is dependent on the $\Delta G_f^\circ$ value used for the iron hydroxide $Fe(OH)_3$—for better crystallized specimens approaching hematite ($Fe_2O_3$) $\Delta G_f^\circ$ is more negative, shifting the line to lower pH values as shown. Thus some of the values can be explained by assuming that the crystallinity of the dispersed iron oxides in the aquifer is between those of $Fe(OH)_3$ and $Fe_2O_3$. An alternative explanation is that the leaching of iron has proceeded to the extent that all iron oxides have dissolved, thus freeing the groundwater from the poising action of the $Fe^{2+}$–$Fe(OH)_3$ couple.

The strong leaching of iron oxides in the aquifer is likely to lead to the release into solution of other heavy metals which had coprecipitated with iron oxides. The possibly low sulphide content of the water consequent on its low original sulphate content implies that the formation of insoluble sulphides will not be as important in controlling heavy metal content in the Atlantic coastal plain aquifers as it was in the Lincolnshire Limestone.

### 9.2.5. *Diffusion*

Opportunities to observe diffusion in groundwater systems are few because diffusion is usually masked by groundwater flow. However, ideal conditions exist in the Chalk of eastern England which has a high porosity but a very low matrix hydraulic conductivity ($2 \times 10^{-3}$ m day$^{-1}$) as a consequence of small pore size (Price, Bird, and Foster 1976). Bulk water movement below the zone of secondary permeability is negligible. At Trunch the thickness of the Chalk is about 470 m, only the top 40 m of which is fissured and developed as an aquifer (Bath and Edmunds 1981). Below this depth, pore water samples from a cored borehole showed that the Chalk water becomes progressively more saline, reaching 20 000 mg l$^{-1}$ $Cl^-$ at the base of the Chalk. As the Chalk is a marine deposit of Cretaceous age and has subsequently been subject to marine transgressions during the Eocene and Pleistocene, there is little doubt that the saline water is of marine origin, but its age is uncertain.

The situation was therefore treated as a diffusion problem by assuming that after the Chalk was uniformly filled with saline water, the top of the Chalk profile was maintained at a constant zero salinity by water movement in the aquifer. For these boundary conditions, the appropriate solution of the diffusion equation (eqn (3.98)) is

$$\frac{C}{C_0} = \operatorname{erf} \frac{z}{2(D't)^{\frac{1}{2}}} \tag{9.9}$$

where $C = C_0$ at $t = 0$, $z$ is the depth below the interface, $t$ is the time since the commencement of flushing, $D'$ is the diffusion coefficient and

$$\operatorname{erf}(x) = \frac{2}{\sqrt{\pi}} \int_0^x \exp(-\alpha^2)\, d\alpha.$$

Substituting data for sodium concentrations in this equation ($D' = 0.016$ m$^2$ year$^{-1}$, and $C/C_0 = 0.5$ at $z = 228$ m) yields a time of $3 \times 10^6$ years. The actual shape of the profile differs from that calculated from eqn (9.9), causing the calculated time to decrease with depth, e.g. using $C/C_0 = 0.9$ at $z = 432$ m gives $t = 1.5 \times 10^6$ years. In view of the complex history of the area during the late Pleistocene glaciations such discrepancies are not surprising. From these time estimates it is concluded that the saline water has Pleistocene and possibly Eocene components.

### 9.2.6. *Mixing*

The most important hydrochemical consequence of transport processes such as diffusion and dispersion is the mixing of bodies of groundwater with different chemistries. Almost all hydrochemical interpretations involve a mixing component. The simplest approach to such problems is a linear model, demonstrated by Figs. 6.15 and 9.10. The ion most suitable for following mixing in most situations is chloride, since it is conservative in most conceivable situations. However, certain parameters by their very nature are not conservative, particularly $E_H$.

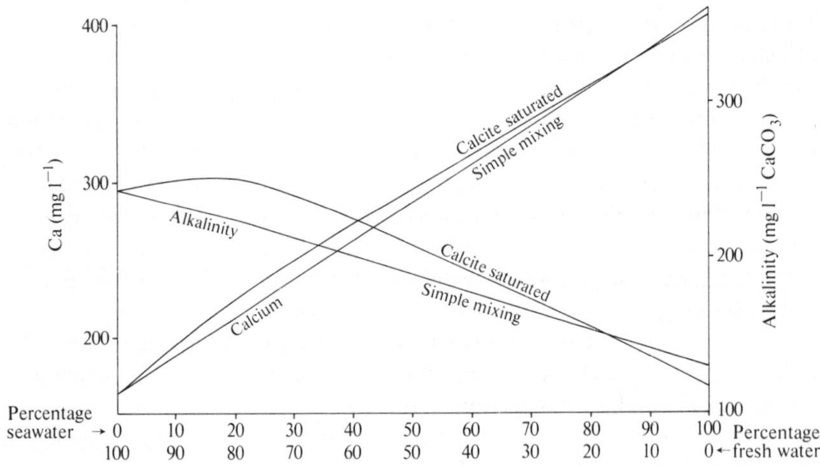


**Fig. 9.10.** Mixing between fresh water and seawater illustrating consequent calcite dissolution.

Where solution equilibria exist in the mixing system, the mixing process disturbs them, as appreciated by Bögli (1964). The chemistry of mixing in the carbonate system is especially important and interesting and has been described in detail by Wigley and Plummer (1976). The mixing of two waters, each separately in equilibrium with calcite, does not generally produce a mixture in equilibrium with calcite. Several effects are shown to be responsible for this result: the algebraic effect produces an increase in $SI_C$ when the waters have different $Ca^{2+}/CO_3^{2-}$ ratios, the $CO_2$ effect produces a decrease in $SI_C$ when the waters have different $pP_{CO_2}$ values, and the ionic strength effect reduces $SI_C$ when the waters differ in ionic strength. Differences in temperature and style of ionic content also cause minor undersaturation.

As a consequence of the $CO_2$ effect, fresh carbonate groundwaters usually produce an undersaturated mixture, a result thought to be important in causing solution below the water table in limestone aquifers (Bögli 1964). When carbonate groundwaters mix with seawater the ionic strength effect reinforces the $CO_2$ effects to produce undersaturation despite the fact that seawater is supersaturated with respect to calcite (Fig. 9.10). Similar effects occur with minerals other than calcite, although for non-carbonate minerals the strong $CO_2$ effect is not present. Since the algebraic and ionic strength effects work in opposite directions, it is possible to produce over-saturation, undersaturation, or no change with these minerals.

## 9.3. Regional studies

### *9.3.1. The classification approach*

A variety of names have been used for classes of water chemistry: the word 'zone' tends to imply a spatially fixed region while the word 'facies' tends to imply the product of an environment. The word 'type' is free from these connotations and is used here to designate a body of water having distinctive characteristics. The various classification parameters which have been used in the past can be divided broadly into those based on chemical composition, e.g. Back (1960), and those based on chemical change, e.g. Downing and Williams (1969). The most successful technique seems to be one which includes both chemical composition (in a broad sense) and chemical change, together with elements of geological and geographical control.

As discussed in Chapter 6, sophisticated methods of classification are seldom satisfactory as they lack the flexibility of other approaches. The Durov diagram is a useful start in correlating major element chemistry. Additional techniques can then be used to separate the waters further, particularly those that plot in the extreme corners of the diagram where subtle variations in the chemistry are swamped by the two predominant ions. Dilution diagrams and *X–Y* plots are useful for this second stage of classification. When the classification is nearing completion, Schoeller diagrams are useful for checking the validity of the scheme.

Almost all classifications suffer from the problems associated with trying to divide a continuous phenomenon into discrete classes, which involves the imposition of arbitrary boundaries. If no clear break between water types is observed, it is often useful to add geographical or geological constraints when determining the boundary in order to produce types with geographical or geological coherence. Even then the random error associated with any individual analysis may require a flexible attitude to classify it in the most suitable type.

More problems are encountered when saline water is present in an aquifer; often a substantial range of composition produced by mixing of fresh and saline waters is present. In such a case it may be appropriate to assign the mixing series to a single water type, with the end member compositions forming further types.

It can be appreciated that the resulting classification will be somewhat subjective although it is usually possible to deduce numerical critera retrospectively to assist others to reproduce the classification. It is important to remember that the classification is a tool to study the hydrochemical and hydrogeological processes, the actual water types having no final significance other than as convenient labels.

These principles are illustrated in the examples below.

### *9.3.2. Chalk aquifer of Essex and Suffolk*

The Chalk† aquifer (Cretaceous) of Essex and Suffolk is partly confined by London Clay (Eocene) and elsewhere is covered by up to 90 m of Pleistocene drift, predominantly argillaceous till (⩽30 m) with sands (0–60 m) at the base. Limited outcrops of Chalk occur only in the larger river valleys. This part of England is an area of low rainfall and consequently little surface water potential and therefore the aquifer is heavily exploited. In view of its restricted recharge potential and the presence in the aquifer of saline water possibly representing intrusion from the North Sea, detailed hydrochemical studies have been undertaken to deduce the recharge mechanism and to investigate the origin of the saline water (Lloyd, Harker, and Baxendale 1981; Heathcote 1981).

The range of groundwater chemistry present in the area is indicated on the Durov diagram shown in Fig. 9.11. A strong mixing trend between calcium–bicarbonate and sodium–chloride waters can be seen, in addition to which there are calcium–bicarbonate waters with sodium–bicarbonate and calcium–sulphate components. In classifying these waters, it was found appropriate to use different techniques for fresh and saline waters, an arbitrary boundary being placed at 300 mg $l^{-1}$ $Cl^-$. (It must be pointed out that Lloyd, Harker, and Baxendale (1981) and Heathcote (1981) used different numbering systems; the system used in this synthesis differs from both.) The fresh waters are easily classified on the basis of the major element chemistry into type I (bicarbonate waters) and type II (sulphate waters). Further subdivision on more subtle differences in chemistry results in four subtypes in type I. The properties of these types are summarized in Table 9.6 and representative analyses are given in Table 9.7.

The saline waters are distinguished by using trace element chemistry, particularly strontium, fluoride, and to some extent iodide, into type III and type IV which have high and low trace element concentrations respectively. Type III can be subdivided on the basis of major element chemistry. Both type III and type IV represent mixing series and therefore have wide ranges of total dissolved

† The Chalk in eastern England is a soft massive coccolith limestone.

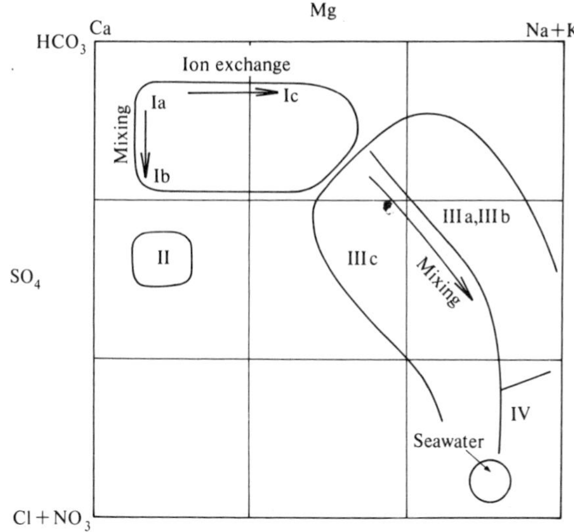


**Fig. 9.11.** Expanded Durov diagram showing the chemistry of the Suffolk and Essex groundwaters.

**Table 9.6** *Fresh water types and their characteristics in Essex and Suffolk*

| Type | Characteristics | Number in | |
|---|---|---|---|
| | | Lloyd, Harker, and Baxendale (1981) | Heathcote (1981) |
| Ia | Calcium–bicarbonate, often with nitrate | I | IA |
| Ib | Calcium–bicarbonate with significant sulphate; nitrate often absent | Ia | IB |
| Ic | Calcium–sodium–bicarbonate; nitrate usually absent | Ib | Not present |
| Id | Calcium–bicarbonate; nitrate absent; significantly lower calcium and bicarbonate content than Ia | Not present | IIC |
| II | Calcium–sulphate–bicarbonate; high magnesium and no nitrate | II | IIB |

solids. Characteristics and representative analyses are given in Tables 9.8 and 9.9 respectively. The geographical distribution of the water types is shown in Fig. 9.12.

The basic chemistry of Type I waters can be explained by calcite dissolution in the soil zone under apparent open-system conditions. Type Ib shows influence from the till and type Ic is partly ion exchanged. This chemical interpretation is

**Table 9.7** *Representative analyses of fresh waters from Essex and Suffolk*

| Type | $Ca^{2+}$ (mg $l^{-1}$) | $Mg^{2+}$ (mg $l^{-1}$) | $Na^{+}$ (mg $l^{-1}$) | $K^{+}$ (mg $l^{-1}$) | $Cl^{-}$ (mg $l^{-1}$) | $SO_4^{2-}$ (mg $l^{-1}$) | Alkalinity (mg $l^{-1}$ $CaCO_3$) | $NO_3^{-}$–N (mg $l^{-1}$) |
|---|---|---|---|---|---|---|---|---|
| Ia | 116 | 28 | 5 | 4.4 | 25 | 44 | 284 | 6.1 |
| Ib | 196 | 12 | 50 | 5.7 | 55 | 170 | 330 | 3.3 |
| Ic | 93 | 26 | 80 | 11 | 112 | 87 | 300 | 0 |
| Id | 83 | 5.9 | 19 | 3.0 | 26 | 23 | 242 | 0 |
| II | 221 | 99.2 | 66 | 12.9 | 101 | 585 | 364 | 0 |

**Table 9.8** *Characteristics of saline water types in Essex and Suffolk*

| Type | Characteristics | | Number in | |
|---|---|---|---|---|
| | | | Lloyd, Harker, and Baxendale (1981) | Heathcote (1981) |
| IIIa | Sodium–chloride | High $Sr^{2+}$, $F^{-}$, $I^{-}$ | Not present | IIIA |
| IIIb | Sodium–chloride with low calcium | | IIIa, IV | Not present |
| IIIc | Sodium–chloride with high magnesium and sulphate | | Not present | IIIB |
| IV | Sodium–chloride | Low $Sr^{2+}$, $F^{-}$ $I^{-}$ very variable | IIIB, V | IV |

**Table 9.9** *Representative analyses of saline waters from Essex and Suffolk*

| Type | $Ca^{2+}$ (mg $l^{-1}$) | $Mg^{2+}$ (mg $l^{-1}$) | $Na^{+}$ (mg $l^{-1}$) | $K^{+}$ (mg $l^{-1}$) | $Cl^{-}$ (mg $l^{-1}$) | $SO_4^{2-}$ (mg $l^{-1}$) | Alkalinity (mg $l^{-1}$ $CaCO_3$) | $Sr^{2+}$ (mg $l^{-1}$) | $F^{-}$ (mg $l^{-1}$) | $I^{-}$ ($\mu$g $l^{-1}$) |
|---|---|---|---|---|---|---|---|---|---|---|
| IIIa | 228 | 190 | 1800 | 64 | 3500 | 384 | 250 | 19 | 3.2 | 370 |
| IIIb | 27 | 29 | 900 | 27 | 1200 | 192 | 310 | — | 3.8 | — |
| IIIc | 109 | 141 | 900 | 40 | 1220 | 675 | 524 | 21.5 | 3.5 | 140 |
| IV | 120 | 180 | 1575 | 34 | 3400 | 240 | 142 | 8.5 | 0.9 | 39 |

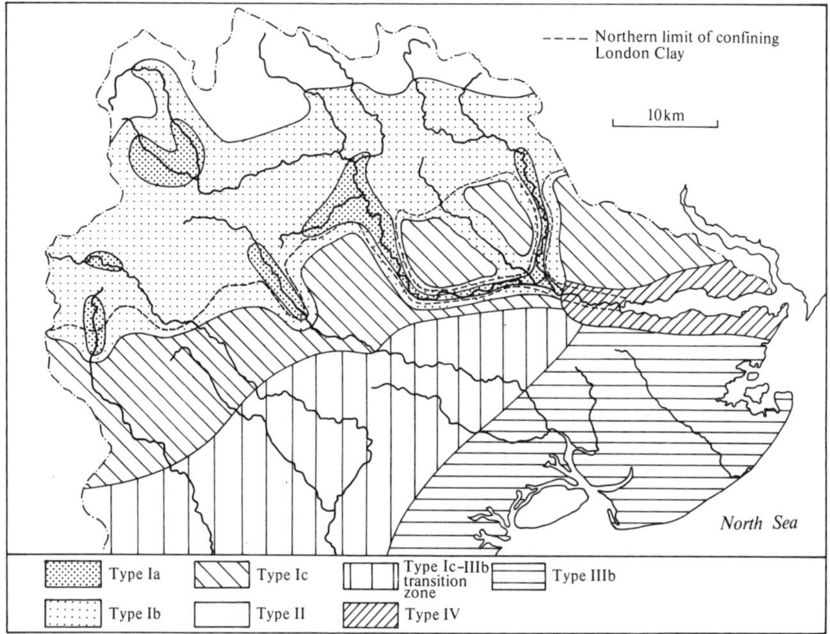


**Fig. 9.12.** Distribution of hydrochemical types in Essex and Suffolk.

compatible with the geographical distribution: type Ia is restricted to the valley bottoms where the Chalk outcrops, type Ib is found near the edges of the till, and type Ic is found beneath the London Clay in the southern part of the area where a basal glauconite sand is developed. Type Id waters are found only in a restricted area and are considered to have formed at the end of the last glaciation when soil $P_{CO_2}$ was lower. This is confirmed by their $^{14}C$ age which is 9000 years (Heathcote 1981). Of the other type I waters, only type Ia waters are definitely modern as shown by nitrate and $^{14}C$ contents, type Ib water ranges from 0 to 2000 years, and type Ic water is about 10 000 years old.

Type II waters are found in the Chalk only beneath the thickest till and are similar in chemistry to pore water extracted from the till, thus showing that the high magnesium and sulphate characteristics derive from the till. Two hypotheses concerning the origin of the sulphate were considered: gypsum dissolution and pyrite oxidation. The presence of either mineral in the till cannot be conclusively proved. Chemical modelling of the processes showed that the chemistry was most satisfactorily produced by normal soil zone calcite dissolution, followed by pyrite oxidation in the presence of calcite and magnesium clays:

$$H_2O + FeS_2 + 3\tfrac{1}{2}O_2 \rightarrow Fe^{2+} + 2H^+ + 2SO_4^{2-} \tag{9.10}$$

$$H^+ + CaCO_3 \rightleftharpoons Ca^{2+} + HCO_3^- \tag{9.11}$$

$$Fe^{2+} + \text{Mg clay} \rightleftharpoons \text{Fe clay} + Mg^{2+}. \tag{9.12}$$

The high trace element concentrations of type III waters could only be explained by aquifer residence, confirmed by infinite $^{14}C$ ages. Mixing studies showed that all three subtypes could be produced from normal seawater subjected to reverse ion exchange and sulphate reduction (see below) followed by dilution by types Id, Ib, and II for types IIIa, IIIb, and IIIc respectively. Ion exchange has continued in type IIIb.

The chemistry of type IV waters can be produced by mixing of type Ia and Ib waters with modern seawater, provided that this has been subject to sulphate reduction and ion exchange. Such processes were demonstrated in undoubted modern saline waters in the area. The high iodide contents of some type IV waters could be closely correlated with the presence of estuarine silts.

The complexity of the flow mechanism in the river valleys is revealed by the changes that take place along the hydraulic gradient from type II via type Ib to type Ia. This is in the direction of decreasing dissolved solids. None of the waters is saturated with any mineral other than calcite and it is therefore impossible to produce the decrease in sulphate and chloride contents other than by dilution. Percolation through the till is slow and produces type II water which dominates the Chalk water chemistry away from the valleys. Water flow also occurs through discontinuities in the till; this is a rapid process and produces type Ib water. This water moves in the sands overlying the Chalk, which it enters only near the valleys. In the broader valleys sufficient direct recharge, uninfluenced by till, takes place to produce type Ia water. Artificial abstraction is important in allowing recharge to occur in the valley bottoms, which are otherwise effluent zones indicated by the presence of old type Ib water. The overall behaviour of the system is illustrated schematically in Fig. 9.13.

Figure 9.14 shows a hydrochemical section along one of the major valleys. When quantitative flow calculations are applied to this, it is found that recharge along the lower part of the valley is insufficient to produce the dilution seen if all the water collected in the upper part of the catchment flows through the aquifer. Hydraulic calculations confirm that a significant proportion of the flow occurs in the river, which the geology shows to be hydraulically isolated from the aquifer in its lower part. Under these circumstances, which were not previously suspected, recharge is adequate to produce the observed dilution. The profile also reveals induced recharge in the lowest part of the catchment.

These two studies contributed greatly to the understanding of the aquifer recharge, thus enabling effective resource management. The realization that groundwater flow along the valleys might be restricted is important in planning the distribution of abstraction wells. Old and modern saline waters, whose response to long-term abstraction differs greatly, were successfully distinguished in areas where they occurred together.

### *9.3.3. Sandstone aquifer of the Lower Mersey Basin*

The aquifer concerned in this study comprises a thick sequence of continental red sandstones with pebble beds and marls of Permo-Triassic age. The base of the

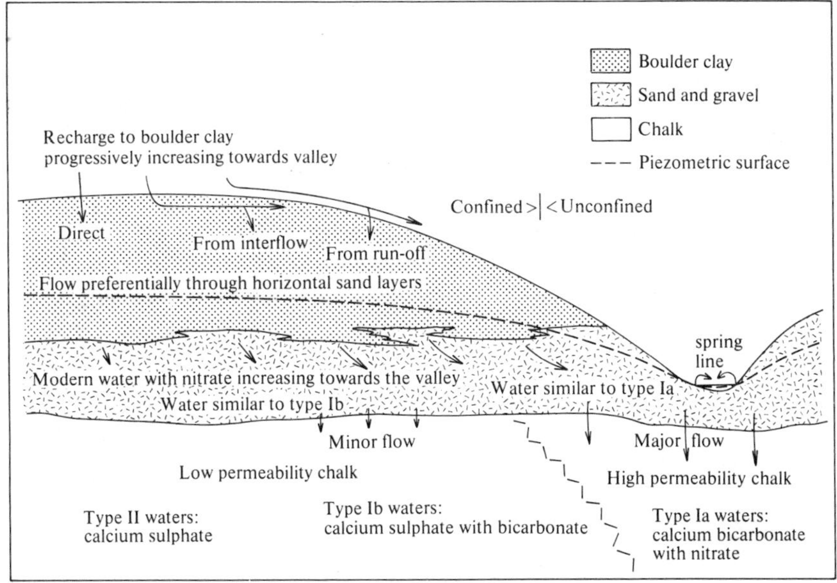


**Fig. 9.13.** Schematic cross-section showing the flow mechanism deduced from hydrochemistry in Essex and Suffolk. (After Lloyd, Harker, and Baxendale 1981.)

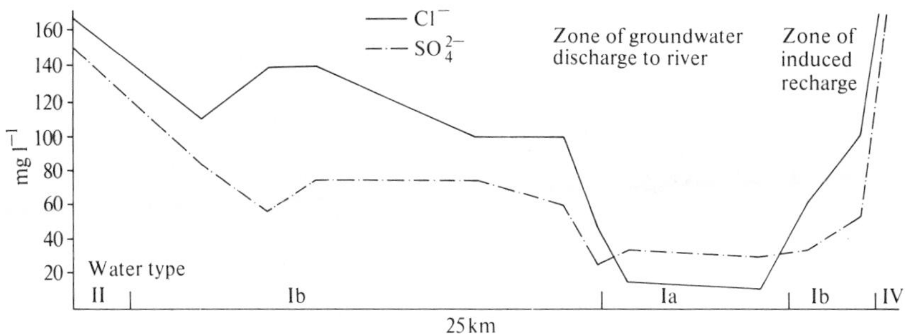


**Fig. 9.14.** Hydrochemical section along the Gipping valley, Suffolk.

aquifer is formed by Carboniferous mudstones and the top of the aquifer is defined, where they are present, by Triassic mudstones and halite deposits. The eetire sequence has been subject to major vertical faulting. A varied sequence of drift deposits including sands, clays, and peat overlies the solid rocks. The area is traversed by the Mersey Estuary and the Manchester Ship Canal (salt water). Heavy groundwater abstraction in this major industrial region has resulted in water table declines of up to 100 m. Future development options for the area have been studied using a digital model (University of Birmingham 1981).

Hydrochemical studies were important in understanding the system to be modelled and in validating the result.

The range of groundwater chemistry present in the area is illustrated in a Piper diagram (Fig. 9.15) and in Table 9.10. The trend is for the lowest dissolved solids waters to be calcium–magnesium–sulphate–chloride in character, whilst the higher dissolved solids waters are calcium-magnesium-bicarbonate and sodium bicarbonate in character. The highest dissolved solids waters in the area, which attain 180 000 mg $l^{-1}$, are sodium chloride in character. Except for these very saline waters, the waters are classified quite well by the Piper diagram. The justification for type 3, which overlaps the other fields, is in its lower dissolved solids content. Its distinctiveness is revealed in the *X–Y* plot shown in Fig. 9.16.

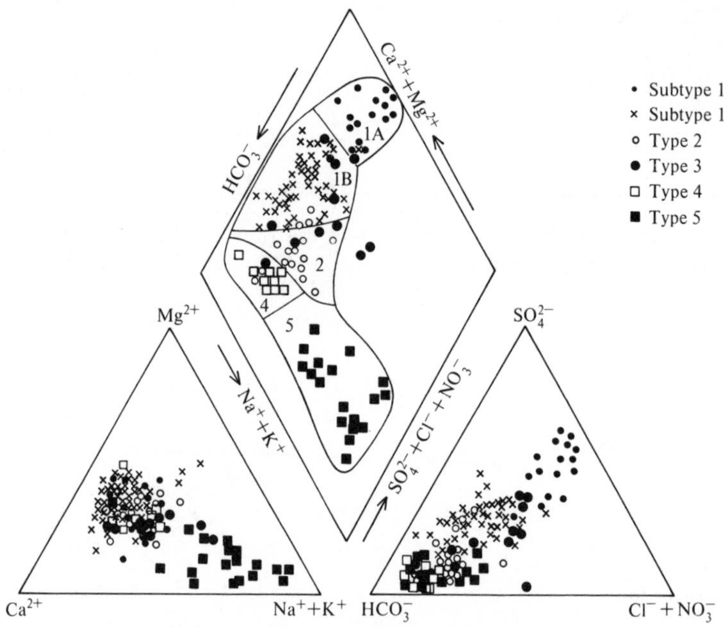


**Fig. 9.15.** Piper diagram showing the variation in groundwater chemistry of the Lower Mersey valley.

The saline waters are best distinguished on the basis of expected concentrations produced by seawater dilution deduced from chloride concentrations. This is considered to be a valid approach since it would seem from their geographical location that at least some of the saline waters represent intrusion from the Mersey Estuary. When this approach is used type A waters are seen to be depleted in potassium, magnesium, and sulphate. Chloride concentrations of some of these waters greatly exceed the seawater value (19 000 mg $l^{-1}$). Type B saline waters have lower chloride contents than seawater, typically only 6000

**Table 9.10** *Representative analyses of groundwater from the Lower Mersey Basin*

| Type | $Ca^{2+}$ (mg $l^{-1}$) | $Mg^{2+}$ (mg $l^{-1}$) | $Na^+$ (mg $l^{-1}$) | $K^+$ (mg $l^{-1}$) | $Cl^-$ (mg $l^{-1}$) | $SO_4^{2-}$ (mg $l^{-1}$) | $HCO_3^-$ (mg $l^{-1}$) | $NO_3^-$ (mg $l^{-1}$) |
|---|---|---|---|---|---|---|---|---|
| 1A | 45 | 21 | 24 | 5.9 | 46 | 125 | 54 | 28 |
| 1B | 66 | 32 | 16 | 4.0 | 31 | 92 | 243 | 18 |
| 2 | 69 | 26 | 38 | 5.3 | 38 | 61 | 313 | 2 |
| 3 | 33 | 13 | 17 | 4.1 | 25 | 36 | 121 | 12 |
| 4 | 68 | 31 | 24 | 4.0 | 22 | 14 | 391 | 0 |
| 5 | 46 | 14 | 152 | 3.7 | 82 | 15 | 455 | 0 |
| A | 1000 | 350 | 24000 | 60 | 37500 | 1800 | 6 | 0 |
| B | 585 | 308 | 727 | 15 | 2772 | 137 | 165 | 2 |

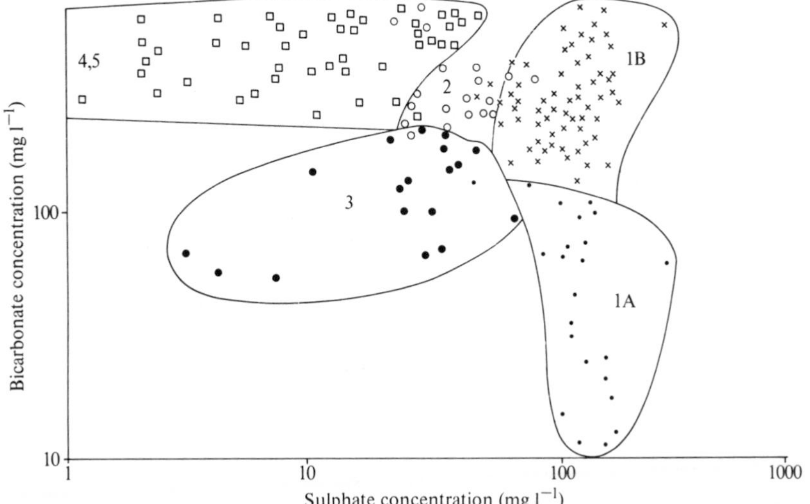


**Fig. 9.16.** *X–Y* plot showing the chemistry of fresh waters from the Lower Mersey valley.

mg $l^{-1}$, and some have reversed ion exchange characteristics (see Fig. 9.5). The iodide concentration of some type B waters is elevated (Lloyd, Howard, Pacey, and Tellam 1982) (see Section 7.2). The distribution of the water types is shown in Fig. 9.17.

The compositional trend shown in Fig. 9.15, which differs so markedly from the trend seen in the Essex example (Section 9.3.2), can be explained as an evolutionary sequence, the chief difference in the processes being the comparative scarcity of carbonate minerals in the sandstone sequences. The high nitrate and slightly elevated chloride concentrations of type I waters are characteristic of modern recharge, which is confirmed by high tritium concentrations. The

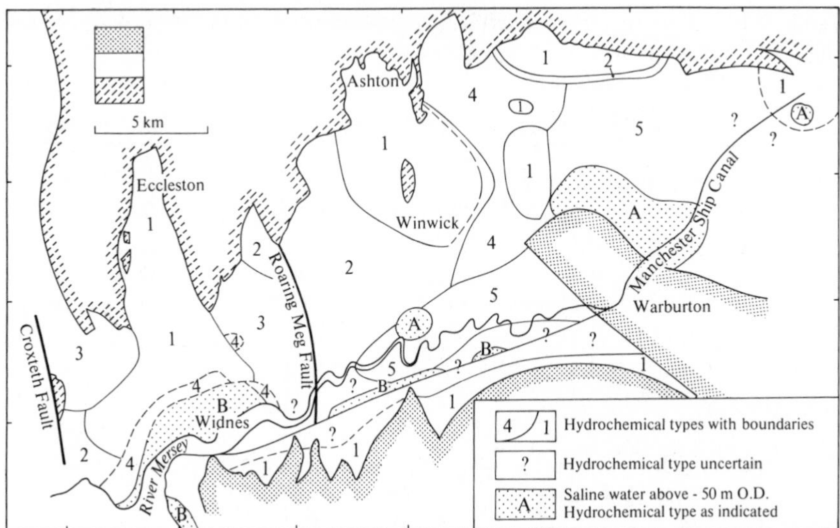


**Fig. 9.17.** Geographical distribution of water types in the Lower Mersey Basin.

relatively high sulphate concentrations are thought to derive in part from pyrite oxidation, with contributions from fertilisers and an above average content of sulphate in rainfall (7 $mg\,l^{-1}$). Type 1A waters are thought to represent the beginning of the evolutionary sequence, the higher calcium and magnesium contents of type 1B representing the effects of slow dolomite dissolution in the upper parts of the aquifer. All type 1 waters are associated with aquifer outcrop or only thin drift cover. As the recharge waters move away from the recharge areas, further dissolution of carbonates occurs, and modelling indicates that this mechanism is dominant in producing type 2 and 4 waters. The low $SO_4^{2-}$, $NO_3^-$, and $Cl^-$ concentrations of type 4 waters are thought to be due in part to low concentrations of these anions at recharge, these waters being old enough to have avoided anthropogenic influences. However, some $SO_4^{2-}$ and $NO_3^-$ reduction is undoubtedly also occurring. Type 2 waters are considered to be equivalents of type 4, but to have suffered mixing with recent recharge water seeping through drift cover as a result of the marked decline in the regional piezometric surfaces. Porewater sampling in the drift deposits indicates high $Cl^-$ and $SO_4^{2-}$ waters are commonly present. Type 5 waters are ion-exchanged equivalents of type 4 waters: they also frequently contain high $Cl^-$ concentrations, and these characteristics strongly suggest that they represent the final stages of the flushing of saline groundwater. $^{14}C$ data suggest that all of the waters present in the region are less than a few thousand years in age, although interpretation of the isotope data is difficult due to the prevalence of mixing. Type 3 waters are associated with relatively inert formations of the aquifer system.

Type A saline waters are characteristically depleted in the heavy water isotopes D and $^{18}O$. The deep location of type A saline waters suggests that they are old ($^{14}C$ dating of these waters is not possible for sampling reasons). Their low magnesium and sulphate contents argue against an origin by reverse osmosis and their low potassium and bromide contents are uncharacteristic of the hypersaline brines in which the overlying mudstones originated. Thus it is suggested that the type A waters originated from dissolution of halite in the overlying beds, which are locally in contact with the aquifer. It would appear hydrogeologically unreasonable that this could have occurred only during the Pleistocene, and it is thus considered that halite dissolution may be a geologically ancient event, and that the depeleted stable isotope signature has resulted from fairly extensive flushing by fresh cool recharge water entering the aquifer during the Pleistocene. From the chemistry of the waters, it is clear that much type B saline water represents modern intrusion from the Mersey Estuary and the Ship Canal, although a proportion undoubtedly arises by upconing of type A waters from depth.

The main hydrogeological conclusions drawn are the distribution of recharge areas, revealed by type 1 and type B saline waters, and the flow directions from type 1 to type 5. Type A saline water is considered to represent *in situ* halite dissolution and is not a result of inflow elsewhere. The coincidence of a water type boundary with the Roaring Meg Fault suggests that this is impermeable. These conclusions, together with hydrogeological data, were used to design and calibrate the aquifer model. The calibration confirmed the hydrochemical conclusions, in particular the location of the recharge areas and the behaviour of the intrusion regions. These latter regions were modelled as fixed-head nodes. Inflow at these nodes during the last half century reproduces the pattern of type B (saline intrusion) waters very closely. Thus the hydrochemical evidence and the model results complement each other.

## 9.4. Concluding statement

These two regional studies, together with the preceding process examples, show that hydrochemical techniques can contribute significantly to the knowledge of the behaviour of complex aquifer systems. They are particularly useful when multi-aquifer situations or strong transmissivity variations make the straightforward interpretation of piezometric data impossible. Hydrochemical and hydraulic techniques each have their uses; together they form an extremely powerful tool. It is essential to have a flexible approach; no one procedure is universally successful, but an integrated study of the chemical data, together with the controls imposed by geological and hydrogeological information, will usually produce results.

## References

Back, W. (1960). Origin of hydrochemical facies of groundwater in the Atlantic Coastal Plain. *Proc. 21st Int. Geological Congr., Part 1*, pp. 87–95.

Barnes, I. and Back, W. (1964). Geochemistry of iron rich groundwater of Southern Maryland. *J. Geol.* **72**, 435–47.

Bath, A. H. and Edmunds, W. M. (1981). Identification of connate water in interstitial solution of chalk sediment. *Geochim. cosmochim. Acta* **45**, 1449–61.

Bögli, A. (1964). Mischungkorrosion, ein Beitrag zum Verkartungsproblem. *Erkunde, Arch. Wisse. Geogr.* **13**, H1/2.

Champ, D. R., Gulkens, J., and Jackson, R. E. (1979). Oxidation–reduction sequences in groundwater systems. *Can. J. earth Sci.* **16**, 12–23.

Cleaves, E. T., Godfrey, A. E., and Bricker, O. P. (1970). Geochemical balance of a small watershed and its geomorphic implications. *Geol. Soc. Am. Bull.* **81**, 3015–32.

Downing, R. A. and Williams, B. P. J. (1969). *The groundwater hydrology of the Lincolnshire Limestone*. Water Resources Board, Reading.

Edmunds, W. M. (1973). Trace element variations across an oxidation–reduction barrier in a limestone aquifer. *Proc. Symp. on Hydrochemistry and Biogeochemistry, Tokyo, 1970*, pp. 500–26.

Freeze, R. A. and Cherry, J. A. (1979). *Groundwater*. Prentice-Hall, Englewood Cliffs, NJ.

Garrels, R. M. and MacKenzie, F. T. (1967). Origin of the chemical composition of some springs and lakes. In *Equilibrium concepts in natural water systems* (ed. R. F. Gould), *Adv. Chem. Ser.* **67**, pp. 336.

Heathcote, J. A. (1981). Hydrochemical aspects of the Gipping Chalk salinity investigation. Ph.D. Thesis. Department of Geological Sciences, University of Birmingham.

Lawrence, A. R., Lloyd, J. W., and Marsh, J. M. (1976). Hydrochemistry and groundwater mixing in part of the Lincolnshire Limestone aquifer, England. *Ground Water* **14**, 12–20.

Lloyd, J, W., Harker, D., and Baxendale, R. A. (1981a). Recharge mechanisms and groundwater flow in the chalk and drift deposits of southern East Anglia. *Q. J. eng. Geol.* **14**, 87–96.

—, Howard, K. W. F., Pacey, N., and Tellam, J. H. (1982). The value of iodide as a parameter in the chemical characterisation of groundwaters. *J. Hydrol.* **57**, 247–65.

Marsh, J. M. (1978). The hydrochemistry of some British aquifers with special reference to trace elements. Ph.D. Thesis. Department of Geological Sciences, University of Birmingham.

Morris, J. C. and Stumm, W. (1967). Redox equilibria and measurements of potentials in the aquatic environment. In *Equilibrium concepts in natural water systems* (ed. R. F. Gould). *Adv. Chem. Ser.* **67**, 270–85.

Price, M., Bird, M. J., and Foster, S. S. D. (1976). Chalk pore size measurements and their significance. *Water Serv.* **80**, 596–600.

Schoff, S. L. (1972). Origin of mineralized water in Precambrian rocks of the upper Paraiba Basin, Paraiba, Brazil. *U.S. Geol. Surv. Water-supply Pap. 1662-H.*

Tardy, Y. (1971). Characterization of the principal weathering types by the geochemistry of waters from some European and African crystalline massifs. *Chem. Geol.* **7**, 253–71.

University of Birmingham (1981). Saline groundwater investigation, Phase 1. Lower Mersey Basin. *Rep. to North West Water Authority, Warrington, England.*

Wigley, T. M. L. and Plummer, L. N. (1976). Mixing of carbonate waters. *Geochim. cosmochim. Acta* **40**, 989–995.

# 10 Water quality criteria

## 10.1. Introduction

Interest in the quality of groundwater for various uses was for many years the only purpose for which hydrochemical analyses were made. The past emphasis on quality criteria can be seen in the derivation of artificial parameters such as alkalinity and hardness which have limited value in studies of groundwater hydrochemical environment and evolution. Although today quality is still a vital factor in assessing a groundwater resource, surprisingly little is known, for example, about the medical effects on humans of many of the constituents found in groundwaters. Further, it is frequently difficult to obtain reliable standards for groundwater quality use in many industries, while standards for irrigation, although available, would appear to rely upon a too simplistic approach.

## 10.2. Groundwater potability

The problems of potability related to inorganic constituents in groundwaters are largely being overwhelmed by contamination resulting from organic pollution; nevertheless important standards are imposed for inorganic concentrations in drinking waters and are better understood than most organic standards. Acceptable standards vary from country to country depending upon economic prosperity, experience, climate, and geographical location. The main standards used are those recommended by the World Health Organization, the European Economic Community, and the United States Environmental Protection Agency. The standards tend to change with time as medical information becomes available so that any set of standards can only be considered as a guide. Where concentrations of a particular determinand are found to be close to the limit of the recommended standard it is advisable to ensure that the criteria being applied are the most up to date.

In Table 10.1 a guide to drinking standards for inorganic parameters is given. While the guide relates to European countries it is clearly applicable world wide.

The guide in some instances shows that a particular parameter may be beneficial at certain concentrations but toxic at other concentrations. In certain cases the optimum concentration range may be very small and open to controversy. If a parameter has a two-fold importance, dose-response curves based upon case history and laboratory studies data are compiled. A general dose-response curve is shown in Fig. 10.1(a) with the curve for fluoride in Fig. 10.1(b).

While it is important that strict attention is paid to drinking water standards, conflicting evidence does occur. For example although the sulphate MAC is

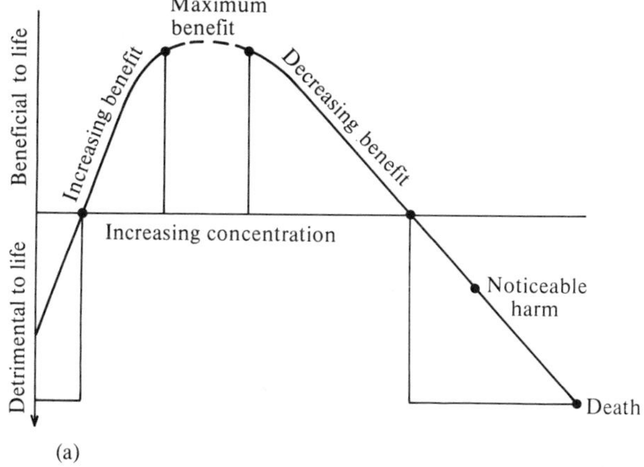


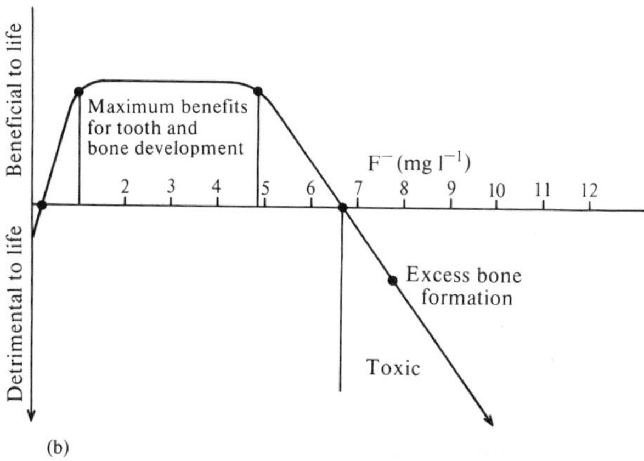


**Fig. 10.1.** Dose–response curves for human consumption: (a) general dose-response curve: (b) fluoride dose-response curve.

250 mg $l^{-1}$, studies in the southern United States have shown that an excess of deaths from hypertension and hypertensive heart disease always occurred in areas where sulphate in water is less than 350 mg $l^{-1}$. Further, the same study showed that although the MAC for magnesium is 50 mg $l^{-1}$, increased deaths from cancer and prostate difficulties occurred in areas where groundwaters contained less than 50 mg $l^{-1}$.

In many cases the standards set are well below toxic or problematic concentrations owing to a lack of firm data. A case in point is water hardness for which the MRC is 10 mg $l^{-1}$. Statistical evidence has shown that a decrease in hardness can be related to cardiovascular death rates for men but no correlation has been shown for women (Lacey 1981).

**Table 10.1** *A guide to drinking water standards*†

| Parameter | Units | Guide level (GL) | Maximum admissible concentration (MAC) | Minimum required concentration (MRC) | Comments |
|---|---|---|---|---|---|
| pH | pH units | 6.5–8.5 | 9.5 | 6.0 | |
| Dissolved oxygen | mg $l^{-1}$ $O_2$ | 1 | 5 | | |
| EC | $\mu S\ cm^{-1}$ | 400 | 1250 | | Related to TDS |
| TDS | mg $l^{-1}$ | | 1500 | | Limit set for taste |
| Total hardness | mg $l^{-1}$ $CaCO_3$ | 35 | | 10 | Decrease may cause cardiovascular death |
| $Ca^{2+}$ | mg $l^{-1}$ | 100 | | 10 | |
| $Mg^{2+}$ | mg $l^{-1}$ | 30 | 50 | 5 | |
| $Na^{+}$ | mg $l^{-1}$ | 20 | 100 | | Normally beneficial and required by body |
| $K^{+}$ | mg $l^{-1}$ | 10 | 12 | | |
| Al | mg $l^{-1}$ | | 0.05 | | |
| $HCO_3^-$ | mg $l^{-1}$ | 30 | | | |
| $SO_4^{2-}$ | mg $l^{-1}$ | 5 | 250 | | Excess causes intestinal irritations |
| $Cl^-$ | mg $l^{-1}$ | 25 | 200 | | Limits set for taste |
| $NO_3^-$ | mg $l^{-1}$ | | 50 | | Methemoglobinemia in infants |
| $NO_2^-$ | mg $l^{-1}$ | | 0.1 | | |
| Ag | $\mu g\ l^{-1}$ | | 10 | | |

| | | | | |
|---|---|---|---|---|
| As | $\mu g\ l^{-1}$ | | 50 | Serious cumulative systematic poisoning |
| $Ba^{2+}$ | $\mu g\ l^{-1}$ | | 100 | Muscle and heart stimulant |
| Cd | $\mu g\ l^{-1}$ | | 5 | |
| $CN^-$ | $\mu g\ l^{-1}$ | | 50 | Rapid fatal poison |
| Cr | $\mu g\ l^{-1}$ | | 50 | |
| Cu | $\mu g\ l^{-1}$ | | 50 | Adults need about 1 mg Cu per day |
| $F^-$ | $\mu g\ l^{-1}$ | 700–1200 | 1400–2400 | Beneficial in small amounts |
| Fe | $\mu g\ l^{-1}$ | 100 | 300 | |
| Hg | $\mu g\ l^{-1}$ | | 1 | Large amounts cause brain damage |
| $H_2S$ | $\mu g\ l^{-1}$ | | 0 | |
| Mn | $\mu g\ l^{-1}$ | 20 | 50 | |
| Mo | $\mu g\ l^{-1}$ | | | Excessive intakes may be toxic |
| Ni | $\mu g\ l^{-1}$ | 5 | 50 | Can cause dermatitis |
| P | $\mu g\ l^{-1}$ | 300 | 2000 | |
| Pb | $\mu g\ l^{-1}$ | | 50 | Serious cumulative body poison |
| Sb | $\mu g\ l^{-1}$ | | 10 | Similar to As but less acute |
| Se | $\mu g\ l^{-1}$ | | 10 | Small amounts beneficial; large amounts toxic |
| Zn | $\mu g\ l^{-1}$ | | 100 | Very small amounts beneficial |

† Largely based upon the EEC standards (European Economic Community 1975).

To supplement the guide standards given in Table 10.1, standards for isotopes and radiation limits in groundwaters are given in Tables 10.2 and 10.3 respectively.

**Table 10.2** *Maximum permissible isotope concentration in drinking water*

| Isotope | Half-life (days) | Concentration (mg $l^{-1}$) | Isotope | Half-life (days) | Concentration (mg $l^{-1}$) |
|---|---|---|---|---|---|
| $^{18}F$ | 0.078 | $2.2 \times 10^{-10}$ | $^{64}Cu$ | 0.54 | $1.3 \times 10^{-10}$ |
| $^{24}Na$ | 0.62 | $9.1 \times 10^{-11}$ | $^{65}Zn$ | 250 | $2.5 \times 10^{-8}$ |
| $^{32}P$ | 14.3 | $6.9 \times 10^{-11}$ | $^{89}Sr$ | 53 | $2.5 \times 10^{-11}$ |
| $^{36}Cl$ | $1.6 \times 10^{8}$ | $1.7 \times 10^{-2}$ | $^{90}Sr$ | $9.1 \times 10^{3}$ | $5.0 \times 10^{-10}$ |
| $^{45}Ca$ | 152 | $5.3 \times 10^{-10}$ | $^{131}I$ | 8 | $4.6 \times 10^{-11}$ |
| $^{51}Cr$ | 26.5 | $2.1 \times 10^{-8}$ | $^{198}Au$ | 2.69 | $2.4 \times 10^{-10}$ |
| $^{56}Mn$ | 0.108 | $1.4 \times 10^{-11}$ | $^{210}Pb$ | $9.1 \times 10^{3}$ | $2.9 \times 10^{-10}$ |
| $^{55}Fe$ | $1.06 \times 10^{3}$ | $2.3 \times 10^{-7}$ | | | |
| $^{59}Fe$ | 43.3 | $2.1 \times 10^{-10}$ | | | |
| $^{60}Co$ | $1.9 \times 10^{3}$ | $3.6 \times 10^{-8}$ | | | |

After Collins 1960.

**Table 10.3** *Maximum radiation limits permissible in drinking water*

| Isotope | Limit ($\mu$Ci $l^{-1}$) | Isotope | Limit ($\mu$Ci $l^{-1}$) |
|---|---|---|---|
| $^{24}Na$ | 0.2 | $^{76}As$ | 0.02 |
| $^{32}P$ | 0.02 | $^{90}Sr$ | $4 \times 10^{-4}$ |
| $^{51}Cr$ | 1.6 | $^{122}Sb$ | 0.03 |
| $^{60}Co$ | 0.05 | $^{131}I$ | $2 \times 10^{-3}$ |
| $^{64}Cu$ | 0.3 | $^{137}Cs$ | $1.5 \times 10^{-2}$ |

After International Atomic Energy Agency 1969.

## 10.3. Livestock water standards

Standards for livestock are not readily available, so that for the most part potable standards for humans are applied. Standards vary depending upon stock type and size and feeding habits. In Table 10.4, EC standards based on American data are given, while key quality criteria for toxic substances are listed in Table 10.5.

## 10.4. Quality of water for industrial use

Water quality requirements for industry are impossible to summarize except very superficially. The requirements vary considerably between countries, industries,

**Table 10.4** *Electrical conductivity (EC) guide for drinking water for livestock and poultry*

| Total soluble salts Content of waters (mg $l^{-1}$) | Comment |
|---|---|
| Less than 1000 mg $l^{-1}$ (EC <1.5 mS $cm^{-1}$) | Relatively low level of salinity. Excellent for all classes of livestock and poultry. |
| 1000–3000 mg $l^{-1}$ (EC = 1.5–5 mS $cm^{-1}$) | Very satisfactory for all classes of livestock and poultry. May cause temporary and mild diarrhoea in livestock not accustomed to it or watery droppings in poultry. |
| 3000–5000 mg $l^{-1}$ (EC = 5–8 mS $cm^{-1}$) | Satisfactory for livestock, but may cause temporary diarrhoea or be refused at first by animals not accustomed to it. Poor water for poultry, often causing water faeces, increased mortality, and decreased growth, especially in turkeys. |
| 5000–7000 mg $l^{-1}$ (EC = 8–11 mS $cm^{-1}$) | Can be used with reasonable safety for dairy and beef cattle, sheep, swine, and horses. Avoid use for pregnant or lactating animals. Not acceptable for poultry. |
| 7000–10000 mg $l^{-1}$ (EC = 11–16 mS $cm^{-1}$) | Unfit for poultry and probably for swine. Considerable risk in using for pregnant or lactating cows, horses, or sheep, or for the young of these species. In general, use should be avoided, although older ruminants, horses, and swine may subsist on them under certain conditions. |
| Over 10 000 mg $l^{-1}$ (EC > 16 mS $cm^{-1}$) | Risks with these highly saline waters are so great that they cannot be recommended for use under any conditions. |

After National Academy of Sciences and National Academy of Engineering 1972.

and processes, and will be specific to a particular product. Some suggested water quality tolerances are given in Table 10.6 based on the recommendations of the New England Water Works Association (1940).

One of the main problems of industrial water quality is related to boiler feed waters in which waters are heated up to considerable temperatures normally under high pressures. Under these conditions carbonate hydrochemistry and to a lesser extent sulphate hydrochemistry are particularly important. Changes in chemistry associated with the temperature and pressure changes can be assessed thermodynamically as discussed in Section 3.4. They are very important and have been subject to much study. Garrels, Thompson, and Siever (1960) discuss the problem of the carbonate chemistry at length while Denman (1961) has examined sulphate hydrochemistry at a range of temperatures. Both authors provide charts for the calculation of mineral stability. As a general summary of boiler feed requirements tolerance limits are included in Table 10.7 taken from Walton (1970).

In the construction industry groundwater and soil are normally analysed for sulphates which can cause the deterioration of concrete. To combat the problem, sulphate-resisting cements have been developed to suit a range of sulphate concentrations. For this particular work the sulphate salt is normally quoted as sulphur trioxide ($SO_3$). The requirements for concrete that may be

**Table 10.5** *Recommendations for levels of toxic substances in drinking water for livestock*

| Constituent | Upper limit ($mg\ l^{-1}$) |
|---|---|
| Aluminium (Al) | 5 |
| Arsenic (As) | 0.2 |
| Beryllium (Be) | No data |
| Boron (B) | 5.0 |
| Cadmium (Cd) | 0.05 |
| Chromium (Cr) | 1.0 |
| Cobalt (Co) | 1.0 |
| Copper (Cu) | 0.5 |
| Fluoride (F) | 2.0 |
| Iron (Fe) | No data |
| Lead (Pb) | 0.1[a] |
| Manganese (Mn) | No data |
| Mercury (Hg) | 0.01 |
| Molybdenum (Mo) | No data |
| Nitrate + Nitrite ($NO_3$-N+$NO_2$-N) | 100 |
| Nitrite ($NO_2$-N) | 10 |
| Selenium (Se) | 0.5 |
| Vanadium (V) | 0.10 |
| Zinc (Zn) | 24 |
| TDS | 10 000[b] |

After National Academy of Sciences and National Academy of Engineering 1972.

[a] Lead is cumulative and problems may begin at a threshold value of 0.05 $mg\ l^{-1}$.

[b] See Table 10.4.

subject to sulphate attack are given in Table 10.8. The specifications are for British Standards (BS) 882 and 1047 which detail aggregates used in concrete.

## 10.5. Irrigation water

The most extensive use of groundwater in the world is for the irrigation of crops. As a result a considerable amount of study has gone into the susceptibility of plants and their growth to the quality of the water used for irrigation. Although this book only has the scope to consider the hydrochemical factors pertinent in irrigation water suitability, it must be stressed that other criteria are also important, such as the soil and crops to be irrigated, local climate, and management

## Table 10.6 *Suggested water quality tolerances for industrial uses (allowable limits)*

| Industry or use | Turbidity | Colour | Odour and taste | Iron as Fe ($mg\ l^{-1}$) | Manganese as Mn ($mg\ l^{-1}$) | Total solids ($mg\ l^{-1}$) | Hardness as $CaCO_3$ ($mg\ l^{-1}$) | Alkalinity as $CaCO_3$ ($mg\ l^{-1}$) | Hydrogen sulphide | Health | pH | Other requirements |
|---|---|---|---|---|---|---|---|---|---|---|---|---|
| Air conditioning | — | — | Low | 0.5 | 0.5 | — | — | — | 1.0 | — | — | No corrosiveness or slime formation |
| Baking | 10 | 10 | Low | 0.2 | 0.2 | — | — | — | 0.2 | Potable | — | |
| Brewing and distilling | | | | | | | | | | | | 275 $mg\ l^{-1}$ NaCl |
| Light beer, gin | 10 | — | Low | 0.1 | 0.1 | 500 | — | 75 | 0.2 | Potable | 6.5–7.0 | |
| Dark beer, whisky | 10 | — | Low | 0.1 | 0.1 | 1000 | — | 150 | 0.2 | Potable | 7.0 | |
| Canning | | | | | | | | | | | | |
| Legumes | 10 | — | Low | 0.2 | 0.2 | — | 25–75 | — | 1.0 | Potable | — | |
| General | 10 | — | Low | 0.2 | 0.2 | — | — | — | 1.0 | Potable | — | |
| Carbonated beverages | 2 | 10 | Low | 0.2 | 0.2 | 850 | 250 | 50–100 | 0.2 | Potable | — | Organic matter infinitesimal; oxygen consumed, 1.5 $mg\ l^{-1}$ |
| Confectionery | — | — | Low | 0.2 | 0.2 | 100 | — | — | 0.2 | Potable | 7.0 | |
| Cooling | 50 | — | — | 0.2 | 0.5 | — | 50 | — | 5 | — | — | No corrosiveness or slime formation |
| Food, general | 10 | — | Low | 0.2 | 0.2 | — | — | — | — | Potable | — | |
| Ice | 5 | 5 | Low | 0.2 | 0.2 | 1300 | — | — | — | Potable | — | 10 $mg\ l^{-1}$ $SiO_2$ |
| Laundering | — | — | — | 0.2 | 0.2 | — | 50 | — | — | — | — | |
| Plastics, clear | 2 | 2 | — | 0.02 | 0.02 | 200 | — | — | — | — | — | No grit or corrosiveness |
| Paper and pulp | | | | | | | | | | | | |
| Ground wood | 50 | 20 | — | 1.0 | 0.5 | — | 180 | — | — | — | — | |
| Kraft pulp | 25 | 15 | — | 0.2 | 0.1 | 300 | 100 | — | — | — | — | |
| Soda and sulphide pulp | 15 | 10 | — | 0.1 | 0.05 | 200 | 100 | — | — | — | — | |
| High-grade light papers | 5 | 5 | — | 0.0 | 0.05 | 200 | 50 | — | — | — | — | No slime formation |
| Rayon (viscose) | | | | | | | | | | | | |
| Pulp production | 5 | 5 | — | 0.05 | 0.03 | 100 | 8 | 50 | — | — | — | 8 $mg\ l^{-1}$ $OH^-$, 8 $mg\ l^{-1}$ $Al_2O_3$, 25 $mg\ l^{-1}$ $SiO_2$, 5 $mg\ l^{-1}$ Cu |
| Manufacture | 0.3 | — | — | 0.0 | 0.0 | — | 55 | — | — | — | 7.8–8.3 | |
| Steel manufacture | — | — | — | — | — | — | 50 | — | — | — | 6.8–7.0 | Temperature 24 °C; 175 $mg\ l^{-1}$ $Cl^-$; suspended matter, 25 $\mu g\ l^{-1}$; minimum organic content and corrosiveness |
| Sugar manufacture | — | — | — | 0.1 | — | — | — | — | — | — | — | 20 $mg\ l^{-1}$ $Ca^{2+}$, 10 $mg\ l^{-1}$ $Mg^{2+}$, 20 $mg\ l^{-1}$ $SO_4^{2-}$ 20 $mg\ l^{-1}$ $Cl^-$, 100 $mg\ l^{-1}$ $HCO_3^-$, |
| Synthetic rubber | — | — | — | — | — | — | 50 | — | — | — | — | Oxygen consumed, 3.0 $mg\ l^{-1}$, minimum organic content and corrosiveness |
| Tanning | 20 | 10–100 | — | 0.2 | 0.2 | — | 50–135 | 135 | — | — | — | 8 $mg\ l^{-1}$ $OH^-$ |
| Textiles | | | | | | | | | | | | |
| General | 5 | 20 | — | 0.25 | 0.25 | — | — | — | — | — | — | Constant composition, |
| Dyeing | 5 | 5–20 | — | 0.25 | 0.25 | 200 | — | — | — | — | — | <0.5 $mg\ l^{-1}$ residual alumina |

Modified from Walton 1970.

**Table 10.7** *Tolerance limits for boiler-feed water*[a]

| | Pressure[b] (lbf in$^{-2}$) | | | |
|---|---|---|---|---|
| | 0–150 | 150–250 | 250–400 | Over 400 |
| Turbidity | 20 | 10 | 5 | 1 |
| Colour | 80 | 40 | 5 | 2 |
| Oxygen consumed | 15 | 10 | 4 | 3 |
| Dissolved oxygen | 1.4 | 0.14 | 0 | 0 |
| Hydrogen sulphide ($H_2S$) | 5 | 3 | 0 | 0 |
| Total hardness as $CaCO_3$ | 80 | 40 | 10 | 2 |
| Sulphate carbonate ratio $Na_2SO_4/Na_2CO_3$ | 1/1 | 2/1 | 3/1 | 3/1 |
| Aluminium oxide ($Al_2O_3$) | 5 | 0.5 | 0.5 | 0.01 |
| Silica ($SiO_2$) | 40 | 20 | 5 | 1 |
| Bicarbonate ($HCO_3^-$) | 50 | 30 | 5 | 0 |
| Carbonate ($CO_3^{2-}$) | 200 | 100 | 40 | 20 |
| Hydroxide ($OH^-$) | 50 | 40 | 30 | 15 |
| Total solids | 3000–500 | 2500–500 | 1500–100 | 50 |
| pH value (minimum) | 8.0 | 8.4 | 9.0 | 9.6 |

[a] It is also required that there be no corrosiveness or slime formation.
[b] 1 lbf in$^{-2}$ $\equiv$ 0.07031 kg cm$^{-2}$.

of irrigation and drainage, before any final decisions can be made with respect to water use.

Water quality constraints in irrigation can be examined using a number of empirical indices that have been established on the basis of field experience and experiments. The indices discussed below provide a good practical guide to water suitability, the application of which is comprehensively explained for example by U.S. Department of Interior (1968), Doorenbos and Pruitt (1975), Sayers and Westcot (1976), and Shainberg and Oster (1978).

### *10.5.1. Total salt concentration hazard*

Osmotic processes form the most important life function in plants so that any changes in osmotic conditions in the root zone can automatically change the rate of water flow to a plant. Normally osmotic effects are caused by total salt concentrations and, as shown in Fig. 10.2, the osmotic pressure of soil waters relates directly to electrical conductivity. In the range of EC that will permit plant growth the osmotic pressure (OP in atmospheres) has the relation:

$$OP = 0.36 \text{ EC mS cm}^{-1} \qquad (10.1)$$

Total salt concentration is usually measured as EC in irrigation work and the United States Salinity Laboratory has established guide groupings of waters based on this parameter as shown in Table 10.9.

**Table 10.8** *Requirements for concrete exposed to sulphate attack*

| Concentration of sulphates expressed as $SO_3$ | | | | Type of cement | Requirements for dense, fully compacted concrete made with aggregates meeting the requirements of BS 882 or BS 1047 | | | |
|---|---|---|---|---|---|---|---|---|
| Class | In soil | | In ground water (parts per thousand) | | Minimum cement content (kg $m^{-3}$) Nominal maximum size of aggregate | | | Maximum free water/ cement ratio |
| | Total $SO_3$ (per cent) | $SO_3$ in 2: 1 water: soil extract (g $l^{-1}$) | | | 40 mm | 20 mm | 10 mm | |
| 1 | Less than 0.2 | — | Less than 30 | Ordinary Portland or Portland blast furnace | 240 | 280 | 330 | 0.55 |
| 2 | 0.2–0.5 | — | 30–120 | Ordinary Portland or Portland blast furnace | 290 | 330 | 380 | 0.50 |
| | | | | Sulphate-resisting Portland | 240 | 280 | 330 | 0.55 |
| | | | | Supersulphated | 270 | 310 | 360 | 0.50 |
| 3 | 0.5–1.0 | 1.9–3.1 | 120–250 | Sulphate-resisting Portland or supersulphated | 290 | 330 | 380 | 0.50 |
| 4 | 1.0–2.0 | 3.1–5.6 | 250–500 | Sulphate-resisting Portland or supersulphated | 330 | 370 | 420 | 0.45 |
| 5 | Over 2 | Over 5.6 | Over 500 | As for Class 4, but with the addition of adequate protective coatings of inert material such as asphalt or bituminous emulsions reinforced with fibreglass membranes. | | | | |

After Building Research Establishment 1975.

This table applies only to concrete made with aggregates complying with the requirements of BS 882 or BS 1047 placed in near-neutral groundwaters of pH 6–9, containing naturally occurring sulphates but not contaminants such as ammonium salts. Concrete prepared from ordinary Portland cement would not be recommended in acidic conditions (pH 6 or less); sulphate-resisting Portland cement is slightly more acid resistant but no experience of large-scale use in these conditions is currently available. Supersulphated cement has given an acceptable life, provided that the concrete is dense and prepared with a free water/cement ratio of 0.40 or less, in mineral acids down to pH 3.5.

The cement contents given in Class 2 are the minima recommended by the manufacturers. For $SO_3$ contents near the upper limit of Class 2, cement contents above these minima are advised.

For severe conditions, e.g. thin sections, sections under hydrostatic pressure on one side only, and sections partly immersed, consideration should be given to a further reduction of the water/cement ratio and, if necessary, an increase in cement content to ensure the degree of workability needed for full compaction and thus minimum permeability.

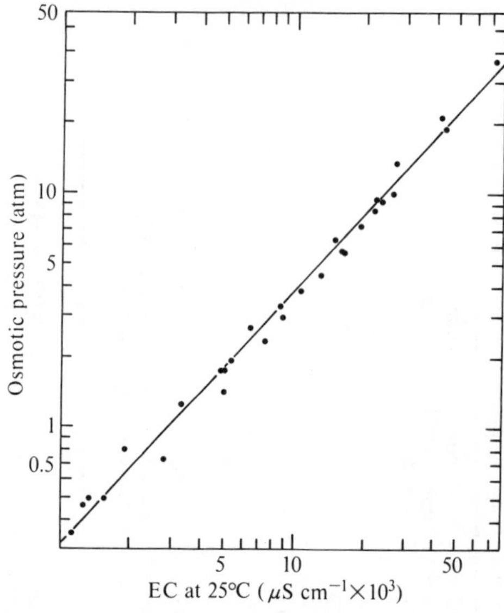


**Fig. 10.2.** Osmotic pressure of soil extracts as related to the electrical conductivity EC. (After Shainberg and Oster 1978.)

**Table 10.9** *Groups of irrigation waters based on electrical conductivity*

| TDS (mg $l^{-1}$) | EC × $10^6$ at 25 °C ($\mu$S $cm^{-1}$) | Class | Remarks |
|---|---|---|---|
| < 200 | <250 | $C_1$ | Low salinity water—can be used for irrigation with most crops on most soils with little likelihood that a salinity problem will develop. Some leaching is required, but this occurs under normal irrigation practices except in soils of extremely low permeability. |
| 200–500 | 250–750 | $C_2$ | Medium salinity water—can be used if a moderate amount of leaching occurs. Plants with moderate salt tolerance can be grown in most instances without special practices for salinity control. |
| 500–1500 | 750–2250 | $C_3$ | High salinity water—cannot be used on soils with restricted drainage, special management for salinity control may be required and plants with good salt tolerance should be selected. |
| 1500–3000 | 2250–5000 | $C_4$ | Very high salinity water—is not suitable for irrigation under ordinary conditions but may be used occasionally under very special circumstances. The soil must be permeable, drainage must be adequate, irrigation water must be applied in excess to provide considerable leaching, and very salt-tolerant crops should be selected. |

After U.S. Salinity Laboratory 1954.

Although the salinity of irrigation water is of major importance, the evolution of salinity within the root zone is the controlling factor on plant growth and crop yield. Shainberg and Oster (1978) represent root zone salinity in terms of a total salt balance:

$$\Delta S_g = D_r C_r + D_g C_g + D_i C_i + S_m - D_d C_d - S_p - S_c \qquad (10.2)$$

where $D$ is the amount of water in terms of depth (e.g. millimetres of irrigation water applied), $C$ is the salt concentration, r denotes the rainfall, g denotes the rising groundwater, i denotes the irrigation application, d denotes the water draining beneath root zone, $S_m$ is the amount of salt dissolved from the soil in the root zone, $S_p$ is the amount of salt precipitated in the root zone, and $S_c$ is the amount of salt removed in the harvested crop. While the balance is composed of simply defined parameters, their accurate determination requires careful field experimentation with the use of lysimeters, for example, before the balance can be applied.

Clearly under simple balance conditions the relationship between the salinity of irrigation application water and the water draining beneath the root zone is a dominant control. A leaching requirement (LR) can be used to assess the relationship:

$$\mathrm{LR} = \mathrm{EC_i}/\mathrm{EC_d^1} = \mathrm{D_d^1}/\mathrm{D_i} \qquad (10.3)$$

where $\mathrm{EC_d^1}$ is the maximum permissible conductivity of drainage water and $\mathrm{D_d^1}$ is the minimum permissible amount (as a depth) of drainage water to maintain high yields.

Maas and Hoffman (1977) in an extensive study have assessed plant tolerance to salinity with respect to leaching requirements and have compiled much of the data shown in Table 10.10 which provides a comprehensive guide.

### *10.5.2. Sodium hazard*

Although account has to be taken of total salt concentrations in groundwaters used for irrigation, specific ion effects are also very important. Owing to its effects on both soil and plants sodium is one of the governing specific ions. A general classification of waters based on sodium concentration is given in Table 10.11. This, however, can only be taken as an initial guide as sodium has to be considered in relation to other parameters.

A number of indices have been proposed to assess sodium effects and the equilibrium between soil chemistry and soil water chemistry. When irrigation water is applied a soil–water hydrochemical equilibrium is eventually reached which will control plant growth. At one time this was assessed by the soluble sodium percentage (SSP) of a water defined as

$$\mathrm{SSP} = \frac{100\,(\mathrm{Na^+} + \mathrm{K^+})}{\mathrm{Ca^{2+}} + \mathrm{Mg^{2+}} + \mathrm{Na^+} + \mathrm{K^+}} \qquad (10.4)$$

where the ionic concentrations are in milliequivalents per litre.

**Table 10.10** *Salt tolerance of agricultural crops and yield decrease to be expected due to salinity of irrigation water*

| Yield decrement | 0% | | | 10% | | | 25% | | | $EC_d$[c] |
|---|---|---|---|---|---|---|---|---|---|---|
| | $EC_e$[a] | $EC_{iw}$[b] | LR[d] | $EC_e$ | $EC_{iw}$ | LR[d] | $EC_e$ | $EC_{iw}$ | LR[d] | |
| *Fruit crops* | | | | | | | | | | |
| Date | 4.0 | 2.7 | 4 | 6.8 | 4.5 | 7 | 11 | 7.3 | 12 | 64 |
| Grapefruit | 1.8 | 1.2 | 8 | 2.4 | 1.6 | 10 | 3 | 2.2 | 14 | 16 |
| Orange | 1.7 | 1.1 | 7 | 2.3 | 1.6 | 10 | 3 | 2.2 | 14 | 16 |
| Apricot | 1.6 | 1.1 | 9 | 2.0 | 1.3 | 12 | 3 | 1.8 | 15 | 12 |
| Peach | 1.7 | 1.1 | 9 | 2.2 | 1.5 | 11 | 3 | 1.9 | 15 | 13 |
| Almond | 1.5 | 1.0 | 7 | 2.0 | 1.4 | 10 | 3 | 1.9 | 14 | 14 |
| Blackberry | 1.5 | 1.0 | 8 | 2.0 | 1.3 | 11 | 3 | 1.8 | 15 | 12 |
| Boysenberry | 1.5 | 1.0 | 8 | 2.0 | 1.3 | 11 | 3 | 1.8 | 15 | 12 |
| Grape | 1.5 | 1.0 | 4 | 2.5 | 1.7 | 7 | 4 | 2.7 | 12 | 24 |
| Plum | 1.5 | 1.0 | 7 | 2.1 | 1.4 | 10 | 3 | 1.9 | 14 | 14 |
| Strawberry | 1.0 | 0.7 | 8 | 1.3 | 0.9 | 11 | 2 | 1.2 | 15 | 8 |
| *Vegetable crops* | | | | | | | | | | |
| Beet | 4.0 | 2.7 | 9 | 5.1 | 3.4 | 11 | 7 | 4.5 | 15 | 30 |
| Broccoli | 2.8 | 1.9 | 7 | 3.9 | 2.6 | 10 | 6 | 3.7 | 14 | 27 |
| Cucumber | 2.5 | 1.7 | 8 | 3.3 | 2.2 | 11 | 4 | 2.9 | 15 | 20 |
| Tomato | 2.5 | 1.7 | 7 | 3.5 | 2.3 | 9 | 5 | 3.4 | 13 | 25 |
| Spinach | 2.0 | 1.3 | 4 | 3.3 | 2.2 | 7 | 5 | 3.5 | 12 | 30 |
| Cabbage | 1.8 | 1.2 | 5 | 2.8 | 1.9 | 8 | 4 | 2.9 | 12 | 24 |
| Potato | 1.7 | 1.1 | 6 | 2.5 | 1.7 | 8 | 4 | 2.5 | 13 | 20 |
| Sweet corn | 1.7 | 1.1 | 6 | 2.5 | 1.7 | 8 | 4 | 2.5 | 13 | 20 |
| Pepper | 1.5 | 1.0 | 6 | 2.2 | 1.5 | 9 | 3 | 2.2 | 13 | 17 |
| Sweet potato | 1.5 | 1.0 | 5 | 2.4 | 1.6 | 8 | 4 | 2.5 | 12 | 21 |
| Lettuce | 1.3 | 0.9 | 5 | 2.1 | 1.4 | 8 | 3 | 2.1 | 12 | 18 |
| Onion | 1.2 | 0.8 | 5 | 1.8 | 1.2 | 8 | 3 | 1.8 | 12 | 15 |
| Radish | 1.2 | 0.8 | 5 | 2.0 | 1.3 | 7 | 3 | 2.1 | 12 | 18 |
| Carrot | 1.0 | 0.7 | 4 | 1.7 | 1.1 | 7 | 3 | 1.9 | 11 | 16 |
| *Forage crops* | | | | | | | | | | |
| Fairway wheatgrass | 7.5 | 5.0 | 11 | 8.9 | 6.0 | 14 | 11 | 7.4 | 17 | 44 |
| Tall wheatgrass | 7.5 | 5.0 | 8 | 9.9 | 6.6 | 11 | 13 | 9.0 | 14 | 63 |
| Bermuda grass | 6.9 | 4.6 | 10 | 8.5 | 5.6 | 13 | 11 | 7.2 | 16 | 45 |
| Barley (hay) | 6.0 | 4.0 | 10 | 7.4 | 4.9 | 12 | 10 | 6.3 | 16 | 40 |
| Perennial ryegrass | 5.6 | 3.7 | 10 | 6.9 | 4.6 | 12 | 9 | 5.9 | 16 | 38 |
| Birdsfoot trefoil | 5.0 | 3.3 | 11 | 6.0 | 4.0 | 13 | 7 | 5.0 | 17 | 30 |
| Harding grass | 4.6 | 3.1 | 9 | 5.9 | 3.9 | 11 | 8 | 5.3 | 15 | 36 |
| Tall fescue | 3.9 | 2.6 | 6 | 5.8 | 3.9 | 9 | 9 | 5.7 | 13 | 46 |
| Crested wheatgrass | 3.5 | 2.3 | 4 | 6.0 | 4.0 | 7 | 10 | 6.5 | 11 | 57 |
| Vetch | 3.0 | 2.0 | 8 | 3.9 | 2.6 | 11 | 5 | 3.5 | 15 | 24 |
| Sudan grass | 2.8 | 1.9 | 4 | 5.1 | 3.4 | 7 | 9 | 5.7 | 11 | 52 |
| Beardless wild rye | 2.7 | 1.8 | 5 | 4.4 | 2.9 | 8 | 7 | 4.6 | 12 | 39 |
| Big trefoil | 2.3 | 1.5 | 10 | 2.8 | 1.9 | 13 | 4 | 2.4 | 16 | 15 |
| Alfalfa | 2.0 | 1.3 | 4 | 3.4 | 2.2 | 7 | 5 | 3.6 | 12 | 31 |
| Love grass | 2.0 | 1.3 | 5 | 3.2 | 2.1 | 8 | 5 | 3.3 | 12 | 28 |
| Orchard grass | 1.5 | 1.0 | 3 | 3.1 | 2.1 | 6 | 5 | 3.7 | 11 | 35 |
| Meadow foxtail | 1.5 | 1.0 | 4 | 2.5 | 1.7 | 7 | 4 | 2.7 | 12 | 24 |
| Alsike clover | 1.5 | 1.0 | 5 | 2.3 | 1.6 | 8 | 4 | 2.4 | 12 | 20 |
| Berseem clover | 1.5 | 1.0 | 3 | 3.3 | 2.2 | 6 | 6 | 3.9 | 10 | 38 |

| Yield decrement | 0% | | | 10% | | | 25% | | | $EC_d$[c] |
|---|---|---|---|---|---|---|---|---|---|---|
| | $EC_e$[a] | $EC_{iw}$[b] | LR[d] | $EC_e$ | $EC_{iw}$ | LR[d] | $EC_e$ | $EC_{iw}$ | LR[d] | |
| *Field crops* | | | | | | | | | | |
| Barley | 8.0 | 5.3 | 10 | 10.0 | 6.7 | 12 | 13 | 8.7 | 11 | 56 |
| Cotton | 7.7 | 5.1 | 10 | 9.6 | 6.4 | 12 | 13 | 8.3 | 16 | 54 |
| Sugarbeet | 7.0 | 4.7 | 10 | 8.7 | 5.8 | 12 | 11 | 7.5 | 16 | 48 |
| Wheat | 6.0 | 4.0 | 10 | 7.4 | 4.9 | 12 | 10 | 6.3 | 16 | 40 |
| Soybean | 5.0 | 3.3 | 17 | 5.5 | 3.7 | 18 | 6 | 4.2 | 21 | 13 |
| Peanut | 3.2 | 2.1 | 16 | 3.5 | 2.4 | 18 | 4 | 2.7 | 20 | 13 |
| Rice (paddy) | 3.0 | 2.0 | 9 | 3.8 | 2.6 | 11 | 5 | 3.4 | 15 | 23 |
| Sesbania | 2.3 | 1.5 | 5 | 3.7 | 2.5 | 8 | 6 | 3.9 | 12 | 33 |
| Corn (grain) | 1.7 | 1.1 | 6 | 2.5 | 1.7 | 8 | 4 | 2.5 | 13 | 20 |
| Sugarcane | 1.7 | 1.1 | 3 | 3.4 | 2.3 | 6 | 6 | 4.0 | 11 | 37 |
| Broad bean | 1.6 | 1.1 | 4 | 2.6 | 1.8 | 7 | 4 | 2.8 | 12 | 24 |
| Flax | 1.7 | 1.1 | 6 | 2.5 | 1.7 | 8 | 4 | 2.5 | 13 | 20 |
| Cowpea | 1.3 | 0.9 | 5 | 2.0 | 1.3 | 8 | 3 | 2.1 | 12 | 17 |
| Bean (field) | 1.0 | 0.7 | 5 | 1.5 | 1.0 | 8 | 2 | 1.5 | 12 | 12 |

After Shainberg and Oster (1978)

[a]$EC_e$ is the electrical conductivity of the saturation extract (mS $cm^{-1}$ at 25 °C) in the root zone where about two-thirds of the water uptake occurs. For 0 per cent yield reduction, $EC_e$ is the threshold salinity at which yield is expected to begin to decline.

[b]$EC_{iw}$ is the electrical conductivity of the irrigation water and was calculated from $EC_e$ according to the expression $3EC_{iw} = 2EC_e$; the irrigation water is concentrated threefold in the root zone, which is equal to $2EC_e$.

[c]$EC_d$ is the maximum electrical conductivity that can develop due to water uptake by the crop. At this EC, crop growth ceases.

[d]LR In percentages

The soluble sodium percentage, however, has been largely superseded by the use of the sodium adsorption ration (SAR) of the water which is given by

$$\mathrm{SAR} = \frac{\mathrm{Na}^+}{\{(\mathrm{Ca}^{2+} + \mathrm{Mg}^{2+})/2\}^{1/2}}, \qquad (10.5)$$

with the cations expressed in milliequivalents per litre, and the exchangeable sodium ratio (ESR) of a soil which is given by

$$\mathrm{ESR} = \frac{\mathrm{Na}^+}{\mathrm{Ca}^{2+} + \mathrm{Mg}^{2+}} = \frac{\mathrm{Na}^+}{\mathrm{CEC} - \mathrm{Na}^+} \qquad (10.6)$$

where CEC is the cation exchange capacity (see Section 3.7.2) and all concentrations are in milliequivalents per 100 g of soil. The relationship between SAR and ESR is given by

$$\mathrm{ESR} = a + b\,(\mathrm{SAR}) \qquad (10.7)$$

**Table 10.11** *Irrigation water classification based on sodium content*

| Classification | Comment |
|---|---|
| $S_1$ | Low sodium water can be used for irrigation on almost all soils with little danger of the development of harmful levels of exchangeable sodium. However, sodium-sensitive crops such as stonefruit trees and avocado may accumulate injurious concentrations of sodium. |
| $S_2$ | Medium sodium water will present an appreciable sodium hazard in fine textured soils having high cation exchange capacity, especially under low leaching conditons, unless gypsum is present in the soil. This water can be used on coarse-textured or organic soils with good permeability. |
| $S_3$ | High sodium water may produce harmful levels of exchangeable sodium in most soils and will require special soil management—good drainage, high leaching, and organic matter additions. Gypsiferous soils may not develop harmful levels of exchangeable sodium from such waters. Chemical amendments may be required for replacement of exchangeable sodium, except that amendments may not be feasible with waters of very high salinity. |
| $S_4$ | Very high sodium water is generally unsatisfactory for irrigation purposes except at low and perhaps medium salinity, where the dissolving of calcium from the soil or the use of gypsum or other additives may make the use of these waters feasible. |

After Wilcox 1955.

where *a* and *b* are empirical constants for a particular soil type and need to be locally established.

The United States Department of Agriculture has adopted an irrigation water classification based upon SAR in combination with EC. The classification is defined in Fig. 10.3. The C and S classification defined on the figure broadly adheres to the descriptions given in Tables 10.9 and 10.11. Equation (10.5) can be determined using the nomogram given in Fig. 10.4.

### *10.5.3. Bicarbonate hazard*

The complications of carbonate precipitation and dissolution discussed in Section 3.4 also provide difficulties in irrigation. To quantify the effects an empirical parameter was devised by Eaton (1950) on the assumption that all calcium and magnesium would precipitate. The parameter, which is termed the residual sodium carbonate (RSC), is given by

$$\mathrm{RSC} = (\mathrm{CO_3^{2-}} + \mathrm{HCO_3^-}) - (\mathrm{Ca^{2+}} + \mathrm{Mg^{2+}}) \tag{10.8}$$

and is in units of milliequivalents per litre of water. The guidelines for applying RSC are given in Table 10.12.

RSC, however, is a very inexact measurement of mineral saturation so that SI values as defined in Section 3.2 provide a much better understanding of when precipitation or dissolution can occur. Some comparisons are given in Table 10.17 below.

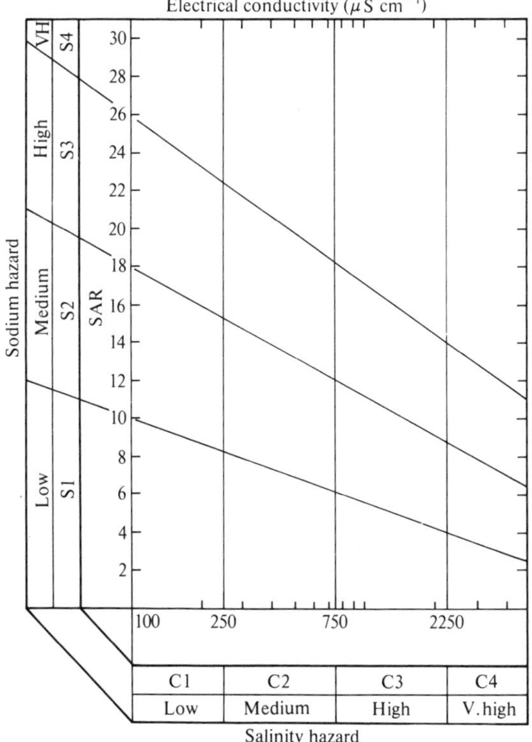


**Fig. 10.3.** Salinity and sodium adsorption classification for irrigation water use.

### *10.5.4. Adjusted SAR*

Because the carbonate dissolution-precipitation conditions clearly relate to sodium concentrations both in the irrigation water and the soil, the sodium adsorption ratio (SAR) has been combined with a carbonate saturation assessment to produce an index referred to as the adjusted SAR (adj. SAR) (Bower, Ogata, and Tucker 1968). This is defined as

$$\text{adj. SAR} = \text{SAR}\{1 + (8.4 - \text{pH}_\text{c})\} \tag{10.9}$$

where SAR is the SAR for irrigation water, 8.4 is the approximate pH of a non-sodic saline soil in equilibrium with $CaCO_3$, and $pH_c$ is the pH for carbonate saturation of the irrigation water and is equivalent to $pH_s$ in Section 11.3.

The adj. SAR now often supercedes SAR and is used in combination with the exchangeable sodium percentage (ESP) of a soil for classification. Soil ESP needs to be determined analytically; however, the theoretical acceptable ESP for a particular irrigation water is shown with respect to SAR in Fig. 10.4.

To calculate adj. SAR the method described in Section 3.2 to determine the carbonate saturation pH is recommended. For a rapid assessment the standard

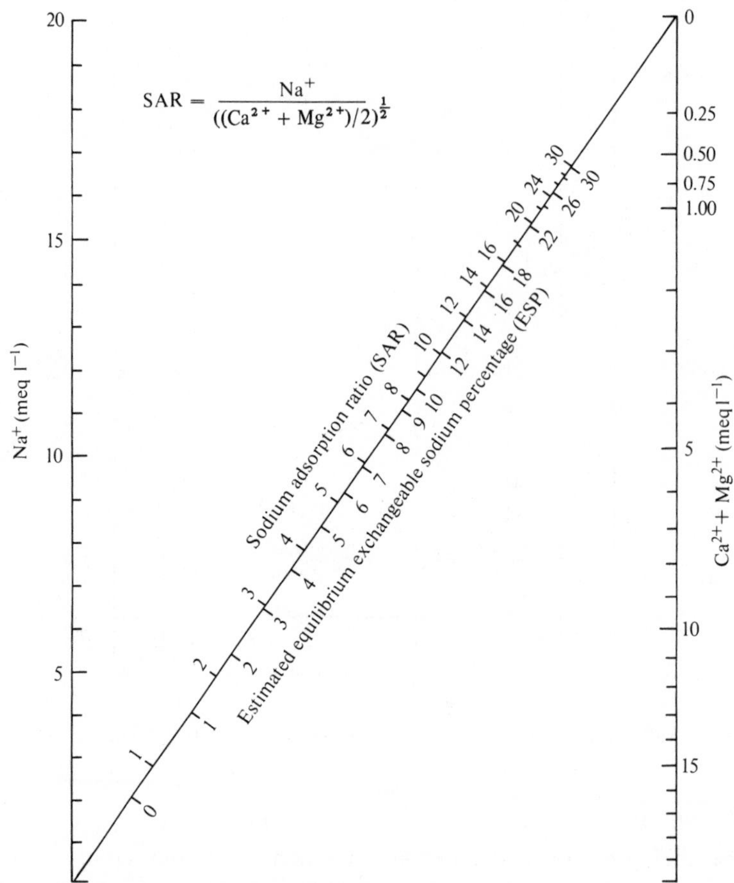


**Fig. 10.4.** Nomogram for determining SAR and for estimating the corresponding ESP of a soil that would be at equilibrium with the irrigation water. (After U.S. Salinity Laboratory 1954.)

**Table 10.12** *Limits for RSC values in irrigation waters*

| Condition | RSC (meq $l^{-1}$) |
|---|---|
| Suitable | <1.25 |
| Marginal | 1.25–2.5 |
| Not suitable | >2.5 |

soils method is given in Appendix III. The soils method is based on saturation calculations given in Section 11.3.

### *10.5.5. Magnesium hazard*

The magnesium hazard (MH) was proposed by Szabolcs and Darab (1964) for irrigation water where

$$\mathrm{MH} = \frac{\mathrm{Mg}^{2+}}{\mathrm{Ca}^{2+} + \mathrm{Mg}^{2+}} \times 100.$$

The units are in milliequivalents per litre, and where MH>50 the effects are considered to be harmful.

### *10.5.6. Chloride hazard*

Chloride adsorbed by plants can adversely affect growth. Adsorption during irrigation can be either through the leaves or through the roots with varying effects. An assessment of the hazard is difficult and needs to be made on a local basis and for specific crops. An example of chloride hazard data is given in Table 10.13.

**Table 10.13** *The chloride hazard (Cl) for citrus in the central coastal areas of Israel*

| $Cl_{iw}$ (meq $l^{-1}$) | Soil texture | | |
|---|---|---|---|
| | Sandy | Loamy | Clayey |
| 6 | $Cl_1$ | $Cl_1$ | $Cl_1$ |
| 6–7.5 | $Cl_1$ | $Cl_1$ | $Cl_2$ |
| 7.5–9 | $Cl_1$ | $Cl_1$ | $Cl_3$ |
| 9–15 | $Cl_1$ | $Cl_2$ | $Cl_4$ |

After Food and Agriculture Organization 1973.
$Cl_{iw}$, chloride concentration in irrigation water.
$Cl_1$ no danger under normal irrigation regime; $Cl_2$, low risk; $Cl_3$, medium risk; $Cl_4$, dangerous.

### *10.5.7. Boron hazard*

Boron is essential for plant growth in very small concentrations; however, it can become extremely toxic at concentrations slightly above optimum. The criteria adopted for boron are given in Tables 10.14 and 10.15 and are based on work by Wilcox (1960).

### *10.5.8. Other specific trace elements*

Branson, Pratt, Rhoades, and Oster (1975) have carried out an exhaustive study of plant tolerance to trace elements and provide a useful guide which is summarized in Table 10.16.

**Table 10.14** *Permissible limits of boron in irrigation waters (mg $l^{-1}$)*

| Class of water | Limit (mg $l^{-1}$) | | |
|---|---|---|---|
| | Sensitive crops | Semi-tolerant crops | Tolerant crops |
| Excellent | <0.33 | <0.67 | <1.00 |
| Good | 0.33–0.67 | 0.67–1.33 | 1.00–2.00 |
| Permissible | 0.67–1.00 | 1.33–2.00 | 2.00–3.00 |
| Doubtful | 1.00–1.25 | 2.00–2.50 | 3.00–3.75 |
| Unsuitable | >1.25 | >2.50 | >3.75 |

**Table 10.15** *Relative tolerance of plants to boron*

| Sensitive | Semi-tolerant | Tolerant |
|---|---|---|
| Lemon | Lima bean | Carrot |
| Grapefruit | Sweet potato | Lettuce |
| Advocado | Bell pepper | Cabbage |
| Orange | Pumpkin | Turnip |
| Apricot | Zinnia | Onion |
| Peach | Oat | Broadbean |
| Cherry | Milo | Gladiolus |
| Persimmon | Corn | Alfalfa |
| Kadota fig | Wheat | Gardenbeet |
| Grape | Barley | Mangel |
| Apple | Olive | Sugarbeet |
| Pear | Field pea | Date palm |
| Plum | Radish | Asparagus |
| Navy bean | Sweetpea | Athel (*Tamarix aphylla*) |
| Jerusalem artichoke | Tomato | |
| Walnut | Cotton | |
| Pecan | Potato | |
| | Sunflower | |

**Table 10.16** *Recommended maximum concentrations of trace elements in irrigation water*[a]

| Elements | For waters used continuously on all soil ($mg\ l^{-1}$) | For use up to 20 years on fine-textured soils at pH 6.0 to 8.5 ($mg\ l^{-1}$) |
|---|---|---|
| Aluminium | 5.0 | 20.0 |
| Arsenic | 0.10 | 2.0 |
| Beryllium | 0.10 | 0.50 |
| Boron | 0.75 | 2.0–10.0 |
| Cadmium | 0.010 | 0.050 |
| Chromium | 0.10 | 1.0 |
| Cobalt | 0.050 | 5.0 |
| Copper | 0.20 | 5.00 |
| Fluorine | 1.0 | 15.0 |
| Iron | 5.0 | 20.0 |
| Lead | 5.0 | 10.0 |
| Lithium | 2.5 | 2.5[b] |
| Manganese | 0.20 | 10.0 |
| Molybdenum | 0.010 | 0.050[c] |
| Nickel | 0.20 | 2.0 |
| Selenium | 0.020 | 0.020 |
| Vanadium | 0.10 | 1.0 |
| Zinc | 2.0 | 10.0 |

After Branson *et al.* 1975.

[a] These levels will not normally have an adverse effect on plants or soils. No data are available for mercury, silver, tin, titanium, or tungsten.

[b] Recommended maximum concentration for citrus is 0.75 $mg\ l^{-1}$

[c] Only for fine-textured acid soils, or acid soils with a relatively high content of iron oxide.

### *10.5.9. General assessment*

To collate the various parameters used in assessing the quality of irrigation water examples of three differing water types are given in Table 10.17 with a summary of their suitability. A feature of the results is the conflicting conclusions from the saturation indices and the adjusted SAR in waters 2 and 3. Both waters are likely to allow carbonate precipitation and permeability reduction; however, this is not reflected in the adj. SAR term.

As can be seen from Table 10.17, if all of the indices are considered then conflicting conclusions may be drawn or, on the other hand, specific hazards may indicate the need for soil management or cropping of particular plant species. Certain of the hazard parameters do not pose major difficulties, for example the magnesium hazard, while others are less reliable such as the residual sodium carbonate index. The indices that are used will depend upon the water analyses available or may be specified because a particular soil type predominates or cropping pattern is required. For general assessment, however, the Food and Agriculture Organization have emphasized that EC, adj. SAR, and certain specific ions should be considered. The guideline values are given in Table 10.18.

**Table 10.17** *Examples of irrigation indices for various potential irrigation waters*

| Parameter | Water 1 | | Water 2 | | Water 3 | |
|---|---|---|---|---|---|---|
| | Value | Hazard | Value | Hazard | Value | Hazard |
| $Ca^{2+}$ (mg $l^{-1}$) | 63.0 | — | 104.0 | — | 61.0 | — |
| $Mg^{2+}$ (mg $l^{-1}$) | 45.0 | — | 39.0 | — | 19.0 | |
| $Na^{+}$ (mg $l^{-1}$) | 161.0 | — | 60.0 | — | 15.0 | |
| $K^{+}$ (mg $l^{-1}$) | 11.0 | — | 7.0 | — | 0.2 | |
| $Cl^{-}$ (mg $l^{-1}$) | 176.0 | High | 112.0 | Low | 64.0 | Low |
| $SO_4^{2-}$ (mg $l^{-1}$) | 250.0 | — | 170.0 | — | 130.0 | |
| $HCO_3^{-}$ (mg $l^{-1}$) | 152.5 | — | 217.2 | — | 209.9 | |
| B | 1.0 | High | 0.1 | Low | 0.2 | Low |
| EC ($\mu$S $cm^{-1}$) | 1460 | High | 1110 | High | 700 | Moderate |
| SAR | 3.79 | Low | 1.27 | Low | 2.69 | Low |
| SSP | 51.55 | Low | 24.90 | Low | 47.02 | Low |
| RSC | −3.34 | Low | −4.82 | Low | −1.11 | Low |
| MH | 54.08 | V. High | 38.21 | Low | 33.93 | Low |
| $SI_C$ | −0.26 | Marginal | 0.37 | High | 0.59 | High |
| $SI_D$ | −0.07 | Marginal | 0.90 | High | 1.23 | High |
| pH | 7.30 | — | 7.68 | — | 8.10 | — |
| $pH_C$ | 6.64 | — | 7.18 | — | 7.67 | — |
| adj. SAR | 10.43 | High | 2.83 | Low | 4.66 | Low |
| ESP | 12.30 | — | 2.80 | — | 5.10 | — |

**Table 10.18** *Guidelines for interpretation of water quality for irrigation*

| Irrigation problem | Degree of problem | | |
|---|---|---|---|
| | No problem | Increasing problem | Severe problem |
| Salinity (affects crop water availability) | | | |
| $EC_W$ (mS cm$^{-1}$) | <0.75 | 0.75–3.0 | >3.0 |
| Permeability (affects infiltration rate into soil) | | | |
| $EC_W$ (mS cm$^{-1}$) | >0.5 | 0.5–0.2 | <0.2 |
| adj. SAR[a, b] | | | |
| Montmorillonite (2:1 crystal lattice) | <6 | 6–9[c] | >9 |
| Illite–vermiculite (2:1 crystal lattice) | <8 | 8–16[c] | >16 |
| Kaolinite–sesquioxides (1:1 crystal lattice) | <16 | 16–24[c] | >24 |
| Specific ion toxicity (affects sensitive crops) | | | |
| Sodium[d, e] (adj. SAR) | <3 | 3–9 | >9 |
| Chloride[d, e] (meq l$^{-1}$) | <4 | 4–10 | >10 |
| Boron (mg l$^{-1}$) | <0.75 | 0.75–2.0 | >2.0 |
| Miscellaneous effects (affects susceptible crops) | | | |
| $NO_3^-$-N (or) $NH_4^+$-N (mg l$^{-1}$) | <5 | 5–30 | >30 |
| $HCO_3^-$ (meq l$^{-1}$) (overhead sprinkling) | <1.5 | 1.5–8.5 | >8.5 |
| pH | (normal range 6.5–8.4) | | |

After Sayers and Westcot 1976.

a adj. SAR means adjusted sodium adsorption ratio and can be calculated using the procedure given in Appendix III.

b Values presented are for the dominant type of clay mineral in the soil since structural stability varies between the various clay types. Problems are less likely to develop if water salinity is high, and are more likely to develop if water salinity is low.

c Use the lower range if $EC_W < 0.4$mS cm$^{-1}$

Use the intermediate range if $EC_W = 0.4 - 1.6$mS cm$^{-1}$

Use the upper limit if $EC_W > 1.6$mS cm$^{-1}$

d Most tree crops and woody ornamentals are sensitive to sodium and chloride (use values shown). Most annual crops are not sensitive (use the salinity tolerance tables, Table 10.10).

e With sprinkler irrigation on sensitive crops, sodium or chloride in excess of 3 meq l$^{-1}$ under certain conditions has resulted in excessive leaf absorption and crop damage.

## References

Bower, C. A., Ogata, G., and Tucker, J. M. (1968). Sodium hazard of waters as influenced by leaching fraction and by precipitation or solution of calcium carbonate. *Soil Sci.* **106**, 29–34.

Branson, R. L., Pratt, P. F., Rhoades, J. D., and Oster, J. D. (1975). Water quality in irrigated watersheds. *J. environ. Quality* **4**, 33–40.

*British Standard 882* (1973). Specification for aggregates from natural sources for concrete, Part 2. British Standards Institution, London.

*British Standard 1047* (1974). Specification for air-cooled blast furnace slag coarse aggregate for concrete, Part 2. British Standards Institution, London.

Building Research Establishment (1975). Concrete in sulphate-bearing soils and groundwater. *Cement Digest* 174. Department of the Environment, London.

Collins, J. C. (1960). *Radioactive wastes, their treatment and disposal.* Spon, London.

Denman, W. L., (1961). Maximum re-use of cooling water. *Ind. eng. Chem.* **53**, 817–22.

Doorenbos, J. and Pruitt, W. O. (1975). Crop water requirements. *Irrigation and Drainage Pap.* 24. Food and Agriculture Organization, Rome.

Eaton, F. M. (1950). Significance of carbonates in irrigation waters. *Soil Sci.* **69**, 123–33.

European Economic Community (1975). Proposal for a council directive, relating to the quality of water for human consumption. *Off. J. European Communities* no. C214.

Food and Agriculture Organization (1973). Quality of irrigation water. *UNESCO–FAO. Pub.*

Garrels, R. M., Thompson, M. E., and Siever, R. (1960). Stability of some carbonates at 25 °C and one atmosphere total pressure. *Am. J. Sci.* **258**, 402–18.

International Atomic Energy Agency. (1969). *Basic safety standards for radiation protection.* International Atomic Energy Agency, Vienna.

Lacey, R. F. (1981). Changes in water hardness and cardiovascular deathrates. *Tech. Rep. 171* (Water Research Centre, England).

Maas, E. V. and Hoffman, G. I. (1977). Crop salt tolerance–current assessment. *Am. Soc. civ. Eng., Irrigation and drainage Div.* **IR2**, 115–34.

National Academy of Sciences and National Academy of Engineering (1972). Water quality criteria. *Rep. EPA-R3-73-033* (National Academy of Sciences and National Academy of Engineering, Washington).

New England Water Works Association (1940). Progress report of the committee on quality tolerances of water for industrial uses. *J. New England water works Ass.* **54**, 261–72.

Sayers, R. S. and Westcot, D. W. (1976). Water quality for agriculture. *Irrigation and Drainage Pap.* 29. Food and Agriculture Organization, Rome.

Shainberg, I. and Oster, J. D. (1978). Quality of irrigation water. *International Irrigation Center, Israel, Publ. No. 2.*

Szabolcs, I. and Darab, C. (1964). The influence of irrigation water of high sodium carbonate content on soils. *Proc. 8th Int. Congr. of ISSS, Trans. II*, pp. 803–12.

U.S. Salinity Laboratory (1954). Diagnosis and improvement of saline and alkali soils. *U.S. Dep. Agric., Agric. Handb. 60.* 160pp.

U.S. Department of the Interior (1968). *Report of the committee on water quality criteria.* U.S. Department of the Interior, Washington, D.C.

Walton, W. C. (1970). *Groundwater resource evaluation.* McGraw-Hill, New York.

Wilcox, L. V. (1955). Classification and use of irrigation waters. *U.S. Dep. Agric., Circ. 969.*

— (1960). Boron injury to plants. *U.S. Dep. Agric.., Agric. Inf. Bull. 211.*

# 11 Corrosion and incrustation

## 11.1. Introduction

Although not necessarily related to the same chemical phenomena, corrosion and incrustation are generally considered together in view of their collective effect upon well installations and water distribution systems. In this chapter features of the two processes are discussed separately; however, possible interdependence is stressed as are the difficulties of assessment.

## 11.2. Corrosion

Refined metals have a tendency to revert to a more natural and stable thermodynamic form; as a result, in the right environment oxidation of the metal occurs, metal is lost from the fabric, and usually passes into solution and may be redeposited. The removal of the metal is termed corrosion.

### *11.2.1. Corrosion of bimetallic couples (galvanic corrosion)*

It has been established that the mechanisms of the corrosion processes in aqueous solutions are electrochemical. If two metals are immersed in conducting solutions an electrochemical cell is created as illustrated in Fig. 11.1(a). At each electrode the electrode creates a 'half-cell' potential $E_H$ with the solution given by

$$E_H = E^\circ + \frac{RT}{nF} \ln \left( \frac{\text{(oxidized)}}{\text{(reduced)}} \right). \tag{11.1}$$

As discussed in Section 3.5 and with reference to Appendix II, the standard potential at the zinc electrode for 25 °C is −0.76 V while the potential at the copper electrode is +0.34 V with respect to the ions in solution. When the zinc and the copper electrode are joined externally a current flows under the potential

$$-0.76 \text{ V} - (+0.34 \text{ V}) = -1.10 \text{ V} \tag{11.2}$$

the zinc becoming the anode and the copper the cathode of the cell. The zinc rod dissolves or corrodes and releases electrons which pass to the copper electrode where they are taken up by cupric ions precipitated from the solution:

$$\text{Zn} \rightarrow \text{Zn}^{2+} + 2e^- \text{ (anodic reaction—oxidation)} \tag{11.3}$$

$$\text{Cu}^{2+} + 2e^- \rightarrow \text{Cu (cathodic reaction—reduction).} \tag{11.4}$$

If a single corrosion or galvanic cell is considered as shown in Fig. 11.1(b) with a simple electrolyte solution and iron forming the anode, oxidation will

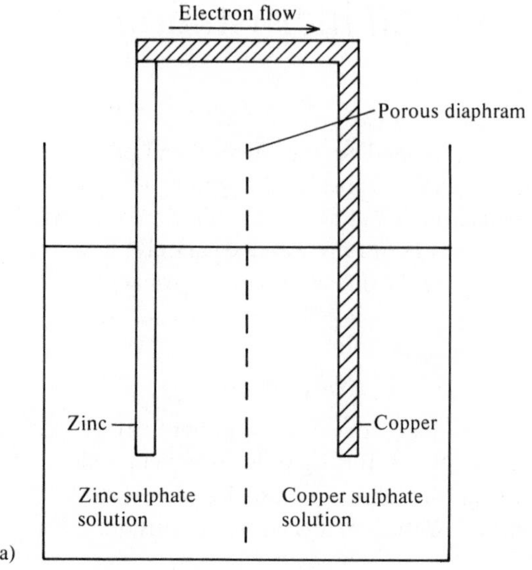


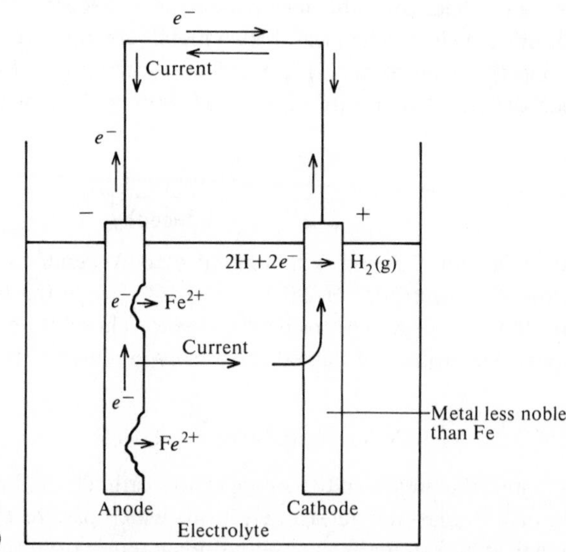


**Fig. 11.1.** (a) Simple electrochemical cell; (b) galvanic cell for the corrosion of iron.

occur at the anode releasing $Fe^{2+}$ ions into solution. An electron imbalance will occur in the anode creating a negative charge and movement of the electrons externally to the cathode:

$$Fe \rightarrow Fe^{2+} + 2e^- . \qquad (11.5)$$

At the cathode the electron may react with hydrogen ions or oxygen if present in solution:

$$2H^+ + 2e^- \rightarrow H_2(g) \qquad (11.6)$$

$$4e^- + 4H^+ + O_2(aq) \rightarrow 2H_2O. \qquad (11.7)$$

In both these reactions $H^+$ is taken up so that the pH rises and significant amounts of $OH^-$ ions appear and move to the anode. As a result a current is conducted through the solution, with a 'positive' current flow to the cathode. The magnitude of the current is dependent on the conductivity, amongst other things, and thence on the total dissolved solids content of the solution.

If oxygen is absent and the pH becomes high then ferrous hydroxide will precipitate at the anode:

$$Fe^{2+} + 2OH^- \rightleftharpoons Fe(OH)_2. \qquad (11.8)$$

In most groundwater situations $Fe_3O_4$ precipitates instead of $Fe(OH)_2$ as it is stable at higher $E_H$ values. Where oxygen is present $Fe^{2+}$ is converted to $Fe^{3+}$ and may precipitate as ferric hydroxide:

$$4\,Fe^{2+} + 4H^+ + O_2(aq) \rightleftharpoons 4Fe^{3+} + 2H_2O \qquad (11.9)$$

$$Fe^{3+} + 3OH^- \rightleftharpoons Fe(OH)_3. \qquad (11.10)$$

Ferrous hydroxide is green to greenish black owing to incipient oxidation, while ferric hydroxide (common rust) is orange to red brown in colour. $Fe(OH)_3$ restricts the rate of corrosion because it forms a tenacious coating while $Fe_3O_4$, which is powdery, tends to flake and provide no protection. Depending upon the carbonate chemistry of the water $Fe^{2+}$ may precipitate as siderite at the cathode:

$$Fe^{2+} + CO_3^{2-} \rightleftharpoons FeCO_3. \qquad (11.11)$$

In the galvanic reactions discussed above to explain corrosion, the important feature is that, although corrosion or dissolution is occurring, analogous precipitation must take place.

In a water well situation galvanic coupling can be very complex so that the theoretical calculation of the rate of corrosion for a single cell using Ohm's law and Faraday's law (Snoeyink and Jenkins 1980) is not realistic in field conditions. The degree of potential galvanic corrosion between two metals can be indicated qualitatively by their position in the galvanic series (Table 11.1) which is an empirical list of metals presented in order of their tendency to corrode.

**Table 11.1** *Galvanic series of metals and alloys*

| | |
|---|---|
| Corroded end *(anode)* | Magnesium |
| | Magnesium alloys |
| | Zinc |
| ↑ | Aluminium 25 |
| | Cadmium |
| | Aluminium 17ST |
| | Steel or Iron |
| | Cast Iron |
| | Cr–Fe (active) |
| Decreasing corrosion resistance | Nickel Resist |
| | 18–8 Cr–Ni–Fe (active) |
| | 18–8–3 Cr–Ni–Mo–Fe (active) |
| | Pb–Sn Solders |
| | Lead |
| | Tin |
| | Nickel (active) |
| | Inconel (active) |
| | Brasses |
| | Copper |
| | Bronzes |
| | Cu–Ni alloys |
| | Silver solder |
| | Nickel (passive) |
| | Inconel (passive) |
| | Cr–Fe (passive) |
| | 18–8 Cr–Ni–Fe (passive) |
| | 18–8–3 Cr–Ni–Mo–Fe (passive) |
| Protected end *(cathode)* | Silver |
| | Gold |
| | Platinum |

After International Nickel Co. Ltd. (quoted by Campbell and Lehr 1973).

The effects of galvanic corrosion are often so serious that joining together different metals or alloys, particularly those far apart in the galvanic series, should be avoided. This type of corrosion is always most intense at the junction of two metals because the currents flowing to the more remote areas of the reactive metal will be reduced owing to the electrolytic resistance of the path traversed.

In the galvanic series the terms 'active' and 'passive' are used to indicate respectively 'more subject' and 'less subject' to corrosion. When oxygen is absent from a water some metals, for example stainless steel, become active, although they are passive in oxygenated water because of the formation of tenacious oxide films.

### *11.2.2. Corrosion of single metals or alloys*

It is not necessary for dissimilar metals to be in contact for corrosive attack to take place. When a metal is exposed to the atmosphere, as for example well casing before emplacement, oxide films develop on the metal surface. Locally an oxide film will be cracked, porous, or very thin, and at such points it will be relatively easier for ions to leave the metal lattice. Irregularities such as couplings and seams will increase the variable oxide film cover which will be accentuated by abrasion during well construction. Once a metal surface has been immersed in a corrosively active water anodic areas develop at the point where the oxide film is thin or weak while cathodic areas occur when the oxide is thicker. The type of reaction envisaged is illustrated in Fig. 11.2.

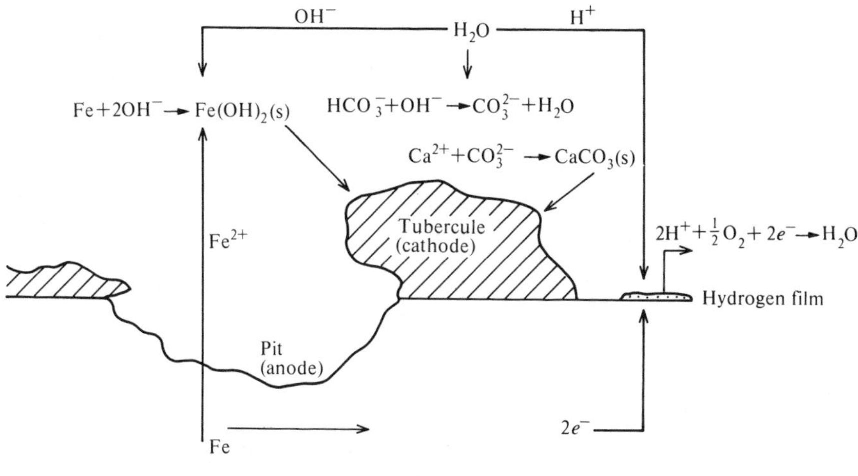


**Fig. 11.2.** Occurrence of local corrosion due to the development of anodic and cathodic areas.

Ferrous and ferric hydroxide will deposit in cathodic areas which may also receive carbonate precipitation depending upon the carbonate chemistry of the water.

Under some circumstances selective corrosion (including dezincification and graphitization) can occur in that one metal of an alloy is selectively removed leaving a spongy and weakened fabric. Dezincification occurs owing to the galvanic cell condition created between different metals in zinc alloys, e.g. brass which is an alloy of zinc and copper. The most favourable environment for dezincification is in saline groundwaters, slightly acid groundwaters, or where oxygen in solution is significant.

Graphitization, for example in low carbon steel and cast iron, is commonly the result of sulphate-reducing conditions. Anaerobic conditions give rise to reactions affecting iron, particularly in the presence of bacteria.

The electrons required for sulphate reduction according to the equation

$$SO_4^{2-} + 8H^+ + 8\bar{e} \rightarrow S^{2-} + 4H_2O \qquad (11.12)$$

can be supplied by the oxidation of iron to $Fe^{2+}$:

$$Fe \rightarrow Fe^{2+} + 2e^-. \qquad (11.13)$$

The high pH produced by the consumption of hydrogen ions in eqn (11.12), together with the sulphide ions produced, precipitate the dissolved iron as $Fe(OH)_2$ and FeS, thus maintaining a low $Fe^{2+}$ concentration which favours the forward direction of eqn (11.13) (le Chatelier's principle). The result is the rapid conversion of iron in the steel to low strength oxides and sulphides according to eqn (11.14):

$$2H_2O + 4Fe + 2H^+ + SO_4^{2-} \rightarrow 3Fe(OH)_2 + FeS \qquad (11.14)$$

As a result of the bacterial activity ferrous hydroxide and black ferrous sulphide are produced.

### *11.2.3. The role of bacteria*

Corrosion can be accentuated by the metabolic process of bacteria under both aerobic and anaerobic conditions.

*Aerobic bacteria.* Many of the common bacteria present in groundwaters such as *Aerobacter aerogenes*, *Escherichia coli*, the *Pseudomonas* family, and *Proteus vulgaris* utilize hydrogen and can therefore take part in the cathodic reaction in a galvanic cell. However, iron bacteria such as *Gallinonella*, *Clonothrix*, *Crenothrix*, *Leptothrin*, etc., which are active in the oxidation of ferrous iron to ferric iron, are probably more problematical and cause more difficulties. While these bacteria are prolific in promoting iron hydroxide incrustation this can only be excessive if iron concentrations are high as result of extensive anodic attack and therefore corrosion. The types of reactions shown in Fig. 11.2 are enhanced considerably by the bacteria. Iron bacteria grow best at low temperatures and are commonly found in groundwaters. They have a wide range of oxygen tolerance and will live in water containing 0.3–9.0 mg $l^{-1}$ dissolved oxygen.

*Anaerobic bacteria.* The action of aerobic bacteria can cause local anaerobic microenvironments and allow the growth of anaerobic bacteria adjacent to aerobic bacteria. More generally, however, such bacteria are prevalent in the truly reducing conditions of confined groundwater systems.

Sulphate-reducing bacteria are the most active participants in the corrosion of well installations. *Desulfovibrio desulfuricans* and *Clostridium nigrificans* are examples that ultilize sulphate as their primary oxygen source in the presence of enzymes and other compounds. *D. desulfuricans* will grow in temperatures up to 50 °C and will develop at pH values between 5.5. and 9.0 although it prefers a neutral pH. *C. nigrificans* grows mainly in water with a temperature in excess of 35 °C.

For their development sulphate-reducing bacteria need organic matter as a source of carbon and energy while sulphate acts as an electron acceptor. The organic matter utilized by these micro-organisms seems to be limited to a number of simple carboxylic acids such as lactic acid (Goldhaber and Kaplan 1975) and is based on the conversion

$$\underset{\text{lactate}}{2CH_3CHOHCOOH} + SO_4^{2-} \rightarrow H_2S + \underset{\text{acetate}}{2CH_3COOH} + 2HCO_3^-. \quad (11.15)$$

However, van Beek and van der Kooif (1982), in a discussion of reducing environments, suggest that microbiological reactions are very complex and probably involve other micro-organisms in addition to sulphate-reducing bacteria.

King and Miller (1971) have proposed the following reasons for the promotion of corrosion by sulphate-reducing bacteria:

(i) stimulation of the cathodic part of the galvanic cell by the removal and utilization of oxygen by the bacteria;

(ii) stimulation of the cathodic reaction by solid ferrous sulphides formed by the reaction of ferrous ions with sulphide ions produced by bacteria;

(iii) stimulation of the anodic reaction (metal dissolution) by bacterially produced sulphide;

(iv) local acid cell formation;

(v) formation of iron phosphide by reaction of the metal with bacterially reduced phosphates.

The galvanic corrosion effected by sulphate-reducing bacteria is shown in Fig. 11.3.

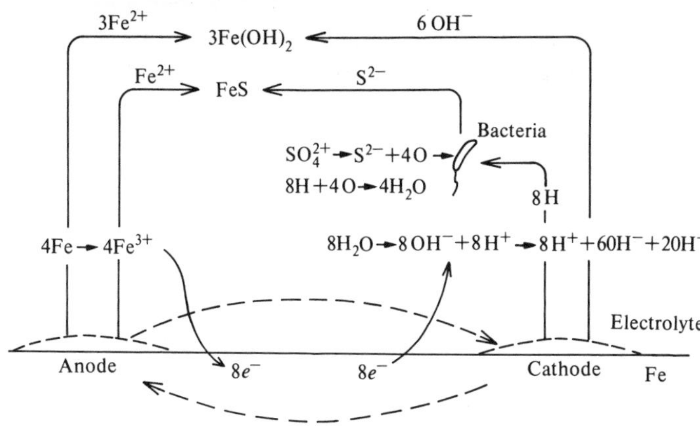


**Fig. 11.3.** Sulphate-reducing bacteria in the corrosion reaction.

Under certain acid conditions reduction can occur as a result of action of methane producing bacteria (Malashenko 1971):

$$8e^- + CO_2 + 8H^+ \rightarrow CH_4 + 2H_2O. \qquad (11.16)$$

Combining this with the anodic corrosion of iron gives

$$4Fe^{2+} + 6H_2O + CO_2 \rightleftharpoons 4Fe(OH)_2 + CH_4. \qquad (11.17)$$

### *11.2.4. Corrosion environments*

Moss (1964) has listed a number of chemical parameters that can be used to indicate corrosion potential. These are pH, dissolved oxygen, hydrogen sulphide, total dissolved solids, carbon dioxide, chloride, and temperature. Several other factors can be added including fluid velocity, bacterial activity, and the presence of heavy metals. While each parameter can be considered separately, the chief problem is the complex interdependence of a number of them.

*The effect of pH.* Invariably the observed pH of a water is not indicative of the pH of the metal surfaces in contact with it. These are usually alkaline owing to the presence of a saturated layer of ferrous hydroxide (pH 9.5). If the pH is between 4 and 10 the corrosion rate depends on how rapidly oxygen diffuses to the metal surface, and this is probably a function of fluid velocity, oxygen saturation, and temperature rather than pH. In acid waters (pH $<4$) corrosion is accelerated as the increased hydrogen ion concentration favours hydrogen evolution and oxygen depolarization. The formation of ferrous hydroxide continues until the pH rises above 9.3 or the electrolytic cell is destroyed. This is a continuous process in flowing water. With pH above 10 corrosion virtually ceases owing to the passivity of metals produced by oxide films favoured by the presence of excess dissolved oxygen.

*Dissolved oxygen effects.* Groundwaters are not normally saturated with dissolved oxygen so that concentrations are usually less than 9 mg $l^{-1}$ at normal pressures and temperatures. In the galvanic cell reactions oxygen removes electrons at the cathode, particularly in aerated flowing water. Oxygen forms a protective covering at the anode but will also obviously cause the oxidation of ferrous ions to ferric ions at a metal surface. At a fresh metal surface corrosion is initially very rapid but is quickly reduced by the formation of the hydroxide film. The corrosion rate of iron is proportional to the dissolved oxygen concentration, although passivity due to excess oxygen occurs at concentrations above 12 mg $l^{-1}$ where waters are under pressure.

High dissolved oxygen concentrations tend to promote passivation of aluminium and stainless steel. In the case of waters with a pH less than 5, direct attack is possible and oxygen is not required.

*Hydrogen sulphide effects.* The generation of hydrogen sulphide in groundwaters has been discussed in Section 11.2.3. It is normally present in confined systems in a down-gradient position to the redox barrier (Section 9.2.4). As the concentration of hydrogen sulphide increases, so the corrosion potential increases.

The hydrogen sulphide can produce destructive intergranular networks of low strength metallic sulphides. There is also a tendency for hydrogen, resulting from corrosion, to penetrate and embrittle steel beneath sulphide deposits particularly in hard alloy steels and cold-worked areas. As there is a considerable electrical potential difference between iron sulphide and steel, galvanic action is pronounced.

*Effect of total dissolved solids.* As salinity increases, corrosion increases particularly with salinities greater than 1000 mg $l^{-1}$ because of the increased conductivity of the electrolyte in the galvanic cells.

*Carbon dioxide effects.* Carbon dioxide is instrumental in corrosion in that its presence accentuates the acidity of a groundwater. $CO_2$ concentrations in groundwaters are usually low; however, in many deep wells $CO_2$ is released from the groundwater at the well face owing to pressure reduction between the well and the aquifer.

The type of reaction envisaged is as follows:

$$HCO_3^- + H^+ \rightarrow H_2CO_3 \qquad (11.18)$$

$$H_2CO_3 \rightarrow H_2O + CO_2(g) \qquad (11.19)$$

$$Ca^{2+} + HCO_3^- \rightarrow CaCO_3(s) + H^+ \qquad (11.20)$$

Calcite is precipitated somewhere in the well installation and $CO_2$ gas is evolved. However, reaction rates are such that precipitation is unlikely to occur within the area of $CO_2$ generation and will not protect against localized acid attack adjacent to bubbles, the reverse of eqns (11.18, 11.19). Water flow inhibits precipitation.

In addition to deep aquifer pressure effects, certain aquifer environments containing organic matter, notably where carbonaceous shales or coals are present, can pose $CO_2$ problems. Under oxidizing conditions the following reaction can occur:

$$O_2 + CH_2O \rightarrow H_2CO_3 \qquad (11.21)$$

with $CO_2$ evolving and the pH decreasing as dictated by eqns (11.18) and (11.19).

In reducing conditions the reaction can include hydrogen sulphide which further enhances the corrosion potential:

$$2CH_2O + SO_4^{2-} \rightarrow HCO_3^- + HS^- + CO_2 + H_2O \qquad (11.22)$$

$$\rightarrow HCO_3^- + HS^- + H_2CO_3. \qquad (11.23)$$

*Chloride effects.* Chloride is most effective at concentrations greater than about 300 mg $l^{-1}$. Its action is to break down protective films of copper and aluminium alloys. However, Kelly and Kemp (1975) have shown that absolute chloride content and pump corrosion do not necessarily relate and have

**Table 11.2** *Estimated pump life based on $Cl^-/CO_3^{2-}$ ratios*

| log ($Cl^-/CO_3^{2-}$) | Cast iron bowls<br>Bronze impellers | Bronze bowls<br>Bronze impellers |
|---|---|---|
| 8 | | UNSAFE |
| 7 | | |
| 6 | UNSAFE | |
| 5 | | |
| 4 | | UNCERTAIN |
| 3 | UNCERTAIN | |
| 2 | | |
| 1 | SAFE | SAFE |

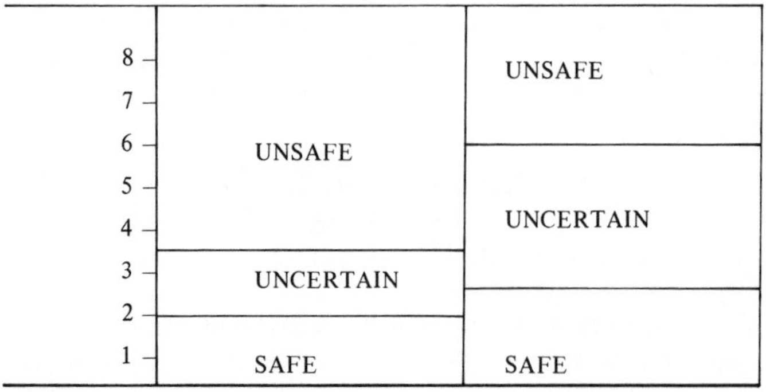

After Kelly and Kemp 1975

| | |
|---|---|
| SAFE | good chance of life exceeding 12 years. |
| UNCERTAIN | Corrosion erratic. |
| UNSAFE | definite risk of life less than 12 years. |

It should be noted that, although precise boundaries have been drawn between different areas, there is a chance that at $Cl^-/CO_3^{2-}$ close to the limits, corrosion may be more or less than predicted. Dissolved oxygen can also influence the degree of attack within each area shown.

suggested that an indicator combining chloride with passivating anions is preferable. The estimated pump life with respect to the $Cl^-/CO_3^{2-}$ ratio is given in Table 11.2 for cast iron and bronze fittings.

*Temperature effects.* In aerated waters the corrosion of steel increases linearly up to 80 °C and then rapidly decreases.

*Fluid velocity effects.* Corrosion increases with fluid velocity as corrosion products are removed and the electrolyte is supplied at a continual rate. Variations in velocity can produce differential concentrations of ions in the fluid so that a potential difference can occur with corrosion occurring in what is called a concentration cell (Snoeyink and Jenkins 1980).

Corrosion can be aggravated by cavitation if suspended solids, which act as abrasives, are present. Dissolved gases including inert gases have a similar effect. Conversely, passivity can occur at high velocities due to oxygen excess.

### *11.2.5. Assessment of corrosion potential*

*Indications using chemical parameters.* The significant chemical parameters involved in corrosion have been discussed in Section 11.2.4. Their means of measurement and qualitative significance are summarized in Table 11.3. Owing to the instability of the majority of the parameters, measurement of their values must be carried out at the well head.

**Table 11.3** *Chemical parameters indicating corrosion potential, their measurement, and general individual effects in groundwaters (mainly applicable to iron and iron-based materials)*

| Parameter | Measurement Method | Comment |
|---|---|---|
| pH | pH probe | As pH decreases below 7, corrosion increases |
| DO[a] | DO probe | Accelerates corrosion in acid, neutral, and slightly alkaline waters |
| $H_2S$ | S probe | As concentration increases, corrosion increases |
| TDS | Laboratory evaporation | As salinity increases, particularly over 1000 mg $l^{-1}$, corrosion increases |
| $CO_2$ | Field titration or calculation | Accelerates corrosion particularly if concentration greater than 50 mg $l^{-1}$ |
| $Cl^-/CO_3^{2-}$ | Specific ion/field titration or calculation | As logarithm of ratio increases corrosion increases |
| Temperature | Thermometer | Corrosion is accelerated at higher temperatures |
| $E_H$ | $E_H$ probe | Indicator of reducing conditions and an important parameter in metal speciation stability |

[a] Dissolved oxygen.

*$E_H$-pH relationship.* Although the corrosion potential indications for the individual parameters are difficult to interrelate, their assessment is important in that no bulk chemical appraisal is really possible. $E_H$-pH relationships come nearest to providing a bulk assessment in that many of the individual parameters listed in Table 11.3 control or are controlled by the oxidation-reduction environment. Unfortunately, while pH can normally be readily measured $E_H$ is frequently not obtainable and in any case the values of both parameters measured in a groundwater do not necessarily reflect the values at the face of a metal installation. Therefore, even where good measurements are available, $E_H$-pH relationships are only general indicators of potential corrosion.

The $E_H$-pH relationship is described in Section 3.5 in which the construction of stability diagrams is illustrated. In terms of metal corrosion, or indeed incrustation, the diagrams are used to indicate which metal species are stable under the appertaining $E_H$-pH conditions of the groundwaters. Examples for two groundwaters with respect to iron speciation are shown in Fig. 11.4. For water 1 reducing conditions prevail with $Fe^{2+}$ the stable component (Clarke and Barnes 1965). Under these conditions protection films such as $Fe(OH)_3$ are unstable and cannot form; consequently $Fe^{2+}$ can be taken into solution and corrosion can proceed under galvanic cell effects. In drawing conclusions from $E_H$-pH diagrams, saturation calculations with respect to various solid phases (Section 3.5) also need to be considered (e.g. $Fe(OH)_2$, $Fe(OH)_3$, $Fe_2O_3$, $Fe_3O_4$,

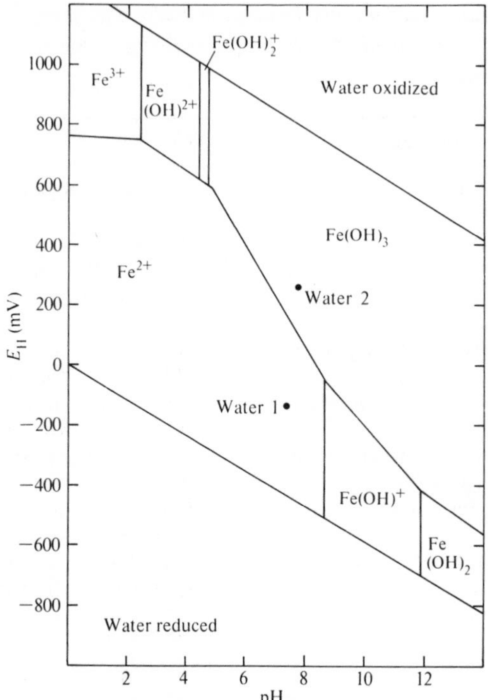


**Fig. 11.4.** Iron speciation and stability in groundwaters based on $E_H$ and pH control.

$FeCO_3$, and others). Water 1 is supersaturated with respect to $Fe_2O_3$; however, at low temperatures and in dilute solutions $Fe_2O_3$ cannot form until $Fe(OH)_3$ has precipitated, which is not possible in this case, owing to equilibrium controls. As a result $Fe_2O_3$ will not form a protective film and corrosion will proceed as indicated by the $E_H$-pH diagram. $Fe(OH)_3$ is the stable phase for water 2 so that a film of $Fe(OH)_3$ would be expected to form on iron-based materials and would provide protection from extensive corrosion.

*Corrosion probes and coupons*. Owing to the difficulties of reliably assessing corrosion by chemical means many industries, including the oil field industry, make direct measurements of metal loss by corrosion probes. As described by Barnes and Clarke (1969) such probes can be extremely valuable in water wells and have been applied to a number of deep-well studies.

A probe consists of a U-shaped loop of 1 mm diameter wire of the material relevant to the installation to be tested. It is normally mounted in the well casing but can be independently suspended. The probe is attached by cable to a monitoring device and power supply (corrosometer) (Fig. 11.5) and acts on an electrical resistance principle. As corrosion occurs, the cross-sectional area of the

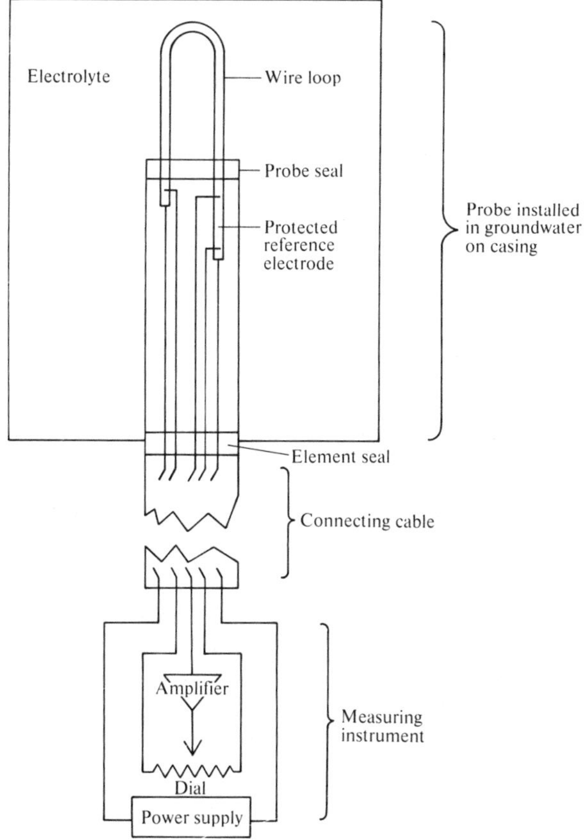


**Fig. 11.5.** Corrosometer probe and measuring equipment.

wire element becomes smaller, thus increasing its electrical resistance. The change in the electrical resistance of the wire element can be equated to the corrosion during the period of exposure by

$$\text{corrosion rate} = \frac{\mathrm{d}D}{\mathrm{d}T} n \times 9.271 \text{ mm per year}$$

where d$D$ is the change on the corrosometer dial reading between data points in units of electrical resistance (the corrosometer contains the meter and a reference element (see Fig. 11.5)), d$T$ is the exposure time in days between the data points, $n$ is a probe multiplier which relates the resistance change to the change in diameter of the probe wire, and 9.271 is a factor for converting the change in the diameter of the wire to the corrosion penetration (in millimetres) per year.

Probe results correlate very well with actual metal losses. They cannot, however, provide quantitative estimates of installation life although they can clearly be used as a guide. Their real value in groundwater work is the identification of suitable materials for water well installations. The qualitative metal loss index given in Table 11.4 is used for this purpose.

**Table 11.4** *Qualitative index of metal loss based on corrosometer readings*

| Corrosion | Loss of metal (mm year$^{-1}$) |
|---|---|
| Insignificant | 0.0–0.05 |
| Moderate | 0.05–0.5 |
| Severe | 0.5–1.25 |
| Extreme | 1.25 or more |

The results of corrosometer readings are shown in Fig. 11.6 for three different metal probes. The results clearly indicate that of the three, stainless steel is the preferred metal for installation.

Samples or coupons of test metal can be suspended in wells for periods to assess corrosion. Weight losses of the sample provide a good indication of corrosion rates. Care should be exercised, however, to ensure that galvanic couples with adjacent materials are not developed. Coupons have the advantage that they can be visually and chemically inspected for corrosion and corrosion products.

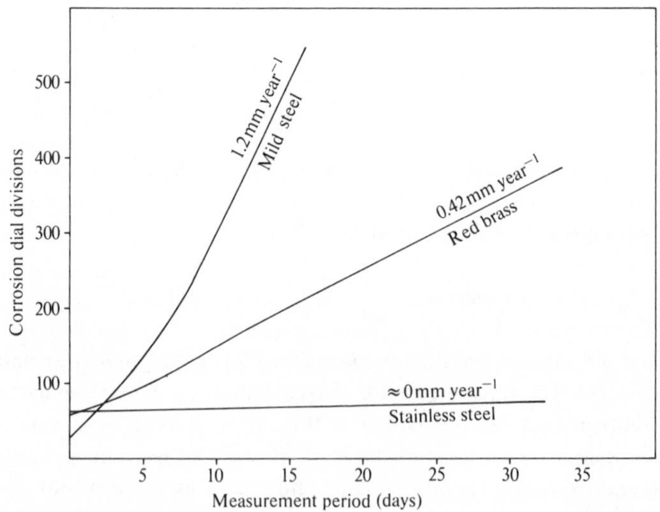


**Fig. 11.6.** Comparative corrosion rates determined using a corrosometer.

### *11.2.6. Prevention and control of corrosion*

Once an assessment of the corrosion potential of a groundwater has shown that it is detrimental, then either corrosion-resistant materials have to be used in well installations or measures have to be taken to combat corrosion in materials prone to corrosion. Where possible it is preferable to use corrosion-resistant materials at the outset of construction as these will reduce the ongoing costs usually associated with control techniques.

*Corrosion resistant-materials.* Corrosion-resistant metals and alloys can prove very expensive so that at their use tends to be restricted to well screens and pump fittings, although stainless steel and copper alloy casing installations are not unknown. Campbell and Lehr (1973) have graded the commonly used metals and alloys in order of decreasing corrosion resistance (Table 11.5).

**Table 11.5** *Corrosion resistance of metals and alloys in decreasing order*

| Metal or alloy | Composition |
|---|---|
| Monel metal | 70 per cent Ni, 30 per cent Cu |
| Stainless steel[a] | 74 per cent low carbon steel, 18 per cent Cr, 8 per cent Ni |
| Everdur metal | 96 per cent Cu, 3 per cent Si, 1 per cent Mn |
| Silicon red brass | 83 per cent Cu, 1 per cent Si, 16 per cent Zn |
| Anaconda red brass | 85 per cent Cu, 15 per cent Zn |
| Common yellow brass | 67 per cent Cu, 33 per cent Zn |
| Armco iron | |
| Low carbon steel | |

[a] Provided that groundwater conditions and screen slots allow the retention of a protective film

Many different types of stainless alloys are available. Normal stainless steel with its chromium and nickel content has an excellent corrosion resistance. Its low carbon content is important with respect to its good weldability. The cost of stainless steel is about six times that of ordinary steel plus galvanizing while a total stainless steel well completion costs only a little over twice an ordinary galvanized-steel completion (Johnson 1966). Everdur is normally more corrosion resistant to sea water than is stainless steel. However, Swan (1982) indicates that the copper-based steels are substantially more expensive than carbon steels and that their anticorrosive performance is no better.

In many groundwater environments non-metallic materials such as epoxy resins (fibre glass) and various plastics (i.e. polyvinyl chloride) are replacing the corrosion-resistant metals and alloys. At shallow depths (up to about 100 m) non-metallic materials are available with sufficient collapse strengths to be used for production purposes (MacDonald and Partners 1965). Problems of material handling, exothermic reactions during cementing, and collapse during well development, however, can be disadvantages in the use of non-metallic materials.

Barring unusual circumstances, such as saline water abstraction, metals or alloys are probably more dependable for long-time installation and major abstraction sources.

*Control of corrosion*. Under circumstances where corrosion resistant materials are not used certain methods can be adopted to combat corrosion. The general methods can be listed as follows:

(i) the use of coated metals or alloys;
(ii) the injection of chemicals into the well;
(iii) cathodic protection.

In addition to the various protective films and anodizing surfaces that can be developed naturally in a well installation during the corrosion process, many metals are artifically coated. Inorganic coatings include cements and silicates which are excellent barriers to attack provided that they remain uncracked. In the water industry organic coatings are probably more familiar and include bitumen, oil paints, plastics, and epoxy resins. While coatings are undoubtedly effective, their retention in an undamaged state during construction usually poses problems.

The injection of chemicals such as calcium hydroxide to neutralize groundwaters in a distribution system poses no problems (Chilton 1973). However, effective injection down wells into zones of corroding groundwaters is not always readily practical so that alternative methods to injection are recommended.

One of the most extensively used corrosion-protection procedures is cathodic protection which is illustrated in Fig. 11.7. In this method the metallic structure is converted into a cathode relative to an introduced material called a sacrificial anode. Figure 11.7(a) shows the introduction of a current such that the anode material is anodic to the casing and will corrode preferentially. Figure 11.7(b) shows the use of magnesium, which is much higher in the galvanic series than iron (Table 11.1), to induce the required potentials naturally. The protective action of galvanizing is partly cathodic in action.

*Control of bacteria*. As indicated in Section 11.2.3 bacterial action in wells is likely to be complex and is outside the scope of this book. However, before bacterial action can be curbed in a well, the types of bacteria present need to be determined. Individual species can if necessary be identified, although in many cases with some difficulty. Normally it is sufficient to identify whether, for example, iron, sulphate, or perhaps nitrate bacteria are active. In view of the need to use a sterile container sampling poses a problem if water specifically adjacent to a corrosion zone is required. Depth samples or discharge samples are usually analysed, although glass slides on which culture will develop can be suspended in a well.

Cathodic protection is probably the most effective protection against bacterial action in well installations, particularly where sulphate-reducing bacteria are

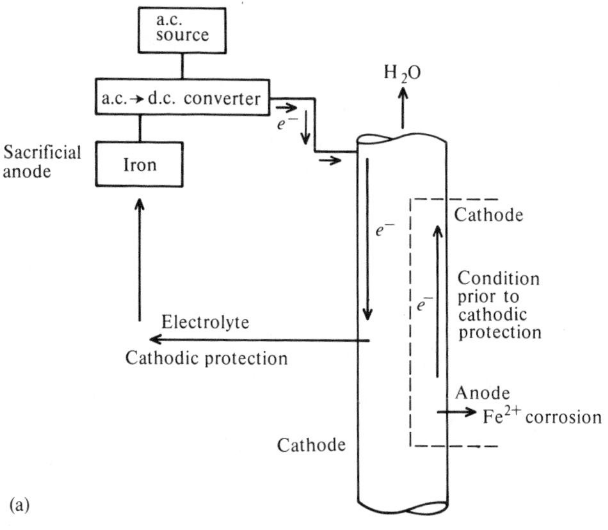


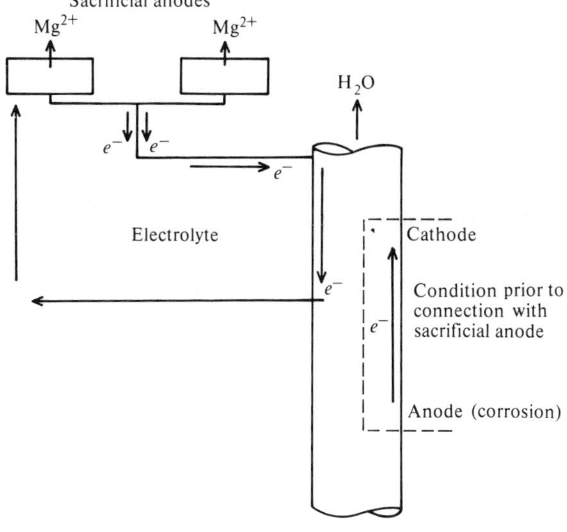


**Fig. 11.7.** Cathodic protection for well casing provided by (a) impressed current and (b) sacrificial magnesium anodes.

present. The development of alkalinity around the cathodically protected installation will help to destroy the bacteria and to reduce corrosion to negligible amounts.

Bacteria may evolve naturally in a well or be introduced during drilling or as a later pollutant. They can be active both on the well materials and in the aquifer where they can reduce permeability. Control of bacteria when they are found to be present is therefore required both at the casing or well screen face and around the well. As a result controls such as cathodic protection may be insufficient and chemicals are required which are best injected under pressure. This can be carried out by the natural build-up of pressure due to chemical reaction in the well, but is normally done by using packer techniques in injection sections. The sections may be located from construction details or by using closed-circuit television surveys. Once packer injection has been completed the packers are immediately removed if acids are being used. Following a suitable period of contact the fluids are removed by well-cleaning techniques.

Sulphate-reducing and iron-oxidizing bacteria prove the most serious problems and are generally susceptable in part to chlorine, hydrochloric acid, or some form of quaternary ammonium compound. None of the procedures are totally effective so that where chemical control is necessary it tends to be repetitious.

Recommended chemical control disinfectants are shown in Table 11.6. Where acids are used in carbonate rocks injection may not be necessary and acidization methods can be used (Stow and Renner 1965).

**Table 11.6** *Recommended disinfectants for bacterial control*

| | |
|---|---|
| Chlorine | Residual chlorine, 0.2 mg $l^{-1}$<br>Hypochlorite, 5 per cent<br>Chlorine dioxide<br>Continuous chlorination |
| Acids | Hydrochloric, 30 per cent<br>Hydroxyacetic, 10 per cent<br>Sulphamic, 10 per cent |
| Surfactants | Rexol |
| Quaternary ammonium compounds | Hyamine |
| Iodine | |

The limited effectiveness of chlorination in controlling sulphate-reducing bacteria is discussed by Updegraff (1955) and Lewis (1965), while Cullimore (1980) has assessed the effectiveness of certain of the disinfectants with respect to iron bacteria as summarized in Table 11.7.

**Table 11.7** *Relative effectiveness of some disinfectants with respect to iron bacteria*

| Compound | Assimilated rating[a] |
|---|---|
| Chlorine dioxide | 82 |
| Hypochlorite | 74 |
| Quaternary ammonium | 68–69 |
| Surfactant | 30–39 |
| Iodine | 33 |

After Cullimore 1980.

[a] Rating of chemical to kill iron bacteria over a concentration range from 0.01 to 1.0 per cent.

## 11.3. Incrustation

Incrustation is the deposition of precipitate from groundwater onto well installation materials or in the surrounding aquifer. The most serious results of incrustation are the reduction in the well-screen open area and the reduction of the aquifer permeability adjacent to the well. Incrustation relates directly to the supersaturation of a groundwater with respect to a certain compound. Therefore if, for example, a groundwater is supersaturated with respect to calcite the mineral will deposit in the well.

The compounds that are of most importance in incrustation studies are carbonates, iron compounds, and manganese compounds. As discussed in Section 3.4 and 11.2.5, supersaturation with respect to a compound can be calculated thermodynamically; however, the limitations of field and laboratory measurements and the permutations of the equations involved tend to give inexact results so that probably with the exception of the better known carbonates saturation indices for the other compounds can only be taken as a guide (Barnes and Clarke 1969).

For practical engineering purposes carbonate incrustation has been extensively examined. An index developed by Langelier (1936) for calcium carbonate is in common use for interpretations. The equation used is

$$\mathrm{pH_s} = (\mathrm{p}K_2 - \mathrm{p}K_\mathrm{s}) + \mathrm{pCa} + \mathrm{pAlk}$$

where $\mathrm{pH_s}$ is the pH at which calcite is in equilibrium with the solution (applicable in the range pH 6.5-9.5), $K_2$ and $K_\mathrm{s}$ are equilibrium constants, pCa is the negative logarithm of the calcium ion concentration in moles per litre, and pAlk is the negative logarithm of the total alkalinity in milliequivalents per litre. The saturation index (SI) determined is

$$\mathrm{SI} = \text{actual pH} - \mathrm{pH_s}.$$

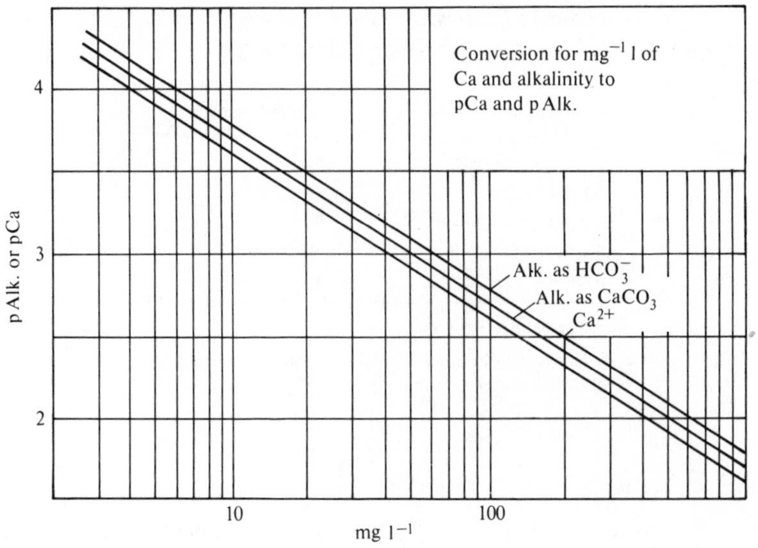


| TDS | $pK_2 - pK_s$ | | | |
|---|---|---|---|---|
| | 0 °C | 10 °C | 20 °C | 25 °C |
| 0 | 2.20 | 2.09 | 1.99 | 1.94 |
| 20 | 2.29 | 2.18 | 2.08 | 2.03 |
| 40 | 2.33 | 2.22 | 2.12 | 2.07 |
| 80 | 2.37 | 2.26 | 2.16 | 2.11 |
| 120 | 2.41 | 2.30 | 2.20 | 2.15 |
| 160 | 2.43 | 2.32 | 2.22 | 2.17 |
| 200 | 2.46 | 2.35 | 2.25 | 2.20 |
| 240 | 2.49 | 2.38 | 2.28 | 2.23 |
| 280 | 2.51 | 2.40 | 2.30 | 2.25 |
| 320 | 2.53 | 2.42 | 2.32 | 2.27 |
| 360 | 2.54 | 2.43 | 2.33 | 2.28 |
| 400 | 2.56 | 2.45 | 2.35 | 2.30 |
| 440 | 2.58 | 2.47 | 2.37 | 2.32 |
| 480 | 2.59 | 2.49 | 2.39 | 2.33 |
| 520 | 2.61 | 2.50 | 2.40 | 2.35 |
| 560 | 3.62 | 2.51 | 2.41 | 2.36 |
| 600 | 2.63 | 2.52 | 2.42 | 2.37 |
| 640 | 2.65 | 2.54 | 2.44 | 2.39 |
| 680 | 2.66 | 2.55 | 2.45 | 2.40 |
| 720 | 2.67 | 2.56 | 2.46 | 2.41 |
| 760 | 2.67 | 2.57 | 2.47 | 2.41 |
| 800 | 2.68 | 2.58 | 2.58 | 2.42 |

Example

pH=7.75, $Ca^{2+}$ =55
Alkalinity as $CaCO_3$=178

TDS=410

From table ($pK_2$-$pK_s$)=2.30

From figure pCa=2.85

From figure pAlk=2.45

Therefore $pH_s$= 7.60

SI=+0.15

The water is incrusting

**Fig. 11.8.** An example of determining SI using the Langelier method.

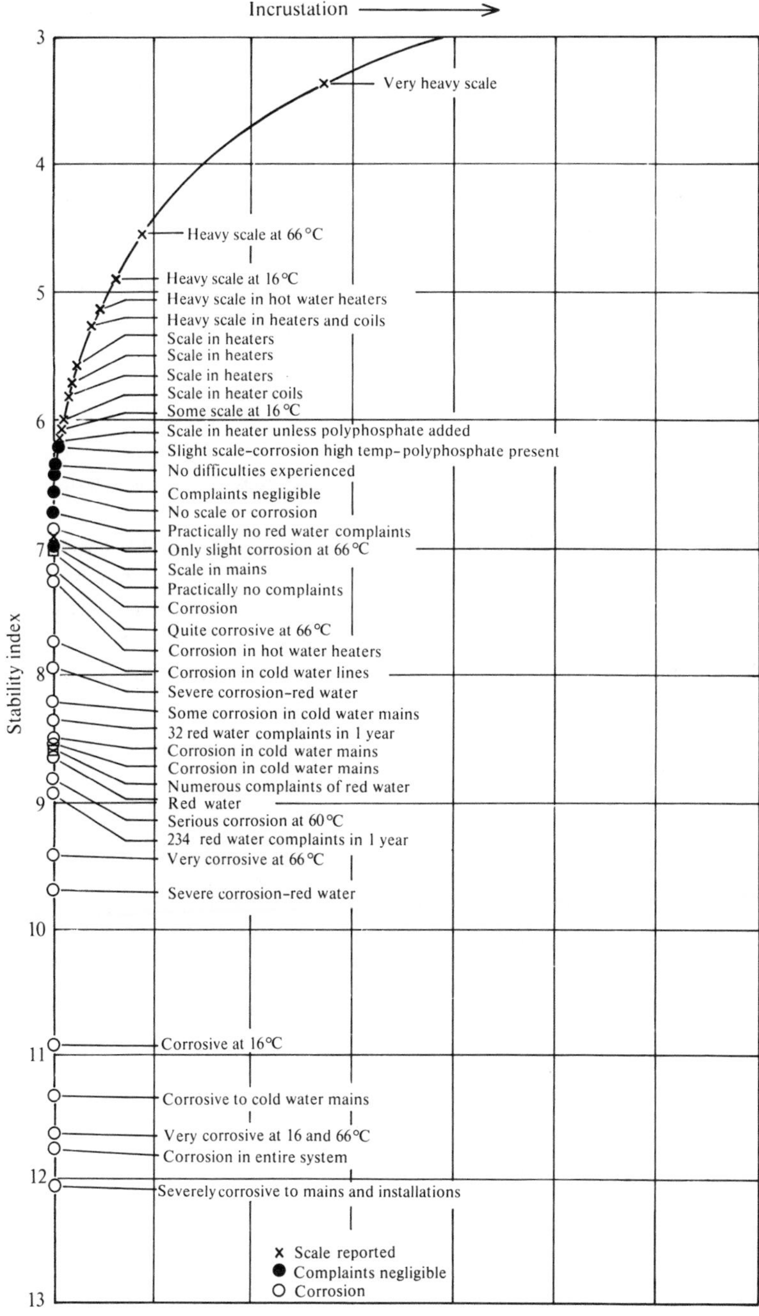


**Fig. 11.9.** The Ryznar index curve based on case history data.

If SI is positive there is a tendency for the water to incrust. The measurements required to determine the index are pH, $Ca^{2+}$, alkalinity and TDS. An example of a calculation is given in Fig. 11.8.

Although in general use, the Langelier index can be misleading in that it is only an approximate method of calculating calcite saturation. Calculations using the saturation index procedures described in Section 3.4 are at variance with the Langelier values because of the use of more accurate equilibrium constants (Section 5.7). The example given in Fig. 11.8 is therefore only for completeness.

Another index which is widely used as an indicator of both incrustation and corrosion is that of Ryznar (1944). Ryznar modified the Langelier index to provide an empirical stability index defined as

$$\text{stability index} = 2\text{pH}_s - \text{pH}.$$

Incrustation is likely to occur, for a stability index below 7, while above 7 corrosion becomes progressively more severe. Ryznar has depicted the index as shown on Fig. 11.9.

Saturation with respect to metal compounds can be calculated thermodynamically (Stumm and Morgan 1981), but provides a less reliable guide to incrustation than do the comparable carbonate calculations. Where $E_H$ and pH data are available, stability field diagrams as shown in Fig. 11.4 should also be used as a guide, as discussed in Section 11.2.5. However, kinetic rate effects must be borne in mind.

## References


Barnes, I. and Clarke, F. E. (1969). Chemical properties of groundwaters and their corrosion and encrustation effects in wells. *U.S. Geol. Surv. Prof. Pap. 498-D.*

van Beek, C. and van der Kooij, D. (1982). Sulphate reducing bacteria in groundwater from clogging and non-clogging shallow wells in the Netherlands River region. *Ground Water* **20**, 298–302.

Campbell, M. D. and Lehr, J. H. (1973). *Water well technology*. McGraw-Hill, New York.

Chilton, J. P. (1973). *Principles of metal corrosion* (2nd edn.) *Chem. Soc. Monogr. for Teachers*, No. 4. Chemical Society, London.

Clarke, F. E. and Barnes, I. (1965). Preliminary study of water well corrosion, Chad Basin, Nigeria. *U.S. Geol. Sur. Open File Rep.*

Cullimore, R. (1980). Iron bacteria—controlling them with disinfectants. *Johnson Drillers J.* July–Aug., 6–8.

Goldhaber, M. B. and Kaplan, I. R. (1975). Controls and consequences of sulphate reduction rates in recent marine sediments. *Soil Sci.* **119**, 42–55.

Johnson, E. E. (1966). *Groundwater and wells.* Edward Johnson, St. Pauls, MN.

Kelly, G. J. and Kemp, R. G. (1975). Guidelines for the selection of turbine pump materials for use in groundwaters. *Aust. Wat. Res. Council. Res. Proj. 71/33. Aust. Gov. Pub. Ser., Canberra.*

King, R. A. and Miller, J. D. (1971). Corrosion by the sulphate reducing bacteria. *Nature* (*Lond.*) **233**, 491–2.

Langelier, W. F. (1936). The analytical control of anti-corrosion water treatment. *J. Am. Waterworks. Assoc.* **28**, 1500–21.

Lewis, R. F. (1965), Control of sulphate reducing bacteria. *J. Am. water works Assoc.* **57**, 1011–15.

MacDonald, Sir M. and Partners (1965). *Lower Indus Report, Physical Resources —Groundwater*, 6. Report to Water and Power Development Authority, West Pakistan.

Malashenko, Y. R. (1971). The isolation of pure cultures of obligate methane oxidising bacteria. *Mikrobiologiya* **40**, 13.

Moss, R. (1964). Design of casings and screens for water production and injection wells. *Am. Petrol. Inst., Pacific Coast District Biennial Symp.* Pub. Amer. Pet. Inst.

Ryznar, J. W. (1944). A new index for determining the amount of $CaCO_3$ scale formed by water. *J. Am. Waterworks Assoc.* **36**, 473–86.

Snoeyink, V. L. and Jenkins, D. (1980). *Water chemistry*, Wiley, New York.

Stow, A. H. and Renner, L. (1965). Acidising boreholes, *J. Inst. water Eng.* **19**, 557–72.

Stumm, W. and Morgan, J. J. (1981). *Aquatic chemistry* (2nd edn.) Wiley, New York.

Swan, J. D. (1982). Relative corrosion resistance of certain steels in soils and waters. *Johnson Drillers J.*, 3rd Quarter, 6–9.

Updegraff, D. M. (1955). Microbiological corrosion of iron and steel. *Corrosion (Houston)* **11**, 442–9.

# Appendices

## Appendix I

*Thermodynamic properties: table of $\Delta G_f^\circ$, $\Delta H_f^\circ$ and $S^\circ$ values for common chemical species in aquatic systems*[a]

Valid at 25 °C, 1 atm pressure, and standard states[b]

| Species | Formation from the elements | | Entropy | Reference[c] |
|---|---|---|---|---|
| | $\Delta G_f^\circ$ (kJ mol$^{-1}$) | $\Delta H_f^\circ$ (kJ mol$^{-1}$) | $S^\circ$ (J mol$^{-1}$ K$^{-1}$) | |
| *Ag (silver)* | | | | |
| Ag (metal) | 0 | 0 | 42.6 | NBS |
| $Ag^+$ (aq) | 77.12 | 105.6 | 73.4 | NBS |
| AgBr | −96.9 | −100.6 | 107 | NBS |
| AgCl | −109.8 | −127.1 | 96 | NBS |
| AgI | −66.2 | −61.84 | 115 | NBS |
| $Ag_2S(\alpha)$ | −40.7 | −29.4 | 14 | NBS |
| AgOH(aq) | −92 | — | — | NBS |
| $Ag(OH)_2^-$(aq) | −260.2 | — | — | NBS |
| AgCl(aq) | −72.8 | −72.8 | 154 | NBS |
| $AgCl_2^-$(aq) | −215.5 | −245.2 | 231 | NBS |
| *Al (aluminium)* | | | | |
| Al | 0 | 0 | 28.3 | R |
| $Al^{3+}$(aq) | −489.4 | −531.0 | −308 | R |
| $AlOH^{2+}$(aq) | −698 | — | — | S |
| $Al(OH)_2^+$(aq) | −911 | — | — | S |
| $Al(OH)_3$(aq) | −1115 | — | — | S |
| $Al(OH)_4^-$(aq) | −1325 | — | — | S |
| $Al(OH)_3$ (amorph) | −1139 | — | — | R |
| $Al_2O_3$(corundum) | −1582 | −1676 | 50.9 | R |

| | | | | |
|---|---|---|---|---|
| AlOOH (boehmite) | −922 | −1000 | 17.8 | R |
| $Al(OH)_3$ (gibbsite) | −1155 | −1293 | 68.4 | R |
| $Al_2Si_2(OH)_4$ (kaolinite) | −3799 | −4120 | 203 | R |
| $KAl_3Si_3O_{10}(OH)_2$ (muscovite) | −1341 | — | — | G |
| $Mg_5Al_2Si_3O_{10}(OH)_8$ (chlorite) | −1962 | — | — | R |
| $CaAl_2Si_2O_8$ (anorthite) | −4017.3 | −4243.0 | 199 | R |
| $NaAlSiO_3O_8$ (albite) | −3711.7 | −3935.1 | — | R |
| *As (arsenic)* | | | | |
| As (α metal) | 0 | 0 | 35.1 | NBS |
| $H_3AsO_4$(aq) | −766.0 | −898.7 | 206 | NBS |
| $H_2AsO_4^-$(aq) | −748.5 | −904.5 | 117 | NBS |
| $HAsO_4^{2-}$ (aq) | −707.1 | −898.7 | 3.8 | NBS |
| $AsO_4^{3-}$ (aq) | −636.0 | −870.3 | −145 | NBS |
| $H_2AsO_3^-$ (aq) | −587.4 | — | — | NBS |
| *Ba (barium)* | | | | |
| $Ba^{2+}$(aq) | −560.7 | −537.6 | 9.6 | R |
| $BaSO_4$ (barite) | −1362 | −1473 | 132 | R |
| $BaCO_3$ (witherite) | −1132 | −1211 | 112 | R |
| *Be (beryllium)* | | | | |
| $Be^{2+}$(aq) | −380 | −382 | −130 | NBS |
| $Be(OH)_2$ (α) | −815.0 | −902 | 51.9 | NBS |
| $Be_3(OH)_3^{3+}$ | −1802 | — | — | NBS |
| *B (boron)* | | | | |
| $H_3BO_3$(aq) | −968.7 | −1072 | 162 | NBS |
| $B(OH)_4^-$(aq) | −1153.3 | −1344 | 102 | NBS |
| *Br (bromide)* | | | | |
| $Br_2$ (l) | 0 | 0 | 152 | NBS |
| $Br_2$(aq) | 3.93 | −2.59 | 130.5 | NBS |
| $Br^-$(aq) | −104.0 | −121.5 | 82.4 | NBS |
| HBrO(aq) | −82.2 | −113.0 | 147 | NBS |
| $BrO^-$(aq) | −33.5 | −94.1 | 42 | NBS |

| Species | Formation from the elements | | Entropy | Reference[c] |
|---|---|---|---|---|
| | $\Delta G_f^\circ$ (kJ mol$^{-1}$) | $\Delta H_f^\circ$ (kJ mol$^{-1}$) | $S^\circ$ (J mol$^{-1}$ K$^{-1}$) | |
| *C (carbon)* | | | | |
| C (graphite) | 0 | 0 | 152 | NBS |
| C (diamond) | 3.93 | −2.59 | 130.5 | NBS |
| $CO_2$ (g) | −394.37 | −393.5 | 213.6 | NBS |
| $H_2CO_3^*$(aq) | −623.2 | −699.6 | 200.8 | NBS[a] |
| $H_2CO_3$(aq) ('true') | ~−607.1 | — | — | S |
| $HCO_3^-$(aq) | −586.8 | −692.0 | 91.2 | S |
| $CO_3^{2-}$(aq) | −527.9 | −677.1 | −56.9 | NBS |
| $CH_4$(g) | −50.75 | −74.80 | 186 | NBS |
| $CH_4$(aq) | −34.39 | −89.04 | 83.7 | NBS |
| $CH_3OH$(aq) | −175.4 | −245.9 | 133 | NBS |
| HCOOH(aq) | −372.3 | −425.4 | 163 | NBS |
| $HCOO^-$(aq) | −351.0 | −425.6 | 92 | NBS |
| HCN(aq) | 119.7 | 107.1 | 124.6 | NBS |
| $CN^-$(aq) | 172.4 | 150.6 | 94.1 | NBS |
| $CH_3COOH$(aq) | −396.6 | −485.8 | 179 | NBS |
| $CH_3COO^-$(aq) | −369.4 | −486.0 | 86.6 | NBS |
| $C_2H_5OH$(aq) | −181.8 | −288.3 | 149 | NBS |
| $NH_2CH_2COOH$(aq) | −370.8 | −514.0 | 158 | NBS |
| $NH_2CH_2COO^-$(aq) | −315.0 | −469.8 | 119 | NBS |
| *Ca (calcium)* | | | | |
| $Ca^{2+}$(aq) | −553.54 | −542.83 | −53 | R |
| $CaOH^+$(aq) | −718.4 | — | — | NBS |
| $Ca(OH)_2$(aq) | −868.1 | −1003 | −74.5 | NBS |
| $Ca(OH)_2$ (portlandite) | −898.4 | −986.0 | 83 | R |
| $CaCO_3$ (calcite) | −1128.8 | −1207.4 | 91.7 | R |
| $CaCO_3$ (aragonite) | −1127.8 | −1207.4 | 88.0 | R |
| $CaMg(CO_3)_2$ (dolomite) | −2161.7 | −2324.5 | 155.2 | R |
| $CaSiO_3$ (wollastonite) | −1549.9 | −1635.2 | 82.0 | R |
| $CaSO_4$ (anhydrite) | −1321.7 | −1434.1 | 106.7 | R |
| $CaSO_4.2H_2O$ (gypsum) | −1797.2 | −2022.6 | 194.1 | R |

| | | | | |
|---|---|---|---|---|
| $Ca_5(PO_4)_3OH$ (hydroxyapatite) | −6338.4 | −6721.6 | 390.4 | R |
| *Cd (cadmium)* | | | | |
| Cd (γ metal) | | | | |
| $Cd^{2+}$(aq) | −77.58 | −75.90 | −73.2 | R |
| $CdOH^-$(aq) | −284.5 | — | — | R |
| $Cd(OH)_3^-$(aq) | −600.8 | — | — | R |
| $Cd(OH)_4^{2-}$(aq) | −758.5 | — | — | R |
| $Cd(OH)_2$ (aq) | −392.2 | — | — | R |
| CdO (s) | −228.4 | −258.1 | 54.8 | |
| $Cd(OH)_2$ (precip.) | −473.6 | −560.6 | 96.2 | R |
| $CdCl^+$(aq) | −224.4 | −240.6 | 43.5 | R |
| $CdCl_2$ (aq) | −340.1 | −410.2 | 39.8 | R |
| $CdCl_3^-$(aq) | −487.0 | −561.0 | 203 | R |
| $CdCO_3$ (s) | −669.4 | −750.6 | 92.5 | R |
| *Cl (chlorine)* | | | | |
| $Cl^-$(aq) | −131.3 | −167.2 | 56.5 | NBS |
| $Cl_2$ (g) | 0 | 0 | 223.0 | NBS |
| $Cl_2$ (aq) | 6.90 | −23.4 | 121 | NBS |
| HClO(aq) | −79.9 | −120.9 | 142 | NBS |
| $ClO^-$(aq) | −36.8 | −107.1 | 42 | NBS |
| $ClO_2$ (aq) | 117.6 | 74.9 | 173 | NBS |
| $ClO_2^-$(aq) | 17.1 | −66.5 | 101 | NBS |
| $ClO_3^-$(aq) | −3.35 | −99.2 | 162 | NBS |
| $ClO_4$ (aq) | −8.62 | −129.3 | 182 | NBS |
| *Co (cobalt)* | | | | |
| Co (metal) | 0 | 0 | 30.04 | R |
| $Co^{2+}$(aq) | −54.4 | −58.2 | −113 | R |
| $Co^{3+}$ | −134 | −92 | −305 | R |
| $HCoO_2^-$(aq) | −407.5 | — | — | NBS |
| $Co(OH)_2$ (aq) | −369 | −518 | 134 | NBS |
| $Co(OH)_2$ (blue precip.) | −450 | — | — | NBS |
| CoO | −214.2 | −237.9 | 53.0 | R |
| $Co_3O_4$ (cobalt spinel) | −725.5 | −891.2 | 102.5 | R |

| Species | Formation from the elements | | Entropy | Reference[c] |
|---|---|---|---|---|
| | $\Delta G_f^\circ$ (kJ mol$^{-1}$) | $\Delta H_f^\circ$ (kJ mol$^{-1}$) | $S^\circ$ (J mol$^{-1}$ K$^{-1}$) | |
| *Cr (chromium)* | | | | |
| Cr (metal) | 0 | 0 | 23.8 | NBS |
| $Cr^{2+}$ (aq) | — | −143.5 | — | NBS |
| $Cr^{3+}$(aq) | −215.5 | −256.0 | 308 | NBS |
| $Cr_2O_3$ (eskolaite) | −1053 | −1135 | 81 | R |
| $HCrO_4^-$(aq) | −764.8 | −878.2 | 184 | R |
| $CrO_4^{2-}$(aq) | −727.9 | −881.1 | 50 | R |
| $Cr_2O_7^{2-}$(aq) | −1301 | −1490 | 262 | R |
| *Cu (copper)* | | | | |
| Cu (metal) | 0 | 0 | 33.1 | NBS |
| $Cu^+$(aq) | 50.0 | 71.7 | 40.6 | NBS |
| $Cu^{2+}$(aq) | 65.5 | 64.8 | −99.6 | NBS |
| $Cu(OH)_2$ (aq) | −249.1 | −395.2 | −121 | NBS |
| $HCuO_2^-$(aq) | −258 | — | — | |
| CuS (covellite) | −53.6 | −53.1 | 66.5 | NBS |
| $Cu_2S$ ($\alpha$) | −86.2 | −79.5 | 121 | NBS |
| CuO (tenorite) | −129.7 | −157.3 | 43 | NBS |
| $CuCO_3.Cu(OH)_2$ (malachite) | −893.7 | −1051.4 | 186 | NBS |
| $2CuCO_3.Cu(OH)_2$ (azurite) | — | −1632 | — | NBS |
| *F (fluorine)* | | | | |
| $F_2$(g) | 0 | 0 | 202 | NBS |
| $F^-$(aq) | −278.8 | −332.6 | −13.8 | NBS |
| HF(aq) | −296.8 | 320.0 | 88.7 | NBS |
| $HF_2^-$(aq) | −578.1 | −650 | 92.5 | NBS |
| *Fe (iron)* | | | | |
| Fe (metal) | 0 | 0 | 27.3 | NBS |
| $Fe^{2+}$(aq) | −78.87 | −89.10 | −138 | NBS |
| $FeOH^+$(aq) | −277.3 | — | — | NBS |

| | | | | |
|---|---|---|---|---|
| $Fe^{3+}(aq)$ | −4.60 | −48.5 | −316 | NBS |
| $FeOH^{2+}(aq)$ | −229.4 | −324.7 | −29.2 | NBS |
| $Fe(OH)_2^+(aq)$ | −438 | — | — | NBS |
| $Fe(OH)_2^-(aq)$ | −659 | — | — | NBS |
| $Fe_2(OH)_2^{4+}(aq)$ | −467.3 | — | — | NBS |
| $FeS_2$ (pyrite) | −160.2 | −171.5 | 52.9 | R |
| $FeS_2$ (marcasite) | −158.4 | −169.4 | 53.9 | R |
| FeO(s) | −251.1 | −272.0 | 59.8 | R |
| $Fe(OH)_2$ (precip.) | −486.6 | −569 | 87.9 | NBS |
| α-$Fe_3O_3$ (hematite)[e] | −742.7 | −824.6 | 87.4 | R |
| $Fe_3O_4$ (magnetite) | −1012.6 | −1115.7 | 146 | R |
| α-FeOOH (goethite)[e] | −488.6 | −559.3 | 60.5 | R |
| FeOOH (amorph)[e] | −462 | — | — | S |
| $Fe(OH)_3$ (amorph)[e] | −699(−712) | — | — | S |
| $FeCO_3$ (siderite) | −666.7 | −737.0 | 105 | R |
| $Fe_2SiO_4$ (fayalite) | −1379.4 | −1479.3 | 148 | R |
| *H (hydrogen)* | | | | |
| $H_2(g)$ | 0 | 0 | 130.6 | NBS |
| $H_2(aq)$ | 17.57 | −4.18 | 57.7 | NBS |
| $H^+(aq)$ | 0 | 0 | 0 | NBS |
| $H_2O(l)$ | −237.18 | −285.83 | 69.91 | NBS |
| $H_2O_2(aq)$ | −134.1 | −191.1 | 144 | NBS |
| $HO_2^-(aq)$ | −67.4 | −160.3 | 23.8 | NBS |
| *Hg (mercury)* | | | | |
| Hg(l) | 0 | 0 | 76.0 | NBS |
| $Hg_2^{2+}(aq)$ | 153.6 | 172.4 | 84.5 | NBS |
| $Hg^{2+}(aq)$ | 164.4 | 171.0 | −32.2 | NBS |
| $Hg_2Cl_2$ (calomel) | −210.8 | 265.2 | 192.4 | NBS |
| HgO (red) | −58.5 | −90.8 | 70.3 | NBS |
| HgS (metacinnabar) | −43.3 | −46.7 | 96.2 | NBS |
| $HgI_2$ (red) | −101.7 | −105.4 | 180 | NBS |
| $HgCl^+(aq)$ | −5.44 | −18.8 | 75.3 | NBS |
| $HgCl_2(aq)$ | −173.2 | −216.3 | 155 | NBS |
| $HgCl_3^-(aq)$ | −309.2 | −388.7 | 209 | NBS |
| $HgCl_4^{2-}(aq)$ | −446.8 | −554.0 | 293 | NBS |
| $HgOH^+(aq)$ | −52.3 | −84.5 | 71 | NBS |

| Species | Formation from the elements | | Entropy | Reference[c] |
|---|---|---|---|---|
| | $\Delta G_f^\circ$ (kJ mol$^{-1}$) | $\Delta H_f^\circ$ (kJ mol$^{-1}$) | $S^\circ$ (J mol$^{-1}$ K$^{-1}$) | |
| $Hg(OH)_2$ (aq) | −274.9 | −355.2 | 142 | NBS |
| $HgO_2^-$(aq) | −190.3 | — | — | NBS |
| *I (iodine)* | | | | |
| $I_2$ (crystal) | 0 | 0 | 116 | NBS |
| $I_2$(aq) | 16.4 | 22.6 | 137 | NBS |
| $I^-$(aq) | −51.59 | −55.19 | 111 | NBS |
| $I_3^-$(aq) | −51.5 | −51.5 | 239 | NBS |
| HIO(aq) | −99.2 | −138 | 95.4 | NBS |
| $IO^-$(aq) | −38.5 | −107.5 | −5.4 | NBS |
| $HIO_3$(aq) | −132.6 | −211.3 | 167 | NBS |
| $IO_3^-$ | −128.0 | −221.3 | 118 | NBS |
| *Mg (magnesium)* | | | | |
| Mg (metal) | 0 | 0 | 32.7 | R |
| $Mg^{2+}$(aq) | −454.8 | −466.8 | −138 | R |
| $MgOH^+$(aq) | −626.8 | — | — | S |
| $Mg(OH)_2$ (aq) | −769.4 | −926.8 | −149 | NBS |
| $Mg(OH)_2$ (brucite) | −833.5 | −924.5 | 63.2 | R |
| *Mn (manganese)* | | | | |
| Mn (metal) | 0 | 0 | 32.0 | R |
| $Mn^{2+}$(aq) | −228.0 | −220.7 | −73.6 | R |
| $Mn(OH)_2$ (precip.) | −616 | — | — | S |
| $Mn_3O_4$ (hausmannite) | −1281 | — | — | S |
| MnOOH (α manganite) | −557.7 | — | — | S |
| $MnO_2$ (pyrolusite) | −465.1 | −520.0 | 53 | R |
| $MnCO_3$ (rhodochrosite) | −816.0 | −889.3 | 100 | R |
| MnS (albandite) | −218.1 | −213.8 | 87 | R |
| $MnSiO_3$ (rhodonite) | −1243 | −1319 | 131 | R |

| | | | | |
|---|---|---|---|---|
| *N (nitrogen)* | | | | |
| $N_2$(g) | 0 | 0 | 191.5 | NBS |
| $N_2O$(g) | 104.2 | 82.0 | 220 | NBS |
| $NH_3$(g) | −16.48 | −46.1 | 192 | NBS |
| $NH_3$(aq) | −26.57 | −80.29 | 111 | NBS |
| $NH_4^+$(aq) | −79.37 | −132.5 | 113.4 | NBS |
| $HNO_2$(aq) | −42.97 | −119.2 | 153 | NBS |
| $NO_2^-$(aq) | −37.2 | −104.6 | 140 | NBS |
| $HNO_3$(aq) | −111.3 | −207.3 | 146 | NBS |
| $NO_3^-$(aq) | −111.3 | −207.3 | 146.4 | NBS |
| *Ni (nickel)* | | | | |
| $Ni^{2+}$(aq) | −45.6 | −54.0 | −129 | R |
| NiO (bunsenite) | −211.6 | −239.7 | 38 | R |
| NiS (millerite) | −86.2 | −84.9 | 66 | R |
| *O (oxygen)* | | | | |
| $O_2$(g) | 0 | 0 | 205 | NBS |
| $O_2$(aq) | 16.32 | −11.71 | 111 | NBS |
| $O_3$(g) | 163.2 | 142.7 | 239 | NBS |
| $OH^-$(aq) | −157.3 | −230.0 | −10.75 | NBS |
| *P (phosphorus)* | | | | |
| P (α, white) | 0 | 0 | 41.1 | |
| $PO_4^{3-}$(aq) | −1018.8 | −1277.4 | −222 | NBS |
| $HPO_4^{2-}$(aq) | −1089.3 | −1292.1 | −33.4 | NBS |
| $H_2PO_4^-$(aq) | −1130.4 | −1296.3 | 90.4 | NBS |
| $H_3PO_4$(aq) | −1142.6 | −1288.3 | 158 | NBS |
| *Pb (lead)* | | | | |
| Pb (metal) | 0 | 0 | 64.8 | NBS |
| $Pb^{2+}$(aq) | −24.39 | −1.67 | 10.5 | NBS |
| $PbOH^+$(aq) | −226.3 | — | — | NBS |
| $Pb(OH)_3^-$(aq) | −575.7 | — | — | NBS |
| $Pb(OH)_2$ (precip.) | −452.2 | — | — | NBS |
| PbO (yellow) | −187.9 | −217.3 | 68.7 | NBS |
| $PbO_2$ | −217.4 | −277.4 | 68.6 | NBS |
| $Pb_3O_4$ | −601.2 | −718.4 | 211 | NBS |
| PbS | −98.7 | −100.4 | 91.2 | NBS |

| Species | Formation from the elements | | Entropy | Reference[c] |
|---|---|---|---|---|
| | $\Delta G_f^\circ$ (kJ mol$^{-1}$) | $\Delta H_f^\circ$ (kJ mol$^{-1}$) | $S^\circ$ (J mol$^{-1}$ K$^{-1}$) | |
| $PbSO_4$ | −813.2 | −920.0 | 149 | NBS |
| $PbCO_3$ (cerussite) | −625.5 | −699.1 | 131 | NBS |
| *S (sulphur)* | | | | |
| S (rhombic) | 0 | 0 | 31.8 | NBS |
| $SO_2$ (g) | −300.2 | −296.8 | 248 | NBS |
| $SO_3$ (g) | −371.1 | −395.7 | 257 | NBS |
| $H_2S$(g) | −33.56 | −20.63 | 205.7 | NBS |
| $H_2S$(aq) | −27.87 | −39.75 | 121.3 | NBS |
| $S^{2-}$(aq) | 85.8 | 33.0 | −14.6 | NBS |
| $HS^-$(aq) | 12.05 | −17.6 | 62.8 | NBS |
| $SO_3^{2-}$(aq) | −486.6 | −635.5 | −29 | NBS |
| $HSO_3^-$(aq) | −527.8 | −626.2 | 140 | NBS |
| $H_2SO_3^*$(aq) | −537.9 | −608.8 | 232 | NBS[f] |
| $H_2SO_3$ (aq) ('true') | ~−534.5 | — | — | S |
| $SO_4^{2-}$(aq) | −744.6 | −909.2 | 20.1 | NBS |
| $HSO_4^-$(aq) | −756.0 | −887.3 | 132 | NBS |
| *Se (selenium)* | | | | |
| Se (black) | 0 | 0 | 42.4 | NBS |
| $SeO_3^{2-}$(aq) | −369.9 | −509.2 | 12.6 | NBS |
| $HSeO_3^-$(aq) | −431.5 | −514.5 | 135 | NBS |
| $H_2SeO_3$ (aq) | −426.2 | −507.5 | 208 | NBS |
| $SeO_4^{2-}$(aq) | −441.4 | −599.1 | 54.0 | NBS |
| $HSeO_4^-$(aq) | −452.3 | −581.6 | 149 | NBS |
| *Si (silicon)* | | | | |
| Si (metal) | 0 | 0 | 18.8 | NBS |
| $SiO_2$ (α, quartz) | −856.67 | −910.94 | 41.8 | NBS |
| $SiO_2$ (α, cristobalite) | −855.88 | −909.48 | 42.7 | NBS |
| $SiO_2$ (α, tridymite) | −855.29 | −909.06 | 43.5 | NBS |
| $SiO_2$ (amorph) | −850.73 | −903.49 | 46.9 | NBS |
| $H_4SiO_4$(aq) | −1316.7 | −1468.6 | 180 | NBS |

| | | | | |
|---|---|---|---|---|
| *Sr (strontium)* | | | | |
| $Sr^{2+}$(aq) | −559.4 | −545.8 | −33 | R |
| $SrOH^{+}$(aq) | −721 | — | — | NBS |
| $SrCO_3$ (strontianite) | −1137.6 | −1218.7 | 97 | R |
| $SrSO_4$ (celestite) | −1341.0 | −1453.2 | 118 | R |
| *Zn (zinc)* | | | | |
| Zn metal | 0 | 0 | 29.3 | NBS |
| $Zn^{2+}$(aq) | −147.0 | −153.9 | 112 | NBS |
| $ZnOH^{+}$(aq) | −330.1 | — | — | NBS |
| $Zn(OH)_2$(aq) | −522.3 | — | — | NBS |
| $Zn(OH)_3^-$(aq) | −694.3 | — | — | NBS |
| $Zn(OH)_4^{2-}$(aq) | −858.7 | — | — | NBS |
| $Zn(OH)_2$ (solid $\beta$) | −553.2 | −641.9 | 81.2 | R |
| $ZnCl^{+}$(aq) | −275.3 | — | — | NBS |
| $ZnCl_2$ (aq) | −403.8 | — | — | NBS |
| $ZnCl_3^-$(aq) | −540.6 | — | — | NBS |
| $ZnCl_4^{2-}$(aq) | −666.1 | — | — | S |
| $ZnCO_3$ (smithsonite) | −731.6 | −812.8 | 82.4 | NBS |

From Stumm and Morgan 1981.

[a] The quality of the data is highly variable: the authors do not claim to have critically selected the 'best' data. For information on precision of the data and for a more complete compendium which includes less common substances, the reader is referred to the references. For research work, the original literature should be consulted.

[b] Thermodynamic properties taken from Robie, Hemingway, and Fisher (1978) are based on a reference state of the elements in their standard states at 1 bar ($10^5$ Pa = 0.987 atm). This change in reference pressure has a negligible effect upon the tabulated vaues for the condensed phases. (For gas phases only data from NBS (reference state = 1 atm) are given).

[c] NBS, Wagman *et al.* (1968, 1969, 1971); R, Robie *et al.* (1978); S, other sources.

[d] $[H_2CO_3^*] = [CO_2]$ (aq) + 'true' $[H_2CO_3]$.

[e] The thermodynamic stability of oxides, hydroxides, or oxyhydroxides of iron(III) depends on the mode of preparation, the age, and the molar surface.

[f] $[H_2SO_3^*] = [SO_2(aq)]$ + 'true' $[H_2SO_3]$.

## Appendix II

*Electrode potentials*

*Potentials in acid solutions*

| Reaction | Potential |
|---|---|
| $K^+ + e^- \rightarrow K$ | −2.93 |
| $Ca^{2+} + 2e^- \rightarrow Ca$ | −2.87 |
| $Na^+ + e^- \rightarrow Na$ | −2.71 |
| $Mg^{2+} + 2e^- \rightarrow Mg$ | −2.37 |
| $Th^{4+} + 4e^- \rightarrow Th$ | −1.90 |
| $U^{3+} + 3e^- \rightarrow U$ | −1.80 |
| $Al^{3+} + 3e^- \rightarrow Al$ | −1.66 |
| $Mn^{2+} + 2e^- \rightarrow Mn$ | −1.18 |
| $V^{2+} + 2e^- \rightarrow V$ | −1.18 |
| $SiO_2 + 4H^+ + 4e^- \rightarrow Si + 2H_2O$ | −0.99 |
| $Zn^{2+} + 2e^- \rightarrow Zn$ | −0.76 |
| $Cr^{3+} + 3e^- \rightarrow Cr$ | −0.74 |
| $Te + 2H^+ + 2e^- \rightarrow H_2Te(aq)$ | −0.74 |
| $U^{4+} + e^- \rightarrow U^{3+}$ | −0.61 |
| $Fe^{2+} + 2e^- \rightarrow Fe$ | −0.41 |
| $Cr^{3+} + e^- \rightarrow Cr^{2+}$ | −0.41 |
| $Se + 2H^+ + 2e^- \rightarrow H_2Se(aq)$ | −0.40 |
| $Co^{2+} + 2e^- \rightarrow Co$ | −0.28 |
| $V^{3+} + e^- \rightarrow V^{2+}$ | −0.26 |
| $Ni^{2+} + 2e^- \rightarrow Ni$ | −0.24 |
| $Sn^{2+} + 2e^- \rightarrow Sn$ | −0.14 |
| $Pb^{2+} + 2e^- \rightarrow Pb$ | −0.13 |
| $2H^+ + 2e^- \rightarrow H_2$ | 0.00 |
| $S + 2H^+ + 2e^- \rightarrow H_2S(aq)$ | +0.14 |
| $Sn^{4+} + 2e^- \rightarrow Sn^{2+}$ | +0.15 |
| $Cu^{2+} + e^- \rightarrow Cu^+$ | +0.16 |
| $SO_4^{2-} + 8H^+ + 8e^- \rightarrow S^- + 4H_2O$ | +0.16 |
| $SO_4^{2-} + 4H^+ + 2e^- \rightarrow H_2SO_3(aq)$ | +0.17 |
| $AgCl + e^- \rightarrow Ag + Cl$ | +0.22 |
| $HAsO_2(aq) + 3H^+ + 3e^- \rightarrow As + 2H_2O$ | +0.25 |
| $UO_2^{2+} + 4H^+ + 2e^- \rightarrow U^{2+} + 2H_2O$ | +0.33 |
| $Cu^{2+} + 2e^- \rightarrow Cu$ | +0.34 |
| $VO^{2+} + 2H^+ + e^- \rightarrow V^{3+} + H_2O$ | +0.34 |
| $H_2SO_3(aq) + 4H^+ + 4e^- \rightarrow S + 3H_2O$ | +0.45 |

| | |
|---|---|
| $Cu^+ + e^- \rightarrow Cu$ | +0.52 |
| $I_2(s) + 2e^- \rightarrow 2I^-$ | +0.54 |
| $I_3^- + 2e^- \rightarrow 3I^-$ | +0.54 |
| $H_2AsO_4(aq) + 2H^+ + 2e^- \rightarrow HAsO_2(aq) + 2H_2O$ | +0.56 |
| $PdCl_4^{2-} + 2e^- \rightarrow Pd + 4Cl^-$ | +0.62 |
| $PtCl_4^{2-} + 2e^- \rightarrow Pt + 4Cl^-$ | +0.73 |
| $H_2SeO_3(aq) + 4H^+ + 4e^- \rightarrow Se + 3H_2O$ | +0.74 |
| $Fe^{3+} + e^- \rightarrow Fe^{2+}$ | +0.77 |
| $Hg_2^{2+} + 2e^- \rightarrow 2Hg$ | +0.79 |
| $Ag^+ + e^- \rightarrow Ag$ | +0.80 |
| $Hg^{2+} + 2e^- \rightarrow Hg$ | +0.85 |
| $Pd^{2+} + 2e^- \rightarrow Pd$ | +0.92 |
| $NO_3^- + 4H^+ + 3e^- \rightarrow NO(g) + 2H_2O$ | +0.96 |
| $Fe(OH)_3 + 3H^+ + e^- \rightarrow Fe^{2+} + 3H_2O$ | +0.98 |
| $AuCl_4^- + 3e^- \rightarrow Au + 4Cl^-$ | +1.00 |
| $VO_2^+ + 2H^+ + e^- \rightarrow VO^{2+} + H_2O$ | +1.00 |
| $Br_2(l) + 2e^- \rightarrow 2Br^-$ | +1.07 |
| $Br_2(aq) + 2e^- \rightarrow 2Br^-$ | +1.09 |
| $S + Hg^{2+} + 2e^- \rightarrow HgS$ | +1.11 |
| $SeO_4^{2-} + 4H^+ + 2e^- \rightarrow H_2SeO_3(aq) + H_2O$ | +1.15 |
| $IO_3^- + 6H^+ + 5e^- \rightarrow \frac{1}{2}I_2(s) + 3H_2O$ | +1.20 |
| $O_2(g) + 4H^+ + 4e^- \rightarrow 2H_2O$ | +1.23 |
| $MnO_2(s) + 4H^+ + 2e^- \rightarrow Mn^{2+} + 2H_2O$ | +1.23 |
| $Cr_2O_7^{2-} + 14H^+ + 6e^- \rightarrow 2Cr^{3+} + 7H_2O$ | +1.33 |
| $Cl_2(g) + 2e^- \rightarrow 2Cl^-$ | +1.36 |
| $PbO_2(s) + 4H^+ + 2e^- \rightarrow Pb^{2+} + 2H_2O$ | +1.46 |
| $Au^{3+} + 3e^- \rightarrow Au$ | +1.50 |
| $Mn^{3+} + e^- \rightarrow Mn^{2+}$ | +1.51 |
| $MnO_4^- + 8H^+ + 5e^- \rightarrow Mn^{2+} + 4H_2O$ | +1.51 |
| $H_5IO_6(aq) + H^+ + 2e^- \rightarrow IO_3^- + 3H_2O$ | +1.6 |
| $Au^+ + e^- \rightarrow Au$ | Approx. +1.68 |
| $Co^{3+} + e^- \rightarrow Co^{2+}$ | +1.82 |
| $F_2(g) + 2e^- \rightarrow 2F^-$ | +2.87 |

*Potentials in basic solutions*

| | |
|---|---|
| $Mg(OH)_2 + 2e^- \rightarrow Mg + 2OH^-$ | −2.69 |
| $UO_2 + 2H_2O + 4e^- \rightarrow U + 4OH^-$ | −2.39 |
| $Al(OH)_4^- + 3e^- \rightarrow Al + 4OH^-$ | −2.32 |

| | |
|---|---|
| $Mn(OH)_2 + 2e^- \rightarrow Mn + 2OH^-$ | −1.55 |
| $Zn(OH)_2 + 2e^- \rightarrow Zn + 2OH^-$ | −1.25 |
| $SO_4^{2-} + H_2O + 2e^- \rightarrow SO_3^{2-} + 2OH^-$ | −0.93 |
| $Se + 2e^- \rightarrow Se^{2-}$ | −0.92 |
| $Sn(OH)_3^- + 2e^- \rightarrow Sn + 3OH^-$ | −0.91 |
| $Sn(OH)_6^{2-} + 2e^- \rightarrow Sn(OH)_3^- + 3OH^-$ | −0.90 |
| $Fe(OH)_2 + 2e^- \rightarrow Fe + 2OH^-$ | −0.89 |
| $2H_2O + 2e^- \rightarrow H_2 + 2OH^-$ | −0.83 |
| $VO(OH)_2 + H_2O + e^- \rightarrow V(OH)_3 + OH^-$ | −0.64 |
| $Fe(OH)_3 + e^- \rightarrow Fe(OH)_2 + OH^-$ | −0.55 |
| $Pb(OH)_3^- + 2e^- \rightarrow Pb + 3OH^-$ | −0.54 |
| $S + 2e^- \rightarrow S^{2-}$ | −0.44 |
| $Cu_2O + H_2O + 2e^- \rightarrow 2Cu + 2OH^-$ | −0.36 |
| $CrO_4^{2-} + 4H_2O + 3e^- \rightarrow Cr(OH)_3 + 5OH^-$ | −0.13 |
| $2Cu(OH)_2 + 2e^- \rightarrow Cu_2O + 2OH^-$ | −0.08 |
| $MnO_2 + 2H_2O + 2e^- \rightarrow Mn(OH)_2 + 2OH^-$ | −0.05 |
| $SeO_4^{2-} + H_2O + 2e^- \rightarrow SeO_3^{2-} + 2OH^-$ | +0.05 |
| $Pd(OH)_2 + 2e^- \rightarrow Pd + 2OH^-$ | +0.07 |
| $HgO(red) + H_2O + 2e^- \rightarrow Hg + 2OH^-$ | +0.10 |
| $Mn(OH)_3 + e^- \rightarrow Mn(OH)_2 + OH^-$ | +0.1 |
| $Co(OH)_3 + e^- \rightarrow Co(OH)_2 + OH^-$ | +0.17 |
| $PbO_2 + H_2O + 2e^- \rightarrow PbO(red) + 2OH^-$ | +0.25 |
| $IO_3^- + 3H_2O + 6e^- \rightarrow I^- + 6OH^-$ | +0.26 |
| $O_2 + 2H_2O + 4e^- \rightarrow 4OH^-$ | +0.40 |

After Krauskopf 1979.

The value of $E_H^\circ$ for each half-reaction is its potential in volts referred to the $H_2$-$H^+$ half-reaction, which is assigned the arbitrary value of zero. The values are given for 25 °C and 1 atm pressure, with all substances at unit activity. All pure substances whose state is not specified in the equations are assumed to be in their standard states at 25 °C and 1 atm.

The equation for each couple is written so that the reducing agent is on the right. Potential differences for complete reactions can be obtained by subtracting potentials for the appropriate half-reactions, provided that formulae of oxidizing and reducing agents are identical in the half-reactions and the complete reaction.

## Appendix III

*Calculation of adj. SAR* (after Sayers and Westcot 1976)

The adjusted sodium adsorption ratio (adj. SAR) is calculated from the following equation:

$$\text{adj. SAR} = \frac{Na^{+}}{\left(\dfrac{(Ca^{2+} + Mg^{2+})}{2}\right)^{\frac{1}{2}}}\{1 + (8.4 - pH_C)\}\ ^{\ddagger\dagger}$$

where $Na^{+}$, $Ca^{2+}$, and $Mg^{2+}$ are in milliequivalents per litre and are obtained from the water analysis. $pH_C$ is calculated using the table given below which relates to the concentration values from the water analysis. The table values are then substituted in the $pH_C$ equation

$$pH_C = (pK'_2 - pK'_c) + p(Ca^{2+} + Mg^{2+}) + p(Alk)$$

$pK'_2 - pK'_c$ is obtained from the sum of $Ca^{2+} + Mg^{2+} + Na^{+}$ in milliequivalents per litre. $p(Ca^{2+} + Mg^{+})$ is obtained from the sum of $Ca^{2+} + Mg^{2+}$ in milliequivalents per litre. p(Alk) is obtained from the sum of $CO_3^{2-} + HCO_3^{-}$ in milliequivalents per litre. All the concentrations are obtained from the water analysis.

| Sum of concentration (meq $l^{-1}$) | $pK'_2 - pK'_c$ | p(Ca + Mg) | p(Alk) |
|---|---|---|---|
| 0.05 | 2.0 | 4.6 | 4.3 |
| 0.10 | 2.0 | 4.3 | 4.0 |
| 0.15 | 2.0 | 4.1 | 3.8 |
| 0.20 | 2.0 | 4.0 | 3.7 |
| 0.25 | 2.0 | 3.9 | 3.6 |
| 0.30 | 2.0 | 3.8 | 3.5 |
| 0.40 | 2.0 | 3.7 | 3.4 |
| 0.50 | 2.1 | 3.6 | 3.3 |
| 0.75 | 2.1 | 3.4 | 3.1 |
| 1.00 | 2.1 | 3.3 | 3.0 |
| 1.25 | 2.1 | 3.2 | 2.9 |
| 1.5 | 2.1 | 3.1 | 2.8 |
| 2.0 | 2.2 | 3.0 | 2.7 |
| 2.5 | 2.2 | 2.9 | 2.6 |
| 3.0 | 2.2 | 2.8 | 2.5 |
| 4.0 | 2.2 | 2.7 | 2.4 |
| 5.0 | 2.2 | 2.6 | 2.3 |
| 6.0 | 2.2 | 2.5 | 2.2 |
| 8.0 | 2.3 | 2.4 | 2.1 |
| 10.0 | 2.3 | 2.3 | 2.0 |
| 12.5 | 2.3 | 2.2 | 1.9 |
| 15.0 | 2.3 | 2.1 | 1.8 |
| 20.0 | 2.4 | 2.0 | 1.7 |
| 30.0 | 2.4 | 1.8 | 1.5 |
| 50.0 | 2.5 | 1.6 | 1.3 |
| 80.0 | 2.5 | 1.4 | 1.1 |

† A nomogram for determining $Na/\{\frac{1}{2}(Ca + Mg)\}^{\frac{1}{2}}$ is presented in Fig. 10.4

‡ $pH_C$ is a theoretical calculated pH of the irrigation water in contact with lime and in equilibrium with soil $CO_2$.

## References to appendices

Krauskopf, K. B. (1979). *Introduction to geochemistry*. McGraw-Hill, New York.

Robie, R. A., Hemingway, B. S., and Fisher, J. R. (1978). Thermodynamic properties of minerals and related substances at 298. 15 K and 1 bar ($10^5$ Pa) pressure and at higher temperatures. *U.S. Geol. Surv. Bull. 1452.*

Sayers, R. S. and Westcot, D. W. (1976). Water quality for agriculture. *Irrigation and Drainage 29.* Food and Agriculture Organization, Rome.

Stumm, W. and Morgan, J. J. (1981). *Aquatic chemistry* (2nd Edn.). Wiley, New York.

Wagman, D. D., *et al.* (1968). Selected values of chemical thermodynamic properties. *U.S. Natl. Bur. Stand. Tech. Note 270-3.*

—(1969). Selected values of chemical thermodynamic properties. *U.S. Natl. Bur. Stand. Tech. Note 270-4.*

— (1971). Selected values of chemical thermodynamic properties. *U.S. Natl. Bur. Stand. Tech. Note 270-5.*

# Index